AN ASSESSMENT OF THE SBIR PROGRAM AT THE NATIONAL INSTITUTES OF HEALTH

Committee for
Capitalizing on Science, Technology, and Innovation:
An Assessment of the Small Business Innovation Research Program

Policy and Global Affairs

Charles W. Wessner, Editor

NATIONAL RESEARCH COUNCIL
OF THE NATIONAL ACADEMIES

THE NATIONAL ACADEMIES PRESS
Washington, D.C.
www.nap.edu

THE NATIONAL ACADEMIES PRESS   500 Fifth Street, N.W.   Washington, DC 20001

NOTICE: The project that is the subject of this report was approved by the Governing Board of the National Research Council, whose members are drawn from the Councils of the National Academy of Sciences, the National Academy of Engineering, and the Institute of Medicine. The members of the committee responsible for the report were chosen for their special competences and with regard for appropriate balance.

This study was supported by Contract/Grant No. DASW01-02-C-0039 between the National Academy of Sciences and the U.S. Department of Defense, NASW-03003 between the National Academy of Sciences and the National Aeronautics and Space Administration, DE-AC02-02ER12259 between the National Academy of Sciences and the U.S. Department of Energy, NSFDMI-0221736 between the National Academy of Sciences and the National Science Foundation, and N01-OD-4-2139 (Task Order #99) between the National Academy of Sciences and the U.S. Department of Health and Human Services. The content of this publication does not necessarily reflect the views or policies of the Department of Health and Human Services, nor does mention of trade names, commercial products, or organizations imply endorsement by the U.S. Government. Any opinions, findings, conclusions, or recommendations expressed in this publication are those of the author(s) and do not necessarily reflect the views of the organizations or agencies that provided support for the project.

International Standard Book Number-13: 978-0-309-10951-2
International Standard Book Number-10: 0-309-10951-5

Limited copies are available from the Policy and Global Affairs, National Research Council, 500 Fifth Street, N.W., Washington, DC 20001; 202-334-1529.

Additional copies of this report are available from the National Academies Press, 500 Fifth Street, N.W., Lockbox 285, Washington, DC 20055; (800) 624-6242 or (202) 334-3313 (in the Washington metropolitan area); Internet, http://www.nap.edu.

Printed in the United States of America

THE NATIONAL ACADEMIES
Advisers to the Nation on Science, Engineering, and Medicine

The **National Academy of Sciences** is a private, nonprofit, self-perpetuating society of distinguished scholars engaged in scientific and engineering research, dedicated to the furtherance of science and technology and to their use for the general welfare. Upon the authority of the charter granted to it by the Congress in 1863, the Academy has a mandate that requires it to advise the federal government on scientific and technical matters. Dr. Ralph J. Cicerone is president of the National Academy of Sciences.

The **National Academy of Engineering** was established in 1964, under the charter of the National Academy of Sciences, as a parallel organization of outstanding engineers. It is autonomous in its administration and in the selection of its members, sharing with the National Academy of Sciences the responsibility for advising the federal government. The National Academy of Engineering also sponsors engineering programs aimed at meeting national needs, encourages education and research, and recognizes the superior achievements of engineers. Dr. Charles M. Vest is president of the National Academy of Engineering.

The **Institute of Medicine** was established in 1970 by the National Academy of Sciences to secure the services of eminent members of appropriate professions in the examination of policy matters pertaining to the health of the public. The Institute acts under the responsibility given to the National Academy of Sciences by its congressional charter to be an adviser to the federal government and, upon its own initiative, to identify issues of medical care, research, and education. Dr. Harvey V. Fineberg is president of the Institute of Medicine.

The **National Research Council** was organized by the National Academy of Sciences in 1916 to associate the broad community of science and technology with the Academy's purposes of furthering knowledge and advising the federal government. Functioning in accordance with general policies determined by the Academy, the Council has become the principal operating agency of both the National Academy of Sciences and the National Academy of Engineering in providing services to the government, the public, and the scientific and engineering communities. The Council is administered jointly by both Academies and the Institute of Medicine. Dr. Ralph J. Cicerone and Dr. Charles M Vest are chair and vice chair, respectively, of the National Research Council.

www.national-academies.org

Committee for
Capitalizing on Science, Technology, and Innovation:
An Assessment of the Small Business Innovation Research Program

Chair
Jacques S. Gansler (NAE)
Roger C. Lipitz Chair in Public Policy and Private Enterprise
and Director of the Center for Public Policy and Private Enterprise
School of Public Policy
University of Maryland

David B. Audretsch
Distinguished Professor and
    Ameritech Chair of Economic
    Development
Director, Institute for Development
    Strategies
Indiana University

Gene Banucci
Executive Chairman
ATMI, Inc.

Jon Baron
Executive Director
Coalition for Evidence-Based Policy

Michael Borrus
Founding General Partner
X/Seed Capital

Gail Cassell (IOM)
Vice President, Scientific Affairs and
Distinguished Lilly Research Scholar
    for Infectious Diseases
Eli Lilly and Company

Elizabeth Downing
CEO
3D Technology Laboratories

M. Christina Gabriel
Director, Innovation Economy
The Heinz Endowments

Trevor O. Jones (NAE)
Founder and Chairman
Electrosonics Medical, Inc.

Charles E. Kolb
President
Aerodyne Research, Inc.

Henry Linsert, Jr.
CEO
Columbia Biosciences Corporation

W. Clark McFadden
Partner
Dewey & LeBoeuf, LLP

Duncan T. Moore (NAE)
Kingslake Professor of Optical
    Engineering
University of Rochester

Kent Murphy
President and CEO
Luna Innovations

Linda F. Powers
Managing Director
Toucan Capital Corporation

Tyrone Taylor
President
Capitol Advisors
    on Technology, LLC

Charles Trimble (NAE)
CEO, *retired*
Trimble Navigation

Patrick Windham
President
Windham Consulting

PROJECT STAFF

Charles W. Wessner
Study Director

Sujai J. Shivakumar
Senior Program Officer

McAlister T. Clabaugh
Program Associate

Adam H. Gertz
Program Associate

David E. Dierksheide
Program Officer

Jeffrey C. McCullough
Program Associate

RESEARCH TEAM

Zoltan Acs
University of Baltimore

David H. Finifter
The College of William and Mary

Alan Anderson
Consultant

Michael Fogarty
University of Portland

Philip A. Auerswald
George Mason University

Robin Gaster
North Atlantic Research

Robert-Allen Baker
Vital Strategies, LLC

Albert N. Link
University of North Carolina

Robert Berger
Robert Berger Consulting, LLC

Rosalie Ruegg
TIA Consulting

Grant Black
University of Indiana South Bend

Donald Siegel
University of California at Riverside

Peter Cahill
BRTRC, Inc.

Paula E. Stephan
Georgia State University

Dirk Czarnitzki
University of Leuven

Andrew Toole
Rutgers University

Julie Ann Elston
Oregon State University

Nicholas Vonortas
George Washington University

Irwin Feller
American Association for the
    Advancement of Science

POLICY AND GLOBAL AFFAIRS

Ad hoc Oversight Board for
Capitalizing on Science, Technology, and Innovation:
An Assessment of the Small Business Innovation Research Program

Robert M. White (NAE), Chair
University Professor Emeritus
Electrical and Computer Engineering
Carnegie Mellon University

Anita K. Jones (NAE)
Lawrence R. Quarles Professor
   of Engineering and Applied
   Science
School of Engineering and Applied
   Science
University of Virginia

Mark B. Myers
Senior Vice President, *retired*
Xerox Corporation

Contents

APPENDIXES

Preface

Today's knowledge-based economy is driven in large part by the nation's capacity to innovate. One of the defining features of the U.S. economy is a high level of entrepreneurial activity. Entrepreneurs in the United States see opportunities and are willing and able to take on risk to bring new welfare-enhancing, wealth-generating technologies to the market. Yet, while innovation in areas such as genomics, bioinformatics, and nanotechnology present new opportunities, converting these ideas into innovations for the market involves substantial challenges.[1] The American capacity for innovation can be strengthened by addressing the challenges faced by entrepreneurs. Public-private partnerships are one means to help entrepreneurs bring new ideas to market.[2]

The Small Business Innovation Research (SBIR) program is one of the largest examples of U.S. public-private partnerships. An underlying thesis of the program is that small businesses can be a strong area for new ideas, but that they likely will need some support in their early stages, thus the desirability for public-private partnerships in the small business, high-technology arena. Founded in 1982, SBIR was designed to encourage small business to develop new processes and products and to provide quality research in support of the many missions of the U.S. government. By including qualified small businesses in the nation's R&D effort, SBIR awards are intended to stimulate innovative new technologies

---

[1]See Lewis M. Branscomb, Kenneth P. Morse, Michael J. Roberts, Darin Boville, *Managing Technical Risk: Understanding Private Sector Decision Making on Early Stage Technology Based Projects*, Gaithersburg, MD: National Institute of Standards and Technology, 2000.

[2]For a summary analysis of best practice among U.S. public-private partnerships, see National Research Council, *Government-Industry Partnerships for the Development of New Technologies: Summary Report*, Charles W. Wessner, ed., Washington, DC: The National Academies Press, 2002.

to help agencies meet the specific research and development needs of the nation in many areas, including health, the environment, and national defense.

As the SBIR program approached its twentieth year of operation, the U.S. Congress asked the National Research Council to conduct a "comprehensive study of how the SBIR program has stimulated technological innovation and used small businesses to meet federal research and development needs" and make recommendations on still further improvements to the program.[3] To guide this study, the National Research Council drew together an expert committee that includes eminent economists, small businessmen and women, and venture capitalists, led by Dr. Jacques Gansler of the University of Maryland (formerly Undersecretary of Defense for Acquisition and Technology.) The membership of this committee is listed in the front matter of this volume. Given the extent of "green-field research" required for this study, the Steering Committee in turn drew on a distinguished team of researchers to—among other tasks—administer surveys and case studies, and to develop statistical information about the program. The membership of this research team is also listed in the front matter to this volume.

This report is one of a series published by the National Academies in response to the Congressional request. The series includes reports on the Small Business Innovation Research Program at the Department of Defense, the Department of Energy, the National Aeronautics and Space Administration, the National Institutes of Health, and the National Science Foundation—the five agencies responsible for 96 percent of the program's operations. It includes, as well, an Overview Report that provides assessment of the program's operations across the federal government. Other reports in the series include a summary of the 2002 conference that launched the study, and a summary of the 2005 conference on *SBIR and the Phase III Challenge of Commercialization* that focused on the Department of Defense and NASA.

PROJECT ANTECEDENTS

The current assessment of the SBIR program follows directly from an earlier analysis of public-private partnerships by the National Research Council's Board on Science, Technology, and Economic Policy (STEP). Under the direction of Gordon Moore, Chairman Emeritus of Intel, the NRC Committee on Government Industry Partnerships prepared eleven volumes reviewing the drivers of cooperation among industry, universities, and government; operational assessments of current programs; emerging needs at the intersection of biotechnology and information technology; the current experience of foreign government partnerships and opportunities for international cooperation; and the changing roles of

---

[3]See SBIR Reauthorization Act of 2000 (H.R. 5667-Section 108).

government laboratories, universities, and other research organizations in the national innovation system.[4]

This analysis of public-private partnerships included two published studies of the SBIR program. Drawing from expert knowledge at a 1998 workshop held at the National Academy of Sciences, the first report, *The Small Business Innovation Research Program: Challenges and Opportunities*, examined the origins of the program and identified some operational challenges critical to the program's future effectiveness.[5] The report also highlighted the relative paucity of research on this program.

Following this initial report, the Department of Defense asked the NRC to assess the Department's Fast Track Initiative in comparison with the operation of its regular SBIR program. The resulting report, *The Small Business Innovation Research Program: An Assessment of the Department of Defense Fast Track Initiative*, was the first comprehensive, external assessment of the Department of Defense's program. The study, which involved substantial case study and survey research, found that the SBIR program was achieving its legislated goals. It also found that DoD's Fast Track Initiative was achieving its objective of greater commercialization and recommended that the program be continued and expanded where appropriate.[6] The report also recommended that the SBIR program overall would benefit from further research and analysis, a perspective adopted by the U.S. Congress.

SBIR REAUTHORIZATION AND CONGRESSIONAL REQUEST FOR REVIEW

As a part of the 2000 reauthorization of the SBIR program, Congress called for a review of the SBIR programs of the agencies that account collectively for 96 percent of program funding. As noted, the five agencies meeting this criterion, by size of program, are the Departments of Defense, the National Institutes of Health, the National Aeronautics and Space Administration, the Department of Energy, and the National Science Foundation.

HR 5667 directed the NRC to evaluate the quality of SBIR research and evaluate the SBIR program's value to the agency mission. It called for an assessment of the extent to which SBIR projects achieve some measure of com-

---

[4]For a summary of the topics covered and main lessons learned from this extensive study, see National Research Council, *Government-Industry Partnerships for the Development of New Technologies: Summary Report*, op. cit.

[5]See National Research Council, *The Small Business Innovation Research Program: Challenges and Opportunities*, Charles W. Wessner, ed., Washington, DC: National Academy Press, 1999.

[6]See National Research Council, *The Small Business Innovation Research Program: An Assessment of the Department of Defense Fast Track Initiative*, Charles W. Wessner, ed., Washington, DC: National Academy Press, 2000. Given that virtually no published analytical literature existed on SBIR, this Fast Track study pioneered research in this area, developing extensive case studies and newly developed surveys.

mercialization, as well as an evaluation of the program's overall economic and noneconomic benefits. It also called for additional analysis as required to support specific recommendations on areas such as measuring outcomes for agency strategy and performance, increasing federal procurement of technologies produced by small business, and overall improvements to the SBIR program.

ACKNOWLEDGMENTS

On behalf of the National Academies, we express our appreciation and recognition for the insights, experiences, and perspectives made available by the participants of the conferences and meetings, as well as survey respondents and case study interviewees who participated over the course of this study. We are also very much in debt to officials from the leading departments and agencies. Among the many who provided assistance to this complex study, we are especially in debt to Kesh Narayanan, Joseph Hennessey, and Ritchie Coryell of the National Science Foundation, Michael Caccuitto of the Department of Defense, Robert Berger and later Larry James of the Department of Energy, Carl Ray and Paul Mexcur of NASA, and—particularly relevant for this volume—Jo Anne Goodnight and Kathleen Shino of the National Institutes of Health.

The Committee's Research Team deserves major recognition for their instrumental role in the Research Team's preparation of this report. Special thanks are due to Robin Gaster who led the NIH assessment of the program for the Committee and made many valuable analytical contributions to the Committee's deliberations. Thanks are also due to Paula Stephan of Georgia State University and to Andrew Toole of Rutgers University, who conducted numerous case studies and were active and effective contributors to the Committee's analysis. Without their collective efforts, amidst many other competing priorities, it would not have been possible to prepare this report.

NATIONAL RESEARCH COUNCIL REVIEW

This report has been reviewed in draft form by individuals chosen for their diverse perspectives and technical expertise, in accordance with procedures approved by the National Academies' Report Review Committee. The purpose of this independent review is to provide candid and critical comments that will assist the institution in making its published report as sound as possible and to ensure that the report meets institutional standards for objectivity, evidence, and responsiveness to the study charge. The review comments and draft manuscript remain confidential to protect the integrity of the process.

We wish to thank the following individuals for their review of this report: Richard Bendis, Innovation Philadelphia; William Bonvillian, Massachusetts Institute of Technology; Michael Gallaher, Research Triangle Institute; Marsha

Schachtel, Johns Hopkins University; and Michael Squillante, Radiation Measurement Devices, Inc.

Although the reviewers listed above have provided many constructive comments and suggestions, they were not asked to endorse the conclusions or recommendations, nor did they see the final draft of the report before its release. The review of this report was overseen by Robert Frosch, Harvard University, and Robert White, Carnegie Mellon University. Appointed by the National Academies, they were responsible for making certain that an independent examination of this report was carried out in accordance with institutional procedures and that all review comments were carefully considered. Responsibility for the final content of this report rests entirely with the authoring committee and the institution.

Jacques S. Gansler   Charles W. Wessner

Summary

I. INTRODUCTION

The Small Business Innovation Research (SBIR) program was created in 1982 through the Small Business Innovation Development Act. As the SBIR program approached its twentieth year of operation, the U.S. Congress requested the National Research Council (NRC) of the National Academies to "conduct a comprehensive study of how the SBIR program has stimulated technological innovation and used small businesses to meet Federal research and development needs" and to make recommendations with respect to the SBIR program. Mandated as a part of SBIR's reauthorization in late 2000, the NRC study has assessed the SBIR program as administered at the five federal agencies that together make up some 96 percent of SBIR program expenditures. The agencies, in order of program size are the Department of Defense, the National Institutes of Health, the National Aeronautics and Space Administration, the Department of Energy, and the National Science Foundation.

Based on that legislation, and after extensive consultations with both Congress and agency officials, the NRC focused its study on two overarching questions.[1] First, how well do the agency SBIR programs meet four societal objectives

---

[1]Three primary documents condition and define the objectives for this study: These are the Legislation—H.R. 5667, the NAS-Agencies *Memorandum of Understanding*, and the NAS contracts accepted by the five agencies. These are reflected the Statement of Task addressed to the Committee by the Academies leadership. Based on these three documents, the NRC Committee developed a comprehensive and agreed set of practical objectives to be reviewed. These are outlined in the Committee's formal Methodology Report, section on "Clarifying Study Objectives." National Research Council, *An Assessment of the Small Business Innovation Research Program—Project Methodology*, Washington, DC: The National Academies Press, 2004, accessed at: *<http://www7.nationalacademies.org/sbir/SBIR_Methodology_Report.pdf>*.

of interest to Congress: (1) to stimulate technological innovation; (2) to increase private sector commercialization of innovations (3) to use small business to meet federal research and development needs; and (4) to foster and encourage participation by minority and disadvantaged persons in technological innovation.[2] Second, can the management of agency SBIR programs be made more effective? Are there best practices in agency SBIR programs that may be extended to other agencies' SBIR programs?

To satisfy the congressional request for an external assessment of the program, the NRC conducted empirical analyses of the operations of SBIR based on commissioned surveys and case studies. Agency-compiled program data, program documents, and the existing literature were reviewed. In addition, extensive interviews and discussions were conducted with program managers, program participants, agency 'users' of the program, as well as program stakeholders.

The study as a whole sought to answer questions of program operation and effectiveness, including the quality of the research projects being conducted under the SBIR program, the commercialization of the research, and the program's contribution to accomplishing agency missions. To the extent possible, the evaluation included estimates of the benefits (both economic and noneconomic) achieved by the SBIR program, as well as broader policy issues associated with public-private collaborations for technology development and government support for high technology innovation.

Taken together, this study is the most comprehensive assessment of SBIR to date. Its empirical, multifaceted approach to evaluation sheds new light on the operation of the SBIR program in the challenging area of early stage finance. As with any assessment, particularly one across five quite different agencies and departments, there are methodological challenges. These are identified and discussed at several points in the text. This important caveat notwithstanding, the scope and diversity of the report's research should contribute significantly to the understanding of the SBIR program's multiple objectives, measurement issues, operational challenges, and achievements. This volume presents the Committee's assessment of the SBIR program at the National Institutes of Health.

---

[2]These congressional objectives are found in the Small Business Innovation Development Act (PL 97-219). In reauthorizing the program in 1992, (PL 102-564) Congress expanded the purposes to "emphasize the program's goal of increasing private sector commercialization developed through Federal research and development and to improve the Federal government's dissemination of information concerning small business innovation, particularly with regard to woman-owned business concerns and by socially and economically disadvantaged small business concerns."

BOX S-1
Special Features of the NIH SBIR Program

A Major Grant-based Program

The NIH SBIR Program is the second largest program after the Department of Defense. In 2005, the program expended approximately $562 million. NIH employs grants for almost all its SBIR awards, unlike the Department of Defense, which relies on contracts.

Highly Decentralized Organization

SBIR operates in 23 different Institutes and Centers (ICs) at the National Institutes of Health. Each is an independent, grant-making authority, with coordination provided by a small central office in the Office of the NIH Director. The program is called on to meet a wide variety of needs ranging from early stage support for drug development to medical diagnostics and devices to health related instructional material.

A Core Research Mission and SBIR

The NIH has a different mission and structure than other agencies with large research budgets. The NIH is focused on the pursuit of fundamental knowledge to extend healthy life and reduce the burdens of illness for the nation's citizens. Most NIH programs generally do not seek to develop products and services for the marketplace. The SBIR program does.

SBIR at NIH

As a government grant program intended to support science and the commercialization of biomedical applications for the public good, NIH SBIR does not focus on a return on investment the way a private sector investor would. Unlike most commercial, venture, or angel investors, NIH SBIR funds research projects, not companies as a whole.

Procurement Is Not a Goal

The NIH SBIR program differs fundamentally from those at DoD and NASA, where the primary objective of the program is to develop technologies for use by the agency, via the procurement process. At NIH, the vast majority of projects have no proposed utilization within the agency. As a result, definitions of and metrics for "commercialization" and "agency mission" are quite different, reflecting these different missions.

II. SUMMARY OF KEY PROGRAM FINDINGS

- **The SBIR program at the National Institutes of Health is meeting most of the four legislative objectives of the program. These are to:**

  1. Stimulate technological innovation;

  2. Use small business to meet federal research and development needs;

  3. Foster and encourage participation by minority and disadvantaged persons in technological innovation; and

  4. Increase private sector commercialization of federal R&D.

  In doing so, the NIH SBIR program is:

- **Expanding knowledge.**

  ○ The NIH SBIR program is contributing to the nation's stock of knowledge and supporting products that contribute to the nation's health.[3]

- **Supporting the NIH Mission.**

  ○ NIH's SBIR activities are aligned with the agency's mission, which "is science in pursuit of fundamental knowledge about the nature and behavior of living systems and the application of that knowledge to extend healthy life and reduce the burdens of illness and disability." The SBIR program funds projects that are aligned with this mission.[4]

- **Supporting small business.** The SBIR program at NIH supports a diverse array of small businesses, which in turn contribute to achieving the NIH mission.[5]

  ○ SBIR-funded research projects appear to help small businesses develop new technologies, processes, and products that support the NIH mission of improving the nation's health.

  ○ Awards to woman-owned businesses have increased, and their share of all awards is trending upward. However, the declining trend in the percentage of Phase I and Phase II awards made to minority-owned firms is a matter of concern. In FY2006, these firms accounted for 5.6 percent of Phase I awards and 3.3 percent of Phase II awards.[6] Data collection on these groups (described below) has been problematic.[7]

---

[3]See Finding H in Chapter 2.
[4]See Finding C in Chapter 2.
[5]See Finding D in Chapter 2.
[6]See Finding E in Chapter 2. See also Figures 4-18 and 2-1.
[7]Note: Following discussions with the NRC staff, the NIH made an effort to recalculate the data for woman and minority owners' participation in the SBIR program. In September 2007, the NIH provided corrected data, which is shown in Appendix A and in several figures in this report. However,

- **Achieving Significant Commercialization.**

    o A variety of metrics shows that a meaningful percentage of NIH SBIR projects enter the commercial market.

    o NRC Phase II Survey data shows that 40 percent of SBIR-funded projects reach the marketplace. This is an impressive figure for such early-stage research. Data from NIH indicates that this figure is likely to rise significantly over time.[8]

    o A smaller number (3-4 percent) of projects each generate more than $5 million in revenues, a skew not atypical of early-stage technology funding.[9]

    o To facilitate commercialization further, NIH has undertaken a series of initiatives to help awardees develop effective commercialization plans.[10]

- **Attracting third-party funding.** SBIR awards help small companies to create products and the expertise needed to attract third-party funding. This additional funding is derived from a variety of sources, including

    o **Angel and venture funding.** SBIR awardees at NIH have attracted the interest of private equity investors. Initial NRC research suggests that some 50 of the 200 NIH SBIR awardees with the highest number of awards have received venture funding totaling more than $1.5 billion.[11]

    o **Acquisition.** In some cases, the technology developed through an SBIR award demonstrated sufficient commercial potential to attract investors in-

---

apparent anomalies in the NIH data on the participation of women and minorities in 2001-2002 could not be resolved by the time of publication of this report.

[8]See Figure 4-1.

[9]See Figure 4-2. Of the 496 projects recently surveyed by the NRC Phase II Survey, one firm generated revenues of more than $50 million. This type of "skew"—in which a majority of projects fail or are modestly successfully while a small proportion earns large revenues—is not atypical of early-stage finance and has been noted in previous research. See National Research Council, *The Small Business Innovation Research Program: An Assessment of the Department of Defense Fast Track Initiative*, Charles W. Wessner, ed., Washington, DC: National Academy Press, 2000. See also Joshua Lerner, "'Public Venture Capital': Rationales and Evaluation," in National Research Council, *The Small Business Innovation Research Program: Challenges and Opportunities*, Charles W. Wessner, ed., Washington, DC: National Academy Press, 1999.

[10]To implement the Commercialization Assistance Program, NIH procured the services of Larta Institute, a business-consulting firm located in Los Angeles, CA. To implement the Niche Assessment Program, NIH procured the services of Foresight Science and Technology of Providence, RI. To implement the 2007 Pilot Manufacturing Assistance Program, NIH has procured the services of Dawnbreaker of Rochester, NY. The NIH Commercialization Assistance Plan (CAP) is described in more detail in Chapter 4.

[11]See Figure 4-7. See the discussion of the relationship between SBIR awardees and venture funds in Finding G in Chapter 2. To better understand the ramifications of the ruling, NIH has commissioned additional NRC research to identify the impact of the 2004 SBA ruling excluding majority venture backed firms from the program.

terested in acquisition of the company receiving the award. For example, in 2000, Philips bought out SBIR recipient Optiva for a reported sum of more than $1 billion.[12]

  o **Other private investment.** A significant number of awardees have received additional funds from a wide range of sources, notably angel investors and non-SBIR government support. Fifty-eight percent of the NRC Phase II Survey respondents attracted additional investment, not including additional SBIR awards.[13]

• **Encouraging commercialization.** NIH is encouraging commercialization through the NIH SBIR Technology Assistance Program, utilizing limited program funds to enhance the commercialization efforts of small businesses.[14]

• **Developing an assessment culture.**

  o Following the congressional mandate for this study, the NIH program management launched its first major assessment of the SBIR program at NIH.[15] The results of this analysis proved useful for the NRC review of the program.

  o The commissioning of this research, coupled with the support and close engagement of the program management with the NRC assessment, suggests the growth of a positive assessment culture at NIH with regard to the SBIR program.

III. SUMMARY OF KEY RECOMMENDATIONS

The Committee's recommendations are designed to improve the operation of an already effective SBIR program at NIH.[16]

A. **The NIH should retain its distributed management structure for the program while increasing evaluation efforts, improving data collection, obtaining additional resources, and encouraging upper management attention.**

---

[12]See Box 4-3 in Chapter 4 for a description of the Optiva case. Paradoxically, the acquisition of a firm can sometimes limit reporting of commercialization success. The acquired firm normally does not respond to surveys even if it previously had a positive sales record. Not all nonrespondents, of course, are successful; many have gone out of business, yet acquisition of successful firms does constrain the ability of the survey to capture what are often significant sales.

[13]See Table 4-9. Data reflects information from the NRC Phase II Survey.

[14]See Finding B in Chapter 2.

[15]National Institutes of Health, Office of Extramural Research, "National Survey to Evaluate the NIH SBIR Program: Final Report," July 2003. Available online at: <*http://grants.nih. gov/grants/funding/sbir_report_2003_07.pdf*>.

[16]The recommendations below are drawn from analysis of the data, review of program operations, and discussions with program participants.

1. **Flexibility.** It is most important that the program retain the flexibility and experimentation that have characterized its recent management. The SBIR program is effective across the agencies because a "one-size fits all" approach has not been imposed.[17] This flexible approach may well be extended, subject to careful monitoring, across the Institutes and Centers of the NIH.

2. **Evaluation.** Much greater effort is required to evaluate current outcomes, collect relevant data, and document the impact of changes to the program.[18]

   i. Efforts to identify outcomes should be improved.

   ii. Regular evaluations should be undertaken to enable managers to assess program performance and the results of management initiatives.

3. **Innovation.** Efforts to initiate program innovation by NIH should be substantially strengthened and encouraged. Pilot programs are one mechanism that allow for the efficient implementation and subsequent assessment of new initiatives.[19]

4. **Annual report.** Program accountability should be improved through the development and publication of a much-expanded annual report on the NIH SBIR program, in order to supplement current reporting to the SBA and to provide a more complete picture of the program for the NIH management, Congress, awardees, and applicants.[20]

B. **The NIH SBIR program is focused on commercialization and has seen meaningful achievement. However, the limited number of highly successful commercial projects suggests that continued management attention and additional efforts to facilitate commercialization are needed.[21]**

   1. **Commercialization programs.** NIH should continue to experiment with commercialization programs, monitor their result, and adopt them for general application when they show signs of success.

   2. **Funding for commercialization programs.** Congress should consider expanding funding, if only to account for inflation, and relaxing the current restrictions on spending for this purpose.

C. **The program should be provided with additional management funding to**

---

[17]See Recommendation H in Chapter 2.
[18]See Recommendations A and I in Chapter 2.
[19]See Recommendation I in Chapter 2.
[20]See Recommendation C in Chapter 2.
[21]See Recommendation B in Chapter 2.

develop and maintain a results-oriented program with a focused evaluation culture.[22]

1. Effective oversight relies on appropriate funding. A data-driven program requires high quality data and systematic assessment. As noted above, sufficient resources are not currently available for these functions.

2. Increased funding is needed to provide effective oversight, including site visits, program review, systematic third-party assessments, and other necessary management activities.

3. To enhance program utilization, management, and evaluation, additional funds should be provided. There are three ways that this might be achieved:

    i. Additional funds might be allocated internally, within the existing budgets of the services and agencies, as the Navy has done.

    ii. Funds might be drawn from the existing set-aside for the program to carry out these activities.

    iii. The set-aside for the program, currently at 2.5 percent of external research budgets, might be increased, with the goal of providing additional resources to maximize the program's return to the nation.

4. These recommended improvements should enable the NIH SBIR managers to address the four mandated congressional objectives in a more efficient and effective manner.

D. **Possible areas of improvement and experimentation.** The NRC study identified a number of areas where improvements in the program would make it significantly better. While some of these may require NIH-wide initiatives, others might be addressed initially through carefully designed and evaluated pilot programs. Such a capability would need to be developed, and could also be used to address some recent developments that have already occurred within the program. Key areas for potential improvement include:

1. **Improving selection procedures.** Chapter 5 of this report outlines a number of areas where the selection process could be improved. These include more attention to possible conflicts of interest and addressing difficulties in evaluating commercialization plans.[23]

2. **Speeding cycle time.** Fairly minimal management changes focused on reducing the cycle time for awards could substantially accelerate innovation.[24]

---

[22]See Recommendation J in Chapter 2.
[23]See Section 5.5.
[24]See Recommendation H in Chapter 2.

3. **Developing a rationale for large awards.**[25]

    i. The NIH program has recently experimented with a limited number of substantially larger awards. In itself, this could be a positive step, reflecting the flexibility in experimentation that characterizes an effective SBIR program.

    ii. Assessing the impact of the larger awards is challenging insofar as NIH has not developed a clearly articulated rationale for these awards, and no systematic effort has been made to determine the impact of extra large awards.

    iii. Thus while flexibility remains a laudable characteristic of the program, deviations from established program boundaries should be based on clear rationales and followed by equally clear assessment programs to determine whether such initiatives have been effective. This is especially important in this case because larger award size necessarily implies a smaller number of awards.

4. **Understanding the impact of program change, e.g., the limits on venture funding.** NIH is the agency most affected by the SBA ruling barring firms with 51 percent venture funding (or other nonindividual) ownership from the program. To better understand the ramifications of the ruling for the NIH SBIR Program, the NIH recently commissioned an empirical analysis by the National Academies. Timely assessment of the impact of major changes in the program should be a standard practice.[26]

5. **Improving monitoring of awards to women and minorities.** Program management resources to do not appear sufficient to permit effective monitoring of the program on a consistent basis, nor the development of appropriate databases to underpin this effort. These difficulties have been most apparent in relation to collecting data and monitoring the participation of women and minorities, one of the four primary congressional mandates for the program.[27]

---

[25]See Recommendation I-4 in Chapter 2.
[26]See Finding G in Chapter 2.
[27]See Recommendation D in Chapter 2.

1

Introduction

1.1 SMALL BUSINESS INNOVATION RESEARCH
PROGRAM CREATION AND ASSESSMENT

Created in 1982 by the Small Business Innovation Development Act, the Small Business Innovation Research (SBIR) program was designed to stimulate technological innovation among small private-sector businesses while providing the government cost-effective new technical and scientific solutions to challenging mission problems. SBIR was also designed to help to stimulate the U.S. economy by encouraging small businesses to market innovative technologies in the private sector.[1]

As the SBIR program approached its twentieth year of existence, the U.S. Congress requested that the National Research Council (NRC) of the National Academies conduct a "comprehensive study of how the SBIR program has stimulated technological innovation and used small businesses to meet Federal research and development needs," and make recommendations on improvements to the program.[2] Mandated as a part of SBIR's renewal in 2000, the NRC study has assessed the SBIR program as administered at the five federal agencies that together make up 96 percent of SBIR program expenditures. The agencies are, in

---

[1]The SBIR legislation drew from a growing body of evidence, starting in the late 1970s and accelerating in the 1980s, which indicated that small businesses were assuming an increasingly important role in both innovation and job creation. This evidence gained new credibility with the Phase I empirical analysis by Zoltan Acs and David Audretsch of the U.S. Small Business Innovation Database, which confirmed the increased importance of small firms in generating technological innovations and their growing contribution to the U.S. economy. See Zoltan Acs and David Audretsch, *Innovation and Small Firms*, Cambridge, MA: The MIT Press, 1990.

[2]See Public Law 106-554, Appendix I—H.R. 5667, Section 108.

decreasing order of program size: the Department of Defense (DoD), the National Institutes of Health (NIH), the National Aeronautics and Space Administration (NASA), the Department of Energy (DoE), and the National Science Foundation (NSF).

The NRC Committee assessing the SBIR program was not asked to consider if SBIR should exist or not—Congress has affirmatively decided this question on three occasions.[3] Rather, the Committee was charged with providing assessment-based findings to improve public understanding of the program as well as recommendations to improve the program's effectiveness.

1.2 SBIR PROGRAM STRUCTURE

Eleven federal agencies are currently required to set aside 2.5 percent of their extramural research and development budget exclusively for SBIR awards. Each year these agencies identify various R&D topics, representing scientific and technical problems requiring innovative solutions, for pursuit by small businesses under the SBIR program. These topics are bundled together into individual agency "solicitations"—publicly announced requests for SBIR proposals from interested small businesses. A small business can identify an appropriate topic it wants to pursue from these solicitations and, in response, propose a project for an SBIR award. The required format for submitting a proposal is different for each agency. Proposal selection also varies, though peer review of proposals on a competitive basis by experts in the field is typical. Each agency then selects the proposals that are found best to meet program selection criteria, and awards contracts or grants to the proposing small businesses.

As conceived in the 1982 Act, SBIR's award-making process is structured in three phases at all agencies:

- Phase I awards essentially fund feasibility studies in which award winners undertake a limited amount of research aimed at establishing an idea's scientific and commercial promise. Today, the legislation anticipates Phase I awards as high as $100,000.[4]
- Phase II awards are larger—typically about $750,000—and fund more extensive R&D to further develop the scientific and commercial promise of research ideas.
- Phase III. During this phase, companies do not receive additional funding from the SBIR program. Instead, award recipients should be obtaining additional funds from a procurement program at the agency that made the

---

[3]These are the 1982 Small Business Development Act, and the subsequent multi-year reauthorizations of the SBIR program in 1992 and 2000.

[4]With the agreement of the Small Business Administration, which plays an oversight role for the program, this amount can be substantially higher in certain circumstances, e.g., drug development at NIH, and is often lower with smaller SBIR programs, e.g., EPA or the Department of Agriculture.

award, from private investors, or from the capital markets. The objective of this phase is to move the technology from the prototype stage to the marketplace.

Obtaining Phase III support is often the most difficult challenge for new firms to overcome. In practice, agencies have developed different approaches to facilitate SBIR grantees' transition to commercial viability; not least among them are additional SBIR awards.

Previous NRC research has shown that firms have different objectives in applying to the program. Some want to demonstrate the potential of promising research but may not seek to commercialize it themselves. Others think they can fulfill agency research requirements more cost-effectively through the SBIR program than through the traditional procurement process. Still others seek a certification of quality (and the investments that can come from such recognition) as they push science-based products towards commercialization.[5]

1.3 SBIR REAUTHORIZATIONS

The SBIR program approached reauthorization in 1992 amidst continued concerns about the U.S. economy's capacity to commercialize inventions. Finding that "U.S. technological performance is challenged less in the creation of new technologies than in their commercialization and adoption," the National Academy of Sciences at the time recommended an increase in SBIR funding as a means to improve the economy's ability to adopt and commercialize new technologies.[6]

Following this report, the Small Business Research and Development Enhancement Act (P.L. 102-564), which reauthorized the SBIR program until September 30, 2000, doubled the set-aside rate to 2.5 percent.[7] This increase in the percentage of R&D funds allocated to the program was accompanied by a stronger emphasis on encouraging the commercialization of SBIR-funded tech-

---

[5]See Reid Cramer, "Patterns of Firm Participation in the Small Business Innovation Research Program in Southwestern and Mountain States," in National Research Council, *The Small Business Innovation Research Program: An Assessment of the Department of Defense Fast Track Initiative*, Charles W. Wessner, ed., Washington, DC: National Academy Press, 2000.

[6]See National Research Council, *The Government Role in Civilian Technology: Building a New Alliance*, Washington, DC: National Academy Press, 1992, p. 29.

[7]For FY2003, this has resulted in a program budget of approximately $1.6 billion across all federal agencies, with the Department of Defense having the largest SBIR program at $834 million, followed by the National Institutes of Health at $525 million. The DoD SBIR program, is made up of 10 participating components: Army, Navy, Air Force, Missile Defense Agency (MDA), Defense Advanced Research Projects Agency (DARPA), Chemical Biological Defense (CBD), Special Operations Command (SOCOM), Defense Threat Reduction Agency (DTRA), National Imagery and MapPhasing Agency (NIMA), and the Office of Secretary of Defense (OSD). NIH counts 23 separate institutes and agencies making SBIR awards, many with multiple programs.

nologies.[8] Legislative language explicitly highlighted commercial potential as a criterion for awarding SBIR awards. For Phase I awards, Congress directed program administrators to assess whether projects have "commercial potential," in addition to scientific and technical merit, when evaluating SBIR applications.

The 1992 legislation mandated that program administrators consider the existence of second-phase funding commitments from the private sector or other non-SBIR sources when judging Phase II applications. Evidence of third-phase follow-on commitments, along with other indicators of commercial potential, was also to be sought. Moreover, the 1992 reauthorization directed that a small business' record of commercialization be taken into account when evaluating its Phase II application.[9]

The Small Business Reauthorization Act of 2000 (P.L. 106-554) extended SBIR until September 30, 2008. It called for this assessment by the National Research Council of the broader impacts of the program, including those on employment, health, national security, and national competitiveness.[10]

1.4 STRUCTURE OF THE NRC STUDY

This NRC assessment of SBIR has been conducted in two phases. In the first phase, at the request of the agencies, a research methodology was developed by the NRC. This methodology was then reviewed and approved by an independent National Academies panel of experts.[11] Information about the program was also gathered through interviews with SBIR program administrators and during two major conferences where SBIR officials were invited to describe program

[8]See Robert Archibald and David Finifter, "Evaluation of the Department of Defense Small Business Innovation Research Program and the Fast Track Initiative: A Balanced Approach," in National Research Council, *The Small Business Innovation Research Program: An Assessment of the Department of Defense Fast Track Initiative*, op. cit., pp. 211-250.

[9]A GAO report had found that agencies had not adopted a uniform method for weighing commercial potential in SBIR applications. See U.S. General Accounting Office, *Federal Research: Evaluations of Small Business Innovation Research Can Be Strengthened*, GAO/RCED-99-114, Washington, DC: U.S. General Accounting Office, 1999.

[10]The current assessment is congruent with the Government Performance and Results Act (GPRA) of 1993, accessed at: *<http://govinfo.library.unt.edu/npr/library/misc/s20.html>*. As characterized by the GAO, GPRA seeks to shift the focus of government decisionmaking and accountability away from a preoccupation with the activities that are undertaken—such as grants dispensed or inspections made—to a focus on the results of those activities. See *<http://www.gao.gov/new.items/gpra/gpra.htm>*.

[11]National Research Council, *An Assessment of the Small Business Innovation Research Program: Project Methodology*, Washington, DC: The National Academies Press, 2004. The methodology report is available on the Web. Access at: *<http://www7.nationalacademies.org/sbir/SBIR_Methodology_Report.pdf>*.

operations, challenges, and accomplishments.[12] These conferences highlighted the important differences in each agency's SBIR program's goals, practices, and evaluations. The conferences also explored the challenges of assessing such a diverse range of program objectives and practices using common metrics.

The second phase of the NRC study implemented the approved research methodology. The Committee deployed multiple survey instruments and its researchers conducted case studies of a wide profile of SBIR firms. The Committee then evaluated the results and developed both agency-specific and overall findings and recommendations for improving the effectiveness of the SBIR program. The final report includes complete assessments for each of the five agencies and an overview of the program as a whole.

1.5 SBIR ASSESSMENT CHALLENGES

At its outset, the NRC's SBIR study identified a series of assessment challenges that must be addressed. As discussed at the October 2002 conference that launched the study, the administrative flexibility found in the SBIR program makes it difficult to make cross-agency assessments. Although each agency's SBIR program shares the common three-phase structure, the SBIR concept is interpreted uniquely at each agency. This flexibility is a positive attribute in that it permits each agency to adapt its SBIR program to the agency's particular mission, scale, and working culture. For example, NSF operates its SBIR program differently than DoD because "research" is often coupled with procurement of goods and services at DoD but rarely at NSF. Programmatic diversity means that each agency's SBIR activities must be understood in terms of their separate missions and operating procedures. This commendable diversity makes an assessment of the program as a whole more challenging.

A second challenge concerns the linear process of commercialization implied by the design of SBIR's three-phase structure.[13] In the linear model, illustrated in Figure 1-1, innovation begins with basic research supplying a steady stream of fresh and new ideas. Among these ideas, those that show technical feasibility become innovations. Such innovations, when further developed by firms, become marketable products driving economic growth.

As NSF's Joseph Bordogna observed at the study's initial conference, innovation almost never takes place through a protracted linear progression from

---

[12]The opening conference on October 24, 2002, examined the program's diversity and assessment challenges. For a published report of this conference, see National Research Council, *SBIR: Program Diversity and Assessment Challenges*, Charles W. Wessner, ed. Washington, DC: The National Academies Press, 2004. The second conference, held on March 28, 2003, was titled, "Identifying Best Practice." The conference provided a forum for the SBIR Program Managers from each of the five agencies in the study's purview to describe their administrative innovations and best practices.

[13]This view was echoed by Duncan Moore: "Innovation does not follow a linear model. It stops and starts." National Research Council, *SBIR: Program Diversity and Assessment Challenges*, op. cit.

FIGURE 1-1 The linear model of innovation.

research to development to market. Research and development drives technological innovation, which, in turn, opens up new frontiers in R&D. True innovation, Bordogna noted, can spur the search for new knowledge and create the context in which the next generation of research identifies new frontiers. This nonlinearity, illustrated in Figure 1-2, makes it difficult to rate the efficiency of SBIR program. Inputs do not match up with outputs according to a simple function.

A third assessment challenge relates to the measurement of outputs and outcomes. Program realities can and often do complicate the task of data gathering. In some cases, for example, SBIR recipients receive a Phase I award from one agency and a Phase II award from another. In other cases, multiple SBIR awards may have been used to help a particular technology become sufficiently mature to reach the market. Also complicating matters is the possibility that for any particular grantee, an SBIR award may be only one among other federal and nonfederal sources of funding. Causality can thus be difficult, if not impossible, to establish. The task of measuring outcomes is made harder because companies that have garnered SBIR awards can also merge, fail, or change their name before a product reaches the market. In addition, principal investigators or other key individuals can change firms, carrying their knowledge of an SBIR project with them. A technology developed using SBIR funds may eventually achieve

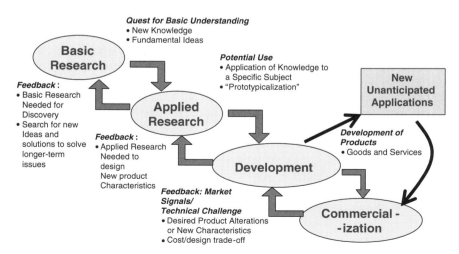

FIGURE 1-2 A feedback model of innovation.

commercial success at an entirely different company than that which received the initial SBIR award.

Complications plague even the apparently straightforward task of assessing commercial success. For example, research enabled by a particular SBIR award may take on commercial relevance in new unanticipated contexts. At the launch conference, Duncan Moore, former Associate Director of Technology at the White House Office of Science and Technology Policy (OSTP), cited the case of SBIR-funded research in gradient index optics that was initially considered a commercial failure when an anticipated market for its application did not emerge. Years later, however, products derived from the research turned out to be a major commercial success.[14] Today's apparent dead end can be a lead to a major achievement tomorrow. Lacking clairvoyance, analysts cannot anticipate or measure such potential SBIR benefits.

Gauging commercialization is also difficult when the product in question is destined for public procurement. The challenge is to develop a satisfactory measure of how useful an SBIR-funded innovation has been to an agency mission. A related challenge is determining how central (or even useful) SBIR awards have proved in developing a particular technology or product. In some cases, the Phase I award can meet the agency's need—completing the research with no further action required. In other cases, surrogate measures are often required. For example, one way of measuring commercialization success is to count the products developed using SBIR funds that are procured by an agency such as DoD. In practice, however, large procurements from major suppliers are typically easier to track than products from small suppliers such as SBIR firms. Moreover, successful development of a technology or product does not always translate into successful "uptake" by the procuring agency. Often, the absence of procurement may have little to do with the product's quality or the potential contribution of SBIR.

Understanding failure is equally challenging. By its very nature, an early-stage program such as SBIR should anticipate a high failure rate. The causes of failure are many. The most straightforward, of course, is *technical failure*, where the research objectives of the award are not achieved. In some cases, the project can be technically successful but a commercial failure. This can occur when a procuring agency changes its mission objectives and hence its procurement priorities. NASA's new Mars Mission is one example of a *mission shift* that may result in the cancellation of programs involving SBIR awards to make room for new agency priorities. Cancelled weapons system programs at the Department of Defense can have similar effects. Technologies procured through SBIR may also *fail in the transition to acquisition*. Some technology developments by small businesses do not survive the long lead times created by complex testing and

---

[14]Duncan Moore, "Turning Failure into Success," in National Research Council, *SBIR: Program Diversity and Assessment Challenges*, op. cit., p. 94.

certification procedures required by the Department of Defense. Indeed, small firms encounter considerable difficulty in penetrating the "procurement thicket" that characterizes defense acquisition.[15] In addition to complex federal acquisition procedures, there are strong disincentives for high-profile projects to adopt untried technologies. Technology transfer in commercial markets can be equally difficult. A *failure to transfer to commercial markets* can occur even when a technology is technically successful if the market is smaller than anticipated, competing technologies emerge or are more competitive than expected, or the product is not adequately marketed. Understanding and accepting the varied sources of project failure in the high-risk, high-reward environment of cutting-edge R&D is a challenge for analysts and policy makers alike.

This raises the issue concerning the standard on which SBIR programs should be evaluated. An assessment of SBIR must take into account the expected distribution of successes and failures in early-stage finance. As a point of comparison, Gail Cassell, Vice President for Scientific Affairs at Eli Lilly, has noted that only 1 in 10 innovative products in the biotechnology industry will turn out to be a commercial success.[16] Similarly, venture capital funds often achieve considerable commercial success on only two or three out of twenty or more investments.[17]

In setting metrics for SBIR projects, therefore, it is important to have a realistic expectation of the success rate for competitive awards to small firms investing in promising but unproven technologies. Similarly, it is important to have some understanding of what can be reasonably expected—that is, what constitutes "success" for an SBIR award, and some understanding of the constraints and opportunities successful SBIR awardees face in bringing new products to market.

---

[15]For a description of the challenges small businesses face in defense procurement, the subject of a June 14, 2005, NRC conference and one element of the congressionally requested assessment of SBIR, see National Research Council, *SBIR and the Phase III Challenge of Commercialization*, Charles W. Wessner, ed., Washington, DC: The National Academies Press, 2007. Relatedly, see remarks by Kenneth Flamm on procurement barriers, including contracting overhead and small firm disadvantages in lobbying in National Research Council, *SBIR: Program Diversity and Assessment Challenges*, op. cit., pp. 63-67.

[16]Gail Cassell, "Setting Realistic Expectations for Success," in National Research Council, *SBIR: Program Diversity and Assessment Challenges*, op. cit., p. 86.

[17]See John H. Cochrane, "The Risk and Return of Venture Capital," *Journal of Financial Economics*, 75(1) 2005:3-52. Drawing on the VentureOne database Cochrane plots a histogram of net venture capital returns on investments that "shows an extraordinary skewness of returns. Most returns are modest, but there is a long right tail of extraordinary good returns. 15% of the firms that go public or are acquired give a return greater than 1,000%! It is also interesting how many modest returns there are. About 15% of returns are less than 0, and 35% are less than 100%. An IPO or acquisition is not a guarantee of a huge return. In fact, the modal or 'most probable' outcome is about a 25% return." See also Paul A. Gompers and Josh Lerner, "Risk and Reward in Private Equity Investments: The Challenge of Performance Assessment," *Journal of Private Equity*, 1 (Winter 1977):5-12. Steven D. Carden and Olive Darragh, "A Halo for Angel Investors" *The McKinsey Quarterly*, 1, 2004 also show a similar skew in the distribution of returns for venture capital portfolios.

From the management perspective, the rate of success also raises the question of appropriate expectations and desired levels of risktaking. A portfolio that always succeeds would not be investing in high-risk, high pay-off projects that push the technology envelope. A very high rate of "success" would, thus, paradoxically suggest an inappropriate use of the program. Understanding the nature of success and the appropriate benchmarks for a program with this focus is therefore important to understanding the SBIR program and the approach of this study.

1.6 STRUCTURE OF THIS REPORT

This report sets out the Committee's assessment of the SBIR program at the National Institutes of Health. The Committee's detailed findings and recommendations are presented in the next chapter. The Committee finds that the NIH SBIR program largely meets it legislative objectives and makes recommendations to improve program outcomes. Chapter 3 reviews awards made by NIH. It analyzes data supplied by NIH, reflecting on both the advantages and disadvantages of NIH data gathering methods. Chapter 4 looks at the outcomes of the NIH SBIR program, including commercial sales and employment effects. Chapter 5 examines how the SBIR program at NIH is managed, including an explanation of the NIH award cycle, outreach efforts to attract the best applicants, and initiatives to support the commercialization of SBIR-funded technologies. Appendix A presents program data collected by NIH, DoD, and the NRC. Appendix B and C provide the template and results of the NRC Firm Survey and surveys of SBIR Phase I and Phase II projects. Appendix D presents illustrative case studies of firms participating in the NIH SBIR program. Finally, Appendix E provides a reference bibliography.

2

Findings and Recommendations

I. NRC STUDY FINDINGS

A. The NIH SBIR program is making significant progress in achieving the congressional goals for the program. The SBIR program is sound in concept and effective in practice at NIH. With the programmatic changes recommended here, the SBIR program should be even more effective in achieving its legislative goals.[1]

1. Overall, the program has made significant progress in achieving its congressional objectives by:

- Stimulating technological innovation;

- Using small business to meet federal research and development needs;

- Fostering and encourage participation by minority and disadvantaged persons in technological innovation; and

- Increasing private sector commercialization of innovations derived from federal research and development.

B. The NIH SBIR program is focused on commercialization and has seen

---

[1]Small Business Innovation Development Act (PL 97-219). In reauthorizing the program in 1992, (PL 102-564) Congress expanded the purposes to "emphasize the program's goal of increasing private sector commercialization developed through Federal research and development and to improve the Federal government's dissemination of information concerning small business innovation, particularly with regard to woman-owned business concerns and by socially and economically disadvantaged small business concerns."

meaningful achievement. There are, nonetheless, opportunities for improvement in commercialization.

1. **A significant percentage of SBIR projects are commercialized to some degree.**

   i. **Reaching the market.** NRC Phase II Survey data suggest that 40 percent[2] of SBIR-funded projects reach the marketplace.[3] Over time, NIH data suggests that this figure will rise significantly; subsequent assessment is required to capture this trend.

   ii. **Revenue skew.** The survey data also show that a much smaller number (7.9 percent of NRC Phase II Survey respondents) of projects generate more than $5 million in revenues.[4] This type of "skew" or concentration—in which a majority of projects are at least modestly successful while a small proportion earns large revenues—is typical of early-stage finance.[5]

   iii. **Licensing revenue.** In some cases, substantial licensing revenues have been generated on the basis of SBIR-funded projects.[6]

   iv. **Additional private investment.** Some companies have received substantial additional investment from the private sector, or have been

---

[2]Forty point seven percent of NRC Phase II Survey respondents reported sales. The NIH Survey found that 30.3 percent of the projects surveyed reached the marketplace. National Institutes of Health, *National Survey to Evaluate the NIH SBIR Program: Final Report*, July 2003.

[3]See Figure 4-1.

[4]See Figure 4-2. One of the 496 projects recently surveyed by the NRC generated revenues of more than $50 million. Case studies identified other projects not included in the survey with similar results (e.g., Optiva, Martek).

[5]As with investments by angel investors or venture capitalists, SBIR awards result in highly concentrated sales, with a few awards accounting for a very large share of the overall sales generated by the program. These are appropriate referent groups, though not an appropriate group for direct comparison, not least because SBIR awards often occur earlier in the technology development cycle than where venture funds normally invest. Nonetheless, returns on venture funding tend to show the same high skew that characterizes commercial returns on the SBIR awards. See John H. Cochrane, "The Risk and Return of Venture Capital," *Journal of Financial Economics*, 75(1):3-52, 2005. Drawing on the VentureOne database Cochrane plots a histogram of net venture capital returns on investments that "shows an extraordinary skewness of returns. Most returns are modest, but there is a long right tail of extraordinary good returns. 15 percent of the firms that go public or are acquired give a return greater than 1,000 percent! It is also interesting how many modest returns there are. About 15 percent of returns are less than 0, and 35 percent are less than 100 percent. An IPO or acquisition is not a guarantee of a huge return. In fact, the modal or 'most probable' outcome is about a 25 percent return." See also Paul A. Gompers and Josh Lerner, "Risk and Reward in Private Equity Investments: The Challenge of Performance Assessment," *Journal of Private Equity*, 1(Winter 1977):5-12. Steven D. Carden and Olive Darragh, "A Halo for Angel Investors," *The McKinsey Quarterly*, 1, 2004 also show a similar skew in the distribution of returns for venture capital portfolios.

[6]See Table 4-7.

bought by other companies, both of which indicate that the company has developed something of value.[7]

2. NIH has increased the significance of the commercialization component of applications over time.

   i. More efforts are now made to ensure that commercialization criteria are applied during Phase II selection.

   ii. NIH has developed several programs under the Technology Assistance Program aimed at helping awardees develop and implement effective commercialization plans. Outside contractors have been hired to implement these programs.[8]

   iii. However, because the focus on commercialization and the deployment of assistance programs are recent, the impact of these efforts on commercialization is not yet clear, although initial results are encouraging, as participant firms have attracted $68 million in third party funding.[9]

3. SBIR-funded research projects enable small businesses to attract third-party interest.

   i. Venture funding. Third parties that identify substantial value in SBIR projects sometimes provided additional funding for the grantee company. At least 50 of the 200 most frequent winners of NIH SBIR awards have received venture funding, and those investments totaled more than $1.5 billion (1992-2005).[10]

   ii. Acquisition. In other cases, the technology developed had sufficient commercial potential that investors bought the grantee company outright. For example, in 2000, Philips bought out SBIR recipient Optiva for a reported sum of more than $1 billion.[11]

   iii. Multiple other sources. Many grantees have found additional funds from a wide range of sources, including angel funding. Fifty-eight percent of NRC Phase II Survey respondents attracted some additional investment (excluding further SBIR awards).[12]

---

[7]See Table 4-11.

[8]See Section 5.8.5.2—Commercialization Assistance Program.

[9]See Section 5.8.5.2.

[10]See Figure 4-7. Other analyses have put the number much higher. See U.S. General Accountability Office, *Small Business Innovation Research: Information on Awards Made by NIH and DoD in Fiscal Years 2002 through 2004*, GAO 06-565, Washington, DC: U.S. Government Accountability Office, 2006.

[11]See Box 4-3 in Chapter 4.

[12]See Table 4-9.

C. The NIH SBIR program is operated in alignment with the agency's mission[13]: awards are made for research that supports improved health within the United States.

1. SBIR funds projects that have a positive impact on public health.

    i. Effective mission alignment. All NIH awards appear to be selected primarily on the basis of their potential to advance knowledge and provide solutions in the field of health care and biomedicine. There is no evidence that NIH awards are made in fields outside those linked to the agency's mission.

    ii. Positive impact on healthcare. SBIR awards have had a substantial impact on many aspects of health care. For example, SBIR awards played an important role in the development of a retractable non-stick needle that makes immunization safer, labor saving advances in the monitoring of epileptics, communication technologies for the disabled, disease specific tests, and improved infant formulas that are sold worldwide. SBIR awards have also helped develop tools that are used by researchers such as an SNP genotyping system, educational CDs and videos, as well as devices with large impacts on small populations—such as the SBIR-supported heart stent—SBIR awards have also helped develop devices with smaller impacts on very large populations, such as the Sonicare electric toothbrush, along with many other improvements in medical technology and practice.

    The impact of an SBIR project on public health is carefully considered during the selection process. Grantees and NIH staff note that impact effects are an important component in every application. In all the cases examined, NIH SBIR funded projects related to public health and biomedical science and technology.

D. The SBIR program at NIH has provided significant support for small business, frequently acting as the impetus for projects and firm creation.

The NRC Phase II Survey and NRC Firm Survey show that the SBIR program has provided substantial benefits for participating small businesses in a number of different ways. Responses indicate that these benefits include:

1. Company creation. Just over 25 percent of companies indicated that they were founded entirely or partly because of an SBIR award;[14]

---

[13]NIH's mission "is science in pursuit of fundamental knowledge about the nature and behavior of living systems and the application of that knowledge to extend healthy life and reduce the burdens of illness and disability." Access at <*http://www.nih.gov/about/*>.

[14]See Table 4-20.

2. **The project initiation decision.** More than 50 percent of SBIR-funded projects reportedly would not have taken place without SBIR funding;

3. **Alternative path development.** Companies often use SBIR to fund alternative development strategies, exploring technological options in parallel with other activities;

4. **Partnering and networking.** SBIR funding pays for outside resources, especially academic consultants and partners, thereby contributing to networking effects and facilitating the transfer of university knowledge to the private sector;

5. **Commercializing academic research.** The partnering between academic institutions and private firms (noted above) and the role of academics in founding firms contribute to the commercialization of university research.[15]

E. **Support for minority- and woman-owned firms.** Data from NIH raise concerns about the shares of awards being made to woman- and minority-owned firms.

1. Awards, applications, and success rates have all declined for minorities, for both Phase I and Phase II (see Figures 2-1 and 2-2), while awards for woman-owned firms have not kept pace with the growth in female Ph.D. recipients in the life scientists.

2. Further research is required to determine whether the pool of potential applicants is not growing fast enough to keep pace with expanded SBIR funding, or whether there are other explanations for these trends.

3. From 2003-2006, average Phase II success rates (awards as a percentage of applications) for minority-owned businesses are almost 10 percentage points lower than those of firms that are neither woman- or minority-owned.

F. **NIH SBIR awards are open to new entrants.**

1. **High proportion of new entrants.** The Phase I share of previous non-winners is quite large, ranging between just under 50 percent in 2000 and just above 35 percent in 2005.[16] As the number of successful participants in the program rises, the proportion of new entrants may be diminishing. Still, the awards are widely distributed, with more than 1,300 companies receiving at least one Phase II award from 1992 to 2002.

2. **Few frequent award winners.** Another measure of openness is the rela-

---

[15]See Table 4-21.
[16]See Figure 3-5.

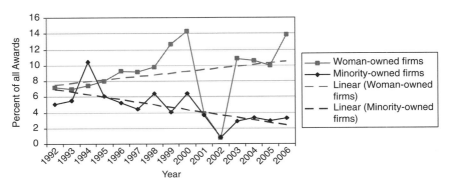

FIGURE 2-1 Share of Phase II awards to woman- and minority-owned firms, 1992-2006.

NOTE: Following discussions with the NRC staff, the NIH made an effort to recalculate the data for woman and minority owners' participation in the SBIR program. In September 2007, the NIH provided corrected data, which is shown in Appendix A and in several figures in this report. However, apparent anomalies in the NIH data on the participation of women and minorities in 2001-2002 could not be resolved by the time of publication of this report.

SOURCE: National Institutes of Health.

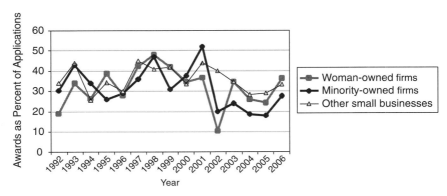

FIGURE 2-2 Success rates for Phase II applications and awards to woman- and minority-owned firms, 1992-2006.

NOTE: Following discussions with the NRC staff, the NIH made an effort to recalculate the data for woman and minority owners' participation in the SBIR program. In September 2007, the NIH provided corrected data, which is shown in Appendix A and in several figures in this report. However, apparent anomalies in the NIH data on the participation of women and minorities in 2001-2002 could not be resolved by the time of publication of this report.

SOURCE: National Institutes of Health.

tively low number of frequent award winners at NIH. Only five companies have been identified as receiving more than 20 Phase II awards between FY1992 and FY2005, and only three received 30 or more, with the maximum being 34.[17,18,19]

3. **Improving access.** The SBIR program at NIH has also made efforts to improve access to the program for researchers outside the "high-award" states. The number of states receiving one or zero Phase II awards declined from 28 in 1995 to 16 in 2003. Similarly, the percentage of Phase II awards going to California fell from 22.8 percent to 13.6 percent in that time period (though the actual number of awards increased in light of the substantial increase in NIH funding during the period).

G. Venture funding and SBIR.

1. **Synergies.** There can often be useful synergies between angel and venture capital investments and SBIR funding; each of these funding sources tends to select highly promising companies.

   i. **Angel investment.** Angel investors often find SBIR awards to be an effective mechanism to bring a company forward in its development to the point where risk is sufficiently diminished to justify investment.[20]

   ii. **Venture investment.** Reflecting this synergy, initial NRC review indicates about 25 percent of the top 200 NIH Phase II award winners

---

[17]See Table 3-6.

[18]NIH has declined to provide company identification data on privacy grounds, so multiple winners are calculated by matching company names. This approach may understate the full distribution of multiple-award winners, even though additional cross-checks of the data were made to reduce the impact of these inaccuracies. The accuracy of these data could be improved by using EINs if they became available.

[19]The top 20 percent of winning companies together received 11.1 percent of awards. This is significantly lower than the Department of Defense.

[20]See Figure 4-7. See the presentation "The Private Equity Continuum" by Steve Weiss, Executive Committee Chair of Coachella Valley Angel Network, at the Executive Seminar on Angel Funding, University of California at Riverside, December 8-9, 2006, Palm Springs, CA. In a personal communication, Weiss points out the critical contributions of SBIR to the development of companies such as CardioPulmonics. The initial Phase I and II SBIR grants allowed the company to demonstrate the potential of their products in animal models of an intravascular oxygenator to treat acute lung infections and thus attract angel investment and subsequently venture funding. Weiss cites this case as an example of how the public and private sectors can collaborate in bringing new technology to markets. Steve Weiss, Personal Communication, December 12, 2006.

(1992-2005) have acquired some venture funding in addition to the SBIR awards.[21]

2. **Program change.** During the first two decades of the program, some venture-backed companies participated in the program, receiving SBIR awards in conjunction with outside equity investments. During this lengthy period, the participation of venture funded firms was not an issue.

In a 2002 directive, the Small Business Administration said that to be eligible for SBIR the small business concern should be "at least 51 percent owned and controlled by one or more individuals who are citizens of, or permanent resident aliens in, the United States, except in the case of a joint venture, where each entity to the venture must be 51 percent owned and controlled by one or more individuals who are citizens of, or permanent resident aliens in, the United States."[22] The effect of this directive has been to exclude companies in which VC firms have a controlling interest.[23]

i. It is important to keep in mind that the innovation process often does not follow a crisp, linear path. Venture capital funds normally (but not always) seek to invest when a firm is sufficiently developed in terms of products to offer an attractive risk-reward ratio.[24] Yet even firms benefiting from venture funding may well seek SBIR awards

---

[21]The GAO report on venture funding within the NIH and DoD SBIR programs used a somewhat different methodology to identify firms with VC funding. As a result of the approach adopted, no conclusions can be drawn from the study as to whether firms identified as VC-funded are in fact excluded from the SBIR program on ownership grounds. In addition, the number of VC-funded firms—reportedly 18 percent of all NIH firms receiving Phase II awards from 2001-2004—is considerably higher than suggested by preliminary NRC analysis. U.S. General Accountability Office, *Small Business Innovation Research: Information on Awards made by NIH and DoD in Fiscal years 2001-2004*, op. cit.

[22]Access the SBA's 2002 SBIR Policy Directive, Section 3(y)(3) at <*http://www.zyn.com/sbir/sbres/sba-pd/pd02-S3.htm*>.

[23]This new interpretation of "individuals" resulted in the denial by the SBA Office of Hearings and Appeals of an SBIR grant in 2003 to Cognetix, a Utah biotech company, because the company was backed by private investment firms in excess of 50 percent in the aggregate. Access this decision at <*http://www.sba.gov/aboutsba/sbaprograms/oha/allcases/sizecases/siz4560.txt*>. The ruling by the Administrative Law Judge stated that VC firms were not "individuals," i.e., "natural persons," and therefore SBIR agencies could not give SBIR grants to companies in which VC firms had a controlling interest. The biotechnology and VC industries have been dismayed by this ruling, seeing it as a new interpretation of the VC-small business relationship by SBA. See, for example, testimony by Thomas Bigger of Paratek Pharmaceuticals before the U.S. Senate Small Business Committee, July 12, 2006.

[24]The last 10 years has seen a decline in venture investments in seed and early stage and a concomitant shift away from higher-risk early-stage funding. See National Science Board, *Science and Engineering Indicators 2006*, Arlington, VA: National Science Foundation, 2006. This decline is reportedly particularly acute in early-stage technology phases of biotechnology where the investment community has moved toward later-stage projects, with the consequence that early-stage projects have greater difficulty raising funds. See the testimony by Jonathan Cohen, founder and CEO of 20/20

as a means of exploring a new concept, or simply as a means of capitalizing on existing research expertise and facilities to address a health-related need or, as one participant firm explained, to explore product-oriented processes not "amenable to review" by academics who review the NIH RO1 grants.[25]

  ii. Some of the most successful NIH SBIR award winning firms—such as Martek—have, according to senior management, been successful only because they were able to attract substantial amounts of venture funding as well as SBIR awards.[26]

  iii. Other participants in the program believe that companies benefiting from venture capital ownership are essentially not small businesses and should therefore not be entitled to access the small percentage of funds set aside for small businesses, i.e., the SBIR Program. They believe further that including venture-backed firms would decrease support for high-risk innovative research in favor of low-risk product development often favored by venture funds.[27]

3. Limits on venture funding. The ultimate impact of the 2004 SBA ruling remains uncertain. What is certain is that no empirical assessment of its impact was made before the ruling was implemented. At the same time, the claims made by proponents and opponents of the change appear overstated.

  i. Preliminary research indicates that approximately 25 percent of the NIH SBIR Phase II winners have received VC funding; that some of these are now graduates of the program (having grown too large or left for other reasons), and some are also not excluded by the ruling because they are still less than 50 percent VC owned. Yet it is important to recognize that these companies may be disproportionately among

---

GeneSystems, at the House Science Committee Hearing on "Small Business Innovation Research: What is the Optimal Role of Venture Capital," July 28, 2005.

[25]See the statements by Ron Cohen, CEO of Acorda Technologies, and Carol Nacy, CEO of Sequella Inc, at the House Science Committee Hearing on "Small Business Innovation Research: What is the Optimal Role of Venture Capital," July 28, 2005. Squella's Dr. Nacy's testimony captures the multiple sources of finance for the 17-person company (June 2005). They included—founder equity investments; angel investments; and multiple, competitive scientific research grants, including SBIR funding for diagnostics devices, vaccines, and drugs. SBIR funding was some $6.5 million out of a total of $18 million in company funding. Dr. Nacy argues that SBIR funding focuses on research to identify new products while venture funding is employed for product development.

[26]See Box 4-4 in Chapter 4.

[27]See the testimony by Jonathan Cohen, founder and CEO of 20/20 GeneSystems, at the House Science Committee Hearing on "Small Business Innovation Research: What is the Optimal Role of Venture Capital," op. cit. In the same hearing Mr. Fredric Abramson, President and CEO of Alpha-Genics, Inc., argues that "any change that permits venture owned small business to compete for SBIR will jeopardize biotechnology innovation as we know it today."

the companies—such as previous highly successful SBIR companies that were also VC funding recipients Invitrogen, MedImmune, and Martek—most likely to generate significant commercial returns.[28] What is not known is how many companies are failing to apply to the program as a result of the ruling.

ii. For firms seeking to capitalize on the progress made with SBIR awards, venture funding may be the only plausible source of funding at the levels required to take a product into the commercial marketplace. Neither SBIR nor other programs at NIH are available to provide the average of $8 million per deal currently characterizing venture funding agreements.[29]

iii. For firms with venture funding, SBIR may allow the pursuit of high risk research or alternative path development that is not in the primary commercialization path, and hence is not budgeted for within the primary development path of the company.[30]

4. **An empirical assessment.** As noted above, the SBA ruling concerning eligibility alters the way the program operated during the period of this review (1992-2002), as it has, presumably, from the program's origin. Anecdotal evidence and initial analysis indicate that a limited number of venture-backed companies have been participating in the program. To better understand the impact of the SBA exclusion of firms receiving venture funding (resulting in majority ownership), the NIH recently commissioned an empirical analysis by the National Academies. This is a further positive step towards an assessment culture and should provide data necessary to illuminate the ramifications of this ruling.[31]

---

[28]For discussion of the factors affecting the returns to venture capital organizations, including incentive and information problems and the role venture funds have played in supporting a limited number of highly successful firms, see P. Gompers and J. Lerner, *The Venture Capital Cycle*, Cambridge: The MIT Press, 2000, Ch. 1.

[29]See National Venture Capital Association, Money Tree Report, November, 2006. The mean venture capital deal size for the first three quarters of 2006 was $8.03 million. This trend has been accelerated by the growth of larger venture firms. See P. Gompers and J. Lerner, *The Venture Capital Cycle*, op. cit., Ch 1.

[30]Firms that have used SBIR in this manner include Neurocrine and Illumina. The latter indicated in interviews that these alternative paths later become critical products that underpinned the success of the company.

[31]This research will address questions such as: which NIH SBIR participating companies have been or are likely to be excluded from the program as a result of the 2002 rule change on Venture Capital Company ownership?; and what is the likely impact of the 2002 ruling had it been applied during the 1992-2006 timeframe and what is its probable current impact? Key variables will include the presence and amount of SBIR support, the receipt of venture capital funding or other outside funding, and output measures including those related to commercialization and knowledge generation.

H. Stimulating technological innovation. The SBIR program at NIH is fulfilling its mission to support the transfer of knowledge into the marketplace. In the process, it is encouraging the general expansion of medical knowledge. The program supports innovation and knowledge transfer in several ways:

    1. Patents and publications. SBIR companies have generated numerous patents and publications, the traditional measures of knowledge transfer activity. Thirty-four percent of projects surveyed by NRC generated at least one patent, and just over half resulted in at least one peer-reviewed article.[32]

    2. Knowledge transfer from universities. The NRC Phase II Survey and NRC Firm Survey also suggest that SBIR awards are supporting the transfer of knowledge, firm creation, and partnerships between universities and the private sector:

        i. In more than 80 percent of responding companies with projects at NIH, at least one founder was previously an academic.[33]

        ii. About 33 percent of founders were most recently employed as academics before the creation of their company.

        iii. About 34 percent of NIH projects had university faculty as contractors on the project, 24 percent used universities as subcontractors, and 15 percent employed graduate students.[34]

    3. Indirect paths. There is strong anecdotal evidence concerning beneficial "indirect path" effects—that projects provide investigators and research staff with knowledge that may later become relevant in a different context—often in another project or even another company. While these effects are not directly measurable, discussion during interviews and case studies suggest they exist.[35]

I. The NIH SBIR program has not benefited from regular evaluation.

    1. Prior to the congressional legislation authorizing this study, no systematic, external program assessment had been undertaken at NIH.

    2. A culture of assessment is now developing. Significant progress has

---

[32]See Table 4-23. Without detailed identifying data on these patents and publications, it is not feasible to apply bibliometric and patent analysis techniques to assess the relative importance of these patents and publications.

[33]See Table 4-21.

[34]See Table 4-25.

[35]For a discussion of the "indirect path" phenomenon with regard to the results of innovation awards, see Rosalie Ruegg, "Taking a Step Back: An Early Results Overview of Fifty ATP Awards," in National Research Council, *The Advanced Technology Program: Assessing Outcomes*, Charles W. Wessner, ed., Washington, DC: National Academy Press, 2001.

already been made in this area. Following the congressional initiative requesting this assessment, NIH commissioned its first large-scale survey of the impact of the SBIR program.[36] The 2003 award recipient survey represents a positive step towards an assessment culture, but a range of issues still need to be addressed and a more systematic approach to evaluation adopted.

3. **In the absence of regular internal and external assessment efforts, the NIH SBIR program is at present not sufficiently evidence based.**

   i. Partly as a result of insufficient resources, data collection, reporting, and analytic capabilities are insufficient, limiting the program's capacity for self-assessment.

   ii. This lack of assessment, together with the decentralized character of the program, means that program management does not have adequate information about how their actions affect outcomes such as commercialization, knowledge generation, and networking.

J. **The SBIR Coordinator's office lacks the funds to manage the program effectively. The lack of resources makes it challenging to manage, monitor, and evaluate the program's performance.**

   1. **Management resources.** If NIH is to take an empirical approach to important program management decisions, sufficient resources are required to collect program data and to analyze it effectively. More resources are required to conduct regular internal and external evaluations of program outcomes.

   2. **Limited monitoring.** Only limited program monitoring is undertaken. For example, there appears to be no mechanism through which an underperforming firm could be excluded from the program, nor is there a formal mechanism through which past performance is integrated into either project review or further selection.[37] Weaknesses in the support of minority- and to a lesser extent woman-owned businesses were not effectively identified and monitored. No site visits to awardees are currently funded.

---

[36]National Institutes of Health, "National Survey to Evaluate the NIH SBIR Program: Final Report," July 2003. Available online at: <*http://grants.nih.gov/grants/funding/sbir_report_2003_07.pdf*>.

[37]For example, it appears that no single staff member is individually responsible to monitor multiple-award winners across ICs, or indeed to consistently track program metrics. Thus the company winning the most Phase I awards at NIH (78) has received only 11 Phase II awards, and has generated no known products and few patents in the course of 10 years of effort. It may be that this firm is working effectively in ways not captured by these data, but the firm has apparently not received a site visit in 10 years, and no one at NIH appears to be charged with assessing whether these funds are being used effectively.

3. **Modest management engagement.** In many cases, SBIR responsibilities are a small part of an Institute and Center (IC) manager's much larger portfolio of projects, and reportedly Institute and Center senior management interest in SBIR is often modest. An absence of management engagement with the program can negatively impact perceptions of the program as well as the resources and staff devoted to its operation.

4. **Limited benchmarking for success.** The SBIR Program Coordinator's office appears to have few formal operational benchmarks for program success, other than compliance—i.e., the full annual disbursement of award funding. This is also true for individual Institutes and Centers that disburse funds and operate the program.

5. **Limited analytic capacity and utilization.** Decisions that affect the character of the program are made and implemented in the absence of data-based analysis, and without clear benchmarks for assessing the success or failure of a given initiative. The recent increase in the mean and median size of Phase I and Phase II awards provides a good example. NIH staff have offered a number of different justifications for the change, but no systematic analysis or review appears to have been made beforehand, and no post hoc assessment of the impact is currently underway.

K. **Selection concerns.** While some interviewees and staff believed that the NIH peer-based selection process is generally equitable and procedurally fair, the selection process generated the most criticisms both internally and externally. Verifying the accuracy of these criticisms is inherently difficult. They are cited here because they were repeatedly raised in interviews and should be reviewed in turn by the management. Key criticisms included:

1. **Limited commercial review.** The commercial potential of projects is often assessed by academic scientists who may have little knowledge of the marketplace.

2. **Conflicts of interest.** Some applicants fear that both academic and nonacademic reviewers may have conflicts of interest with proposals. The challenge, of course, is to find reviewers who are knowledgeable but do not have competing interests.

3. **Timeliness.** Some believe that insufficient effort is made to ensure that the review process is completed as rapidly as possible. This is especially important for small business applicants that need to move forward expeditiously to take advantage of a time-sensitive opportunity.

4. **Resubmission.** The opportunity to resubmit proposals is a major advantage of the NIH program, because it allows applicants to fix minor problems with their proposals and resubmit the applications. It is often cited by NIH staff in response to criticisms of the selection

process. While a very positive mechanism, it should be understood that resubmission can impose real costs on small firms in a commercial environment where delayed funding brings about inefficiencies and lost opportunities. A more timely, targeted response to review mechanism may be required.

II. RECOMMENDATIONS

The recommendations in this section are designed to improve the operation of the NIH SBIR program.[38] It is important to keep in mind that the program is achieving its legislative goals. Meaningful commercialization is occurring and the awards made under the program are making valuable additions to biomedical knowledge and developing products to apply that knowledge to the nation's health. With the programmatic changes recommended here, the NIH SBIR program should be even more effective in achieving its legislative goals.

A. **The NIH should increase commercialization and evaluation efforts, improve data collection, expand outreach, especially for minorities and women, develop a culture of critical evaluation, obtain additional management resources for these tasks, and encourage upper management attention to better exploit the program's potential.**

　　1. **Flexibility.** It is most important that the program retain the flexibility and experimentation that have characterized its recent management. The SBIR program is effective across the agencies because a "one-size fits all" approach has not been imposed.

　　2. **Evaluation.** Much greater effort is required to evaluate current outcomes, collect relevant data, including with regard to participation of minority- and woman-owned firms, and document the impact of changes to the program.

　　　　i. Significant improvement in data collection and assessment is needed.

　　　　ii. Efforts to identify outcomes across a variety of metrics should be improved.

　　　　iii. Regular internal and external evaluations should be undertaken to enable managers to assess program performance and the results of management initiatives.

　　3. **Innovation.** Efforts to initiate program innovation by NIH should be substantially strengthened and encouraged with due regard for best prac-

---

[38]The Committee's recommendations below are drawn from analysis of the NRC survey data, review of program operations, and discussions with program participants.

tice lessons from other programs. Pilot programs, possibly for individual Institutes and Centers are one mechanism that allow for the efficient implementation and subsequent assessment of new initiatives.

4. These recommended improvements should enable the NIH SBIR managers to address the four mandated congressional objectives in a more efficient and effective manner.

B. The NIH SBIR program is focused on commercialization and has seen significant achievement. Nonetheless, there are also clear opportunities for further improvement. Continued management attention and additional efforts and resources to facilitate commercialization are needed.

1. **Commercialization programs.** NIH should continue to experiment with commercialization programs, encouraging general application when they show signs of measurable success. Current data indicate that of the 114 companies participating in the Technology Assistance Program in 2004-2005, 23 had received a total of $22 million in additional funding. Other milestone indicators were also positive.

2. **Funding for commercialization programs.** Congress should consider updating the current limits on spending for this purpose. The current limit of $4,000 per year per awardee imposes considerable constraints on innovative programming in this area. Consideration should be given to substantially increasing this amount, and the flexibility of its use.

C. NIH should adopt a more data-driven culture for its SBIR program, with regular assessment driving policy and program management. The current evaluation efforts at NIH are a good start. Given sufficient additional funding, the Committee recommends:

1. **Annual SBIR Program Report.** The NIH SBIR Program Coordinator should be tasked with preparing a much expanded annual SBIR Program Report for submission to a new Advisory Board (see E, below). The report should summarize all relevant data about awards, outcomes, and program initiatives and activities.

2. **Assessment plan.** The program should review its data collection program, identify improvements and develop a formal plan for evaluation and assessment. The internal assessment program should be supported by systematic, objective outside review and evaluation of the NIH program.

D. NIH should focus greater attention on participation by minority- and woman-owned firms in the program.

1. **Encourage participation.** NIH should encourage woman- and minority-owned businesses to submit SBIR proposals and track their successes in winning Phase I and Phase II awards.

2. Improve data collection and analysis. Data collection efforts, as noted above, need to be substantially improved, particularly with regard to women and minorities.

   i. The absence of effective, timely monitoring of minority and woman participation is troubling. This should be corrected on an urgent basis.

   ii. Further analysis of the data, backed by case interviews, should be undertaken to determine the sources of recent trends and the steps that might be taken to address them.

3. Extend outreach to younger woman and minority students. NIH should encourage and solicit women and underrepresented minorities working at small firms to apply as Principal Investigators and Co-Investigators for SBIR awards and track their success rates.

   i. Encourage emerging talent. The number of women and, to a lesser extent, minorities graduating with advanced scientific and engineering degrees has been increasing significantly over the past decade, especially in the biomedical sciences. This means that many of the woman and minority scientists and engineers with the advanced degrees usually necessary to compete effectively in the SBIR program are relatively young and may not yet have arrived at the point in their careers where they own their own companies. They should be encouraged to serve as principal investigators (PIs) and/or senior co-investigators (Co-Is) on SBIR projects.

   ii. Track success rates. The Committee also strongly encourages NIH to gather and publish the data that would track woman and minority principal investigators (PIs), and to ensure that SBIR is an effective road to opportunity for these PIs as well as for woman- and minority-owned firms. The success rates of woman and minority PIs and Co-Is are a traditional measure of their participation in the non-SBIR research grants funded by nonmission research agencies like NIH and NSF, and should be an appropriate measure of woman and minority participation in the SBIR program. After all, experience as a Principal Investigator or Co-Investigator on a successful SBIR program may well give a woman or minority scientist or engineer the personal confidence and standing with agency program officers that encourage them to apply for SBIR awards and found their own firms.

E. The NIH should consider creating an independent Advisory Board that draws together senior agency managers, outside experts, and other stakeholders to review current operations and recommend changes to the program.

1. **An annual Program Report could be presented to the Board on an annual basis.** The Board would review the report, including program progress, management practices, and make recommendations to senior NIH officials in charge.

2. **The Board might be assembled on the model of the Defense Science Board.** It could include senior NIH staff from the ICs and the Director's Office, on an *ex officio* basis, and bring together, inter alia, representatives from industry (including award recipients), academics, and other experts in early-stage finance and program management.

F. **NIH should support and encourage the use of better tools for quality control and evaluation of the SBIR program.**

1. **Monitor outcomes.** As part of the proposed annual Program Report, the Coordinator should monitor SBIR awards and outcomes across the NIH and each institute should develop a similar and compatible capacity.

2. **Suggestions from surveys.** As part of future surveys, a particular effort should be made to gather suggestions for future program improvement from survey recipients.

3. **Benchmarks.** Operational program benchmarks for both process and outcomes should be developed and used to assess program effectiveness at every IC as well as for the program as a whole.

4. **Public information.** NIH should considerably improve the public distribution of information about the program, including recent data on awards and on outcomes.

5. **Clear responsibilities.** As noted above, the IC management, at the senior level should be responsible for the effective management of each IC-based program and, in cooperation with the SBIR Program Coordinator, share responsibility for serving the needs of both the NIH and the applicants and recipients of SBIR awards.

G. **NIH should consider ways in which the current approach to SBIR award selection might benefit from more program-specific adaptations.** Specifically, there appears to be room for improvements in the following areas:

1. **Conflict of interest.** NIH should explore means of addressing perceived conflicts of interest within the SBIR selection process. While there are inevitable tensions between the need for expertise on selection panels and the interests of those experts, some applicants have expressed concern that the current honor system may not work effectively to deal with those tensions in all cases.

2. **Disclosure.** While disclosure of conflicts is mandatory, NIH could con-

sider mechanisms for ensuring that such disclosure is as effective as possible. NIH might consider spot-checking disclosure statements to improve compliance and to signal that NIH views compliance as important.

3. **Voting.** NIH might consider adjusting the voting mechanism, to help ensure that individual panel members do not exert undue influence on award decisions. Currently, all scores from review panelists are counted; excluding outlier scores might be considered.

4. **Oversight.** The proposed SBIR Advisory Board should be responsible for addressing these and other issues related to award selection, in conjunction with relevant staff at the Center for Scientific Review (CSR—the NIH Center that manages the selection process for the other IC's).

5. **Commercial review.** While the NIH SBIR program has registered substantial commercial success, awardees and agency staff have suggested that there is room for considerable improvement, not least in the way in which selection processes assess commercial potential. The difficulties involved in balancing the need for effective commercial review with the risk of conflicts of interest have not been adequately addressed by NIH. The agency should consider adopting pilot programs that could improve the quality and fairness of commercial reviews.[39] Possible options include:

   i. **Hiring professional commercialization consultants** and attaching them to specific study sections. This option could provide significant additional expertise as a resource for the study sections, without fundamentally changing the review process. It should be evaluated on a test basis and reviewed for enhanced commercialization outcomes.

   ii. **Adding staff with industry experience.** Adding new staff members with significant industry experience in the development and commercialization of new products could bring a new dimension to the review and assessment experience.

   iii. **Separating commercial and scientific review processes**, with commercial review considered by a separate, possibly semi-permanent, panel of commercial experts appointed (or hired) specifically for this purpose.

   iv. **Follow-up assessment.** Best practices might be better identified in the selection process by closer analysis of the connection between award outcomes and selection processes.

H. NIH is to be commended for its flexible, industry-driven approach to the

---

[39]Improving the commercial review process, which this recommendation addresses, is not the same as enhancing commercial potential as a criterion for successful review.

SBIR award process. To improve the program's operation further, NIH should consider mechanisms for substantially shortening the average time between initial application and cash-in-hand for award winners.[40]

1. Strengths of the NIH SBIR award process include:

    i. **Multiple opportunities.** In particular, NIH should be commended for providing three application deadlines, rather than the annual deadline used at some other agencies, encourages timeliness, reduces delay, and therefore facilitates participation by microfirms.

    ii. **Resubmission.** The availability of resubmission is another important and positive aspect of the NIH program, allowing companies to fix problems with their applications rather than simply rejecting them, as is the practice at other SBIR programs.

    iii. **Investigator-driven applications.** NIH's investigator-driven approach to topics also makes it unnecessary for applicants to wait for the "right" topic to be part of a solicitation. This program flexibility is a major advantage of the NIH program.

2. Notwithstanding these strengths, the NIH SBIR program still faces challenges: Even with these advantages, delays still occur. For example, companies sometimes cannot afford to accept the delays involved in resubmission, and, in some cases, they cannot afford the overall time lags inherent in the full cycle from initial application to cash-in-hand. These delays and uncertainties tend to reduce the effectiveness of the program and should be reduced where possible.

3. **Suggested mechanisms for improving the decision cycle include:**

    i. **NIH should develop a selection process that is tuned as much as possible to the specific needs of small business.** The current award process is tightly intertwined with the selection process for other NIH programs, notably R01. This approach may be entirely appropriate for awards to academic institutions and university faculty, but it is often less appropriate for an award program for small business, where delays can in many cases lead firms to abandon promising research.

    ii. **The recent NIH shift to electronic submission is an encouraging development, one that was identified early on in this study.** It should help to reduce cycle delays, especially if NIH uses the new system as an opportunity to improve the process as a whole. The

---

[40]Eighty percent of NIH respondents to the NRC Phase II Survey indicated that they had experienced a gap between Phase I and Phase II.

NASA model and DoE's recent conversion are potential guides to best practice.

   iii. **Quick rebuttal.** Numerous winners and applicants stated in interviews that review panels simply did not understand their applications, or rejected them on questionable grounds.

   ○ NIH should seek ways to use new technology as the basis for new procedures that would allow a more iterative approach within a single review cycle.

   ○ Resubmission is not in itself an adequate response to this problem, in light of the substantial delays it imposes on applicants.[41] One approach would be to have NIH change its selection process to make a short written summary from the lead reviewer available electronically to the applicant *before* the study section meeting. The applicant could then provide a one-page commentary or rebuttal, to be distributed immediately before the meeting. This process might have multiple positive benefits, including improving perceptions of fairness and adding quality control to the selection process.

4. The Committee strongly encourages NIH to experiment with different approaches to selection using the pilot program approach described below.

I. **NIH should develop a formal mechanism for designing, implementing, and evaluating pilot programs.**

1. **Need for experimentation.** Addressing these concerns will require resources and time for experimentation.

   i. **Preserving flexibility.** Making changes initially through pilot programs allows NIH to alter selected areas on a provisional basis. A single approach may not work for a program that funds such highly diverse projects with very different capital requirements and very different product development cycles.

   ii. **Lowering cost.** Pilot programs allow Institutes to investigate program improvements at lower risk and lower cost than through changes to the program as a whole. However, effective pilot programs require rigorous design and evaluation.

2. **Program changes need follow-up assessment.** Some of the most significant changes to the SBIR program at NIH—notably changes in award size—have apparently occurred without any evaluation or a clearly articu-

---

[41]Because rejections are received too late for applicants to resubmit during the next submission cycle, an additional delay of 4 months is widely experienced in addition to the actual time needed to review the proposal again.

lated rationale. Other changes, such as the recent NCI-led commercialization assistance pilot, lack a formal evaluation and assessment component. Performance benchmarks, metrics, and timely evaluation, internal and external, should be included in program modifications.

3. **Improving perceptions of fairness.** Additional improvements to the program to address perceptions of unfairness should be considered. These could include more commercial expertise, the right of rebuttal, enhanced use of resubmission, and measures to address perceptions of conflict of interest.

4. **Suggested pilot programs.** NIH should consider pilot programs designed to shorten the program's award cycle time to be more commercially relevant, refine certain selection processes, and better assess the impact of the trend toward increased award sizes:

    i. **Larger awards.** NIH is unique in the extent to which funding has been made available beyond the standard limits set by SBA. This flexibility is both appropriate and valuable.

       ○ The use of large awards at NIH raises some important questions. NIH staff has offered several different justifications for larger awards.[42] None of these rationales has been based on research and assessment of the program, notwithstanding the possible impact of larger awards on the program.

       ○ At a minimum, NIH should develop a clear justification for these larger awards, based primarily on data drawn from the program or elsewhere, which addresses the range of program risks identified in the program management chapter of this report.

       ○ NIH should also develop a formal program to review the impact of the larger awards that are already being made. This should include developing a clear rationale, identifying selection criteria for larger awards, and a robust assessment component, including third-party review to monitor outcomes. Because the additional resources used to fund these awards are substantial, the awards need to achieve a specific objective and/or yield significantly different or better outcomes than multiple standard-sized awards using equivalent funding.

    ii. **Direct to Phase II.** Some agency staff and recipient companies have suggested that research that is otherwise promising has been excluded

---

[42]These include the need to focus resources on the best applications, the high cost of drug development, the high cost of biomedical research, and the lack of inflation adjustments to the standard award size over the last 10 years.

from receiving adequate Phase II level funding because all award recipients have to garner a Phase I first.[43] As well, some program participants have suggested that consideration be given to changing the requirement that SBIR recipients apply for and receive a Phase I award before applying for Phase II. They suggest the rigid application of this requirement has the potential to exclude promising research that could help agencies meet their congressionally mandated goals.

o However, permitting companies to apply directly to Phase II has the potential to change the program significantly. In particular it could shift the balance of both awards and funding significantly away from Phase I toward Phase II. Every additional Phase II award represents approximately 7.5 Phase I awards. If "direct to Phase II" were as attractive to applicants as proponents suggest, it might became a significant component of the program. This in turn could make a very substantial difference to funding patterns in SBIR to the detriment of Phase I.[44] Moreover, expanded Phase I awards, such as those now used at NIH, can meet the same need without affecting the structure of the program.

o Accordingly, this fundamental change to the program structure should not be made.

iii. **Drug discovery.** Given the large size of the sums required, it would be appropriate for NIH to consider a number of possible approaches to the needs of small companies in this area. Some of these approaches may be appropriately housed within the SBIR program. For example, NIH has already experimented with the Competing Continuation Awards program designed to provide funding during the regulatory review process. However, NIH should also ensure that efforts to address drug development issues do not negatively affect the SBIR program outside drug discovery. Further review of the program's role in drug discovery, and its limitations, should be undertaken.

J. **Additional management resources. To carry out the measures recommended above to improve program utilization, management, and evaluation, the program will require additional funds for management and evaluation.**

---

[43]Discussions with NIH SBIR program managers, June 13, 2006.

[44]Phase I awards may have particular importance in meeting noncommercial objectives of the program, for example, helping academics to transition technologies out of the lab into startup companies.

1. Effective oversight relies on appropriate funding.[45] An evidence-based program requires high quality data and systematic assessment. Sufficient resources are not currently available for these functions.

2. Increased funding is needed to provide effective oversight, including site visits, program review, systematic third-party assessments, and other necessary management activities.

3. To achieve the goal of providing modest amounts of additional funding for management and evaluation, there are three options that might be considered:

    i. Additional funds might be allocated internally, within the existing budgets of NIH, as the Navy has done at DoD.

    ii. Funds might be drawn from the existing set-aside for the program to carry out these activities.

    iii. The set-aside for the program, currently at 2.5 percent of external research budgets, might be marginally increased, with the goal of providing management resources necessary to maximize the program's return to the nation.[46]

The key point is that additional resources for program management and evaluation are necessary to optimize the nation's return on the substantial annual investment in the SBIR program.

---

[45]According to recent OECD analysis, the International Benchmark for program evaluation of large SME and Entrepreneurship Programs is between 3 percent (for small programs) and 1 percent for large-scale programs. See Organisation for Economic Co-operation and Development, "Evaluation of SME Policies and Programs: Draft OECD Handbook," OECD Handbook CFE/SME(2006)17, Paris: Organisation for Economic Co-operation and Development, 2006.

[46]Each of these options has its advantages and disadvantages. For the most part, the Departments, Institutes, and Agencies responsible for the SBIR program have not proved willing or able to make additional management funds available. Without direction from Congress, they are unlikely to do so. With regard to drawing funds from the program for evaluation and management, current legislation does not permit this and would have to be modified, therefore the Congress has clearly intended program funds to be for awards only. The third option, involving a modest increase to the program, would also require legislative action and would perhaps be more easily achievable in the event of an overall increase in the program. In any case, the Committee envisages an increase of the "set-aside" of perhaps 0.03 percent to 0.05 percent on the order of $35 million to $40 million per year or, roughly, double what the Navy currently makes available to manage and augment its program. In the latter case (0.05 percent), this would bring the program "set-aside" to 2.55 percent, providing modest resources to assess and manage a program that is approaching an annual spend of some $2 billion. Whatever modality adopted by the Congress, without additional resources the Committee's call for improved management, data collection, experimentation, and evaluation may prove moot.

3

SBIR Awards at NIH

3.1 INTRODUCTION

This chapter reviews awards made by NIH. The analysis uses data supplied by NIH, and as such reflects some of the advantages and disadvantages of NIH methods of capturing and providing data.

NIH provides data separately for each year of an award, and awards cannot in all cases be connected consistently across award years. Some analysis is therefore presented by award year, rather than by award. Thus, in some cases, our analysis is based on indirect estimates rather than directly on primary data. (These cases are identified below.)

In addition, NIH, citing confidentiality concerns, has not provided NRC with complete access to NIH data. This also means that there may be some areas where NRC analysis is incomplete or not possible.

Finally, NIH has been working since 2005 to correct some problems in the NIH data related to the distribution of awards to woman- and minority-owned businesses that were originally identified by the NRC study.[1]

While about 95 percent of all NIH SBIR awards are grants, a small number of SBIR contracts are awarded each year. These are selected based on the same review criteria but using procedures different than the SBIR award review cycle,

---

[1]Note: Following discussions with the NRC staff, the NIH made an effort to recalculate the data for woman and minority owners' participation in the SBIR program. In September 2007, the NIH provided corrected data, which is shown in Appendix A and in several figures in this report. However, apparent anomalies in the NIH data on the participation of women and minorities in 2001-2002 could not be resolved by the time of publication of this report.

and they are designed to meet specific technical needs of NIH Institutes and Centers (ICs).[2] This chapter focuses primarily on awards.

3.2 PHASE I AWARDS

3.2.1 Number of Phase I, Year One Awards

While funding for NIH and thus for the NIH SBIR program has substantially increased in recent years, the number of Phase I, year one awards awarded has not.[3] In fact, the number of Phase I awards grew by about 25 percent between 1999 and 2004, before falling by 18.5 percent in 2005.[4] It is possible that the decline shown in Figure 3-1 represents an important shift in the NIH SBIR program, as fewer Phase I awards might indicate an effort to concentrate resources of fewer, larger, Phase I awards or on Phase II.

3.2.2 Phase I—Award Size

Unlike almost all other agencies and units, NIH does not strictly cap the size of Phase I and Phase II awards. Instead, NIH has applied for and received a blanket waiver from the SBA SBIR administrator. Figure 3-2 shows that the mean size of Phase I, year one awards has increased substantially at NIH in recent years.

Even though there was no change in the Congressionally mandated maximum award size, the mean size of a Phase I, year one award[5] increased by approximately 70 percent between 1998 and 2005, reaching $171,806 in the latter year.

A comparison of Figures 3-1 and 3-2 shows that the post-1999 increase in NIH SBIR funding has been directed more at increasing the size of Phase I awards than at increasing their number. This is consistent with the opinion expressed by some NIH SBIR staffers that funding should be more heavily concentrated on the highest-quality applications. This effect is more pronounced for Phase II awards, as we shall see below.

In fact, NIH now consistently makes Phase I awards that are substantially

---

[2]ICs are the administrative unit at NIH. There are now 27 ICs—such as the National Cancer Institute—and 23 provide SBIR awards.

[3]Because NIH counts awards separately by award year, it is important to differentiate between the first year of an award and subsequent years, which are treated by NIH for data purposes as separate awards.

[4]All awards data in this chapter are based on data provided privately by NIH to NRC, drawn from NIH awards databases. Provided by NIH SBIR Program Coordinator. Because SBA data is maintained differently, cross-checks against the SBA database are not possible.

[5]NIH differentiates between the first and second years of a Phase I award. In some cases (see below), NIH Phase I awards are supported into a second year; however, this second year of support is not typically at a level comparable to the first year, and is therefore excluded from this analysis.

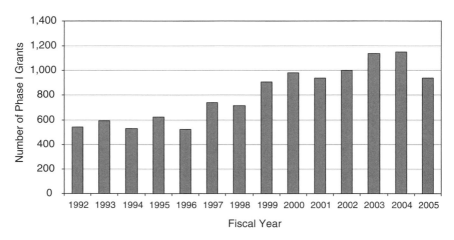

FIGURE 3-1 Number of Phase I awards at NIH, 1992-2005.

SOURCE: National Institutes of Health.

larger than SBA guidelines. The percentage of awards made at or below $100,000 has fallen from 99.7 percent in 1997 to 40.1 percent in 2005.

A comparison of the mean and median size of Phase I awards suggests that the growing mean results from a few large awards. The median award size stayed constant at $100,000 from 1995 to 2002; only in FY2003 did it rise to $106,000. However, by 2003, 16.7 percent of awards were for at least $200,000.

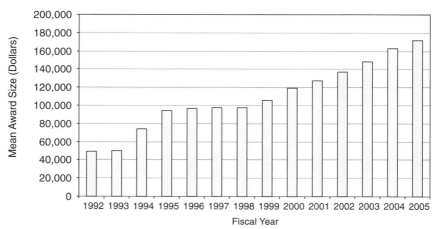

FIGURE 3-2 Phase I, Year One: Mean award size, 1992-2005.

SOURCE: National Institutes of Health.

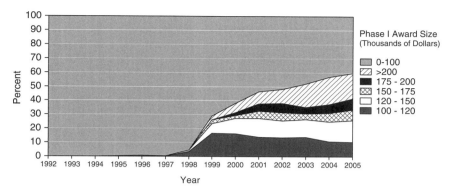

FIGURE 3-3 Oversize Phase I awards at NIH, 1992-2005.

SOURCE: National Institutes of Health

The role and implications of the changing size of NIH awards are discussed separately in Chapter 5—"Program Management."

3.2.3 Phase I New Winners

The percentage of new entrants in the SBIR program is an important indicator of its openness. The figures underscore that the NIH program is not limited to a subset of possible awardees.

Three data sets are especially relevant: the percentage of applications from firms that have previously not won SBIR awards[6]; the percentage of awards going to firms that have not previously won; and the success rate of previous winners vs. new applicants.[7] These data are discussed below.

3.2.3.1 New Applicants

NIH tracks firms that have not previously won SBIR awards at NIH. Some of course will have applied unsuccessfully during previous solicitations; others may have won at other agencies, so the data on previous nonwinners at NIH do not show firms that are necessarily completely new to the SBIR program. However, analysis of previous nonwinners at NIH provides useful metrics for determining the openness or inclusiveness of the program.

Overall, the data show that a very substantial number of applications continue to come in from firms that have not previously won SBIR awards at NIH,

---

[6]While data on applications by new applicants, as opposed to previous nonwinners, would be helpful, these data are not currently available.

[7]The "success rate" here is calculated as winning applications as a percentage of all applications.

and that more than a third of SBIR awards go to previous nonwinners. The NIH SBIR program is substantially open to new entrants.

The data in Figure 3-4 show that since 2000, an average of 61.8 percent of all Phase I applications are from firms that have not previously won SBIR awards at NIH.

This is strong evidence that the opportunities inherent in the NIH SBIR program for the small business biomedical research community are widely understood, and are not limited to a small subset of firms. These data are especially impressive as the number of previous winners has continued to grow sharply during that period, as described immediately below.

3.2.3.2 New Winners

Among all companies winning a Phase I award, an average 41.6 percent are first time winners in the NIH SBIR program.

At least 35 percent of awards went to previous nonwinners in each year since 2000, although that share has declined from 47 percent in 2000. This decline might partly reflect the fact that the number of previous winners has increased sharply during this period, and that many of these new "previous winner" firms continue to apply for more awards.

3.2.3.3 Success Rates

The NIH data permit a comparison of success rates (share of applications that are successful) between new applicants and previous winners. Here there is

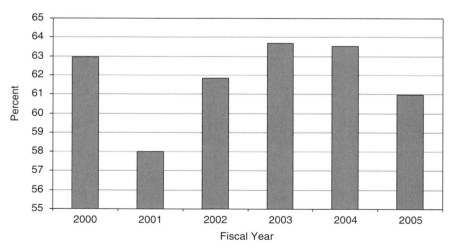

FIGURE 3-4 "New" Phase I applicants (percent of all applicants), 2000-2005.

SOURCE: National Institutes of Health.

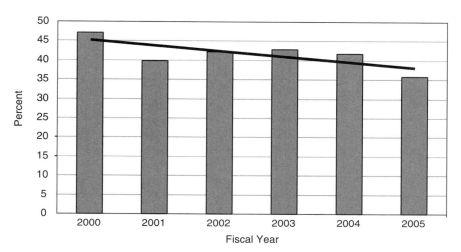

FIGURE 3-5 Percentage of winning companies new to the NIH SBIR program, 2000-2005.

SOURCE: National Institutes of Health.

a clear difference: Previous winners have a success rate more than twice that of previous nonwinners. This may be because previous winners are in a sense already certified as bona fide research companies, while the "previous nonwinners" category includes the entire range of applicants. Moreover, previous winners are likely to have a better understanding of the selection process, and to be able to write applications that better address concerns raised by reviewers.

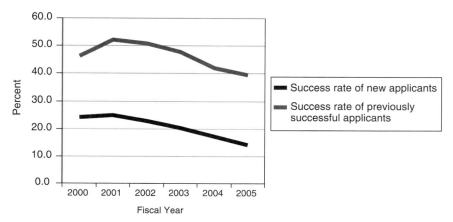

FIGURE 3-6 Phase I success rates of previous winners and nonwinners, 2000-2005.

SOURCE: National Institutes of Health.

The declining rates for both populations since 2001 reflects an increase in the number of applications from 2001-2004, as well as the more recent (FY2005) decline in the absolute number of Phase I awards.

3.2.4 Phase I—Distribution Among the States and Within Them

One of the persistent questions about SBIR is the extent to which awards are distributed among the states. Unsurprisingly, NIH Phase I SBIR awards go disproportionately to states with well-established traditions of life-science entrepreneurship (see Table 3-1). For example, California and Massachusetts together account for 34.7 percent of all Phase I awards between 1992 and 2005.

The top five award-winning states received approximately 51 percent of all awards between 1992 and 2005, ranging from a high of 57.9 percent in 1994 to a low of 47.3 percent in 2002.

Concentration at the top is mirrored in the limited number of awards given to companies in low-award states (see Figure 3-7). The bottom 15 states received 2 percent of all awards 1992-2005, and 1.5 percent in 2005, when six states received zero Phase I awards, and a further five states received only one.

However, outreach efforts by the SBIR program at NIH have supported an increase in the percentage of awards going to the bottom 15 states, which have expanded from barely 0.5 percent of awards in FY1995 to over 3 percent in FY2002 (Figure 3-7).

Further analysis suggests that the raw number of awards conceals other significant differences. Although the National Science Foundation (NSF) does not provide data on the number of life scientists in the workforce, it does offer data on life and physical scientists combined. While not a perfect match for the population of NIH primary investigators, this may be a useful proxy for our purposes here.

The NSF data show that when awards are denominated by the number of life and physical scientists employed, a few states are very successful, but that many are not.

Analysis suggests that the geographical distribution of NIH SBIR awards is understandable. First, Table 3-2 shows that to a considerable extent, awards are made to states which have a high concentration of life scientists. Second, normalizing the data by number of life scientists generates results that place New Hampshire, Vermont, and Oregon among the most successful states, even though these states are not the states that receive the most awards. Finally, and perhaps most persuasively, the selection process (discussed in Chapter 5) is such that the geographical location of applicants is unlikely to play any part in decisions, and awardees interviewed for case studies—even from low award states—indicated that they saw no geographical bias in the selection of awardees.

TABLE 3-1 Phase I Awards—By State, 1992-2005

Number of Phase I Awards by Fiscal Year

State	1992	1993	1994	1995	1996	1997	1998	1999	2000	2001	2002	2003	2004	2005	Grand Total
AK												2		1	3
AL	5	11	6	4	7	5	6	10	8	11	9	8	6	4	100
AR						1		3	1	2	1	3	4	5	22
AZ	14	11	2	10	8	7	10	11	11	15	26	14	12	8	159
CA	110	114	133	144	117	153	136	201	226	240	238	275	232	178	2,497
CO	8	14	8	12	17	22	20	19	36	22	31	36	34	27	306
CT	17	11	13	13	9	19	12	15	13	13	21	26	19	11	212
DC	4	6	6	6	2	3	6	7	11	11	16	10	5	4	97
DE		2	2	2	2	5	1	6	8	7	4	7	6	4	56
FL	10	11	9	14	11	18	8	11	30	18	23	32	20	16	231
GA	5	3	3	6	8	5	4	4	16	12	12	14	19	6	117
HI	8	5	4	3	3	3	3	2	3	6	3	1	1	1	46
IA	1	1			1	1		3	3	4	5	9	4	2	34
ID							1		2	4	1	1	1	1	13
IL	13	21	15	22	11	25	22	19	36	35	28	31	22	29	329
IN	5	5	5	6	5	5	3	7	7	9	7	14	11	12	101
KS	1	2		1	1			4	2	4	5	6	4	1	31
KY	2	3	2	1	2	4	1	3	9	13	10	2	4	4	60
LA	3	2	3	1		2			5	2	3	3	2	1	27
MA	91	114	105	100	105	111	120	175	212	193	189	170	152	106	1,943
MD	68	58	52	54	24	60	38	67	63	84	77	114	85	58	902
ME	2	2	1	6	2		1	3	3	2	5	3	2	2	34
MI	15	18	13	17	10	9	22	20	19	35	24	31	27	24	284
MN	11	13	10	21	11	12	16	7	17	21	24	28	18	19	228
MO	6	8	2	1	3	3	4	5	6	8	12	12	10	8	88
MS			1	1	1	1		2	1	1	2	2	1		13
MT			2		1		3	2	3	4	3	1	4	2	25

continued

TABLE 3-1 Continued

Number of Phase I Awards by Fiscal Year

State	1992	1993	1994	1995	1996	1997	1998	1999	2000	2001	2002	2003	2004	2005	Grand Total
NC	22	10	15	10	14	21	23	21	32	37	39	35	39	28	346
ND	1									1	2	1	3		8
NE	1	3	1	1		4	2	1	3	3	4	3	3	5	34
NH	4	7	6	3	8	13	4	6	12	18	18	20	7	3	129
NJ	12	20	14	15	21	18	17	27	34	34	42	33	29	23	339
NM	7	2	1	3	1	4	2	8	8	10	8	12	11	4	81
NV						3	1		1	2	3	5		3	18
NY	47	38	32	31	25	28	42	43	62	54	71	53	52	45	623
OH	17	19	12	19	12	26	34	26	39	47	47	50	37	35	420
OK	4	3		2	2		3	2	2	3	7	5	4	5	42
OR	9	11	13	19	16	16	14	18	32	27	34	29	19	12	269
PA	22	19	14	24	21	27	28	24	46	57	56	49	47	41	475
PR							1	1	1						3
RI							1	3	8	9	10	7	10	8	58
SC				1	1	1	3	2	3	6	12	8	3	2	41
SD					1	1	1	2		2	3	1	1		13
TN	3	3	4	3	2	8	3	2	6	9	8	9	4	4	68
TX	29	25	23	24	22	35	24	39	64	43	46	65	55	37	531
UT	14	11	10	10	7	14	16	14	23	25	23	11	17	8	203
VA	20	23	21	21	14	29	23	22	36	50	37	57	30	20	403
VT	3	6	3	3	1	1		2	3	2	2	8	4	2	40
WA	16	18	17	22	24	30	28	44	48	47	59	48	37	28	466
WI	4	5	12	9	7	10	15	15	16	29	25	27	18	15	207
WV										1					2
WY						1	2	3	2	1	4	2			15
Grand Total	636	659	596	666	560	764	725	931	1,233	1,293	1,339	1,393	1,135	862	12,792

SOURCE: U.S. Small Business Administration, Tech-Net Database.

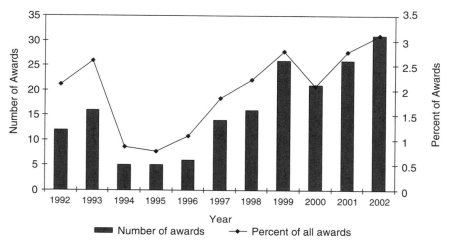

FIGURE 3-7 Phase I awards to the 15 lowest award-receiving states, 1992-2002.
SOURCE: National Institutes of Health.

3.2.4.1 Concentration Within States

Geographic concentration goes considerably further than the state level. Awards are heavily concentrated by zip code.

The data for Massachusetts indicate that 590 zip codes received no awards at all, while the top ten zip codes received more than half of all awards.

The single top winning zip code at NIH, in San Diego, California, where there is a very high concentration of biomedical firms, received more than twice as many Phase I awards as the second most successful zip code.

Concentration can of course be substantially affected by the presence of a single firm: The second most successful zip code in California, Mountain View (94043), received 114 awards, but 69 of those went to a single firm, Panorama Research.

3.2.4.2 Success Rates by State

Table 3-3 shows that success rates vary among states. The range is from 0 percent for Alaska to nearly 31 percent for Massachusetts. Table 3-3 shows the relationship between the ranking of states by number of Phase I applications and Phase I awards.

Variations in success rates across states could be due to a number of factors, such as:

TABLE 3-2 NIH Phase I Awards per 1,000 Life and Physical Scientists Employed

	Life and Physical Scientists, 2003	NIH Phase I, 2003	NIH Phase I per 1,000 Life and Physical Scientists, 2003
New Hampshire	1,480	14.0	9.5
Vermont	850	6.0	7.1
Massachusetts	20,380	140.0	6.9
Maryland	17,910	90.0	5.0
Oregon	5,870	23.0	3.9
Connecticut	5,670	22.0	3.9
California	64,390	248.0	3.9
Virginia	13,030	40.0	3.1
Ohio	15,100	45.0	3.0
Iowa	3,130	9.0	2.9
Colorado	11,710	33.0	2.8
Indiana	4,070	11.0	2.7
Rhode Island	1,580	4.0	2.5
Michigan	9,390	23.0	2.4
Arizona	5,580	13.0	2.3
Nevada	2,510	5.0	2.0
Wyoming	1,510	3	2.0
Delaware	2,020	4.0	2.0
Minnesota	11,200	22.0	2.0
Washington	16,940	33.0	1.9
New Mexico	3,200	6.0	1.9
Wisconsin	11,220	21.0	1.9
New Jersey	17,530	32.0	1.8
Utah	5,060	9.0	1.8
South Carolina	4,610	8.0	1.7
Maine	1,830	3.0	1.6

SOURCE: National Institutes of Health; and National Science Board, *Science and Engineering Indicators 2005*, Arlington, VA: National Science Foundation, 2005.

- Level of entrepreneurial activity.
- University R&D capacities.
- Trained scientists and engineers in the state.
- Access to capital.
- State support activities.[8]

Quantifying the impact of any one of these factors, or of other factors, was be-

---

[8]See U.S. General Accounting Office, *Federal Research: Evaluations of Small Business Innovation Research Can Be Strengthened*, RCED-99-198, Washington, DC: U.S. General Accounting Office, 1999.

	Life and Physical Scientists, 2003	NIH Phase I, 2003	NIH Phase I per 1,000 Life and Physical Scientists, 2003
Pennsylvania	25,080	41.0	1.6
District of Columbia	5,210	8.0	1.5
Oklahoma	3,350	5.0	1.5
North Dakota	1,420	2.0	1.4
New York	30,330	41.0	1.4
North Carolina	17,770	24.0	1.4
Missouri	9,240	12.0	1.3
Kansas	3,910	5.0	1.3
Florida	19,440	24.0	1.2
Alabama	5,170	6.0	1.2
Texas	42,440	49.0	1.2
Illinois	18,300	21.0	1.1
Arkansas	2,700	3.0	1.1
Louisiana	5,540	5.0	0.9
Nebraska	3,920	3.0	0.8
Kentucky	2,660	2.0	0.8
Alaska	2,800	2.0	0.7
South Dakota	1,420	1.0	0.7
Georgia	11,410	8.0	0.7
Tennessee	7,130	4.0	0.6
Hawaii	1,790	1.0	0.6
Montana	2,790	1.0	0.4
Idaho	3,100	1.0	0.3
Mississippi	3,650	1.0	0.3
West Virginia	2,510	0.0	0.0
Average			3.2

yond the scope of this phase of the study, but it would be useful to assess the relative impact of these different factors.

3.2.5 Phase I Awards—By Company

Some companies are very successful in winning Phase I awards. The most successful applicant between FY1992 and FY2003 won 69 Phase I awards. Twenty of the 3,155 companies that received Phase I awards over this period accounted for 776 of the 8,706 Phase I awards awarded (8.9 percent).

However, such individual company success is rare. Only 13 companies re-

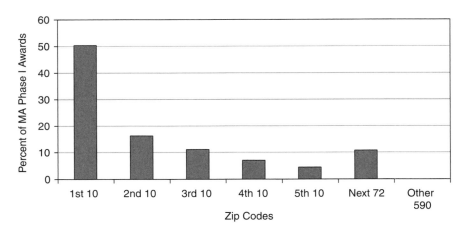

FIGURE 3-8 Concentration of Phase I awards by Zip code in Massachusetts, 1992-2005.

SOURCE: National Institutes of Health.

TABLE 3-3 Phase I Success Rates—By State (Winning applications as percent of total applications)

State	Phase I Success Rate	State	Phase I Success Rate	State	Phase I Success Rate	State	Phase I Success Rate
MA	30.8	WY	25.4	NH	21.2	VA	18.2
UT	30.5	HI	25.1	FL	21.2	MS	18.2
MO	29.7	AZ	24.3	RI	21.0	NY	18.0
WA	29.2	NJ	24.2	MI	21.0	NC	17.8
KY	28.5	LA	24.2	TN	20.8	AR	16.9
TX	27.9	NV	24.2	IL	20.4	MD	16.5
IA	27.2	PR	24.0	IN	20.3	NE	15.4
CT	26.2	PA	24.0	OK	20.2	AL	15.4
MT	25.8	SC	23.8	DC	20.2	WI	14.1
DE	25.8	MN	23.5	OH	19.7	GA	9.8
KS	25.8	ID	23.2	VT	19.3	ND	7.8
CA	25.8	ME	21.6	CO	19.1	OR	7.7
WV	25.7	NM	21.6	SD	18.6	AK	0.0

SOURCE: NRC calculations base on National Institutes of Health data.

ceived more than 30 Phase I awards during from FY1992 to FY2003, and only 33 received at least 20. In contrast, more than 3,000 (3,025) companies received less than 10 awards, and 2,237 received only one or two.[9]

---

[9]NIH data has been received and updated at irregular times during the course of the analysis. While key points have been updated to 2005, where such updates are not critically important, we have utilized older data sets—as in the case above.

TABLE 3-4 Multiple-award Winning Companies at NIH FY1992-2003—Top 20 Winners

Organization	Number of Awards
Panorama Research, Inc.	69
Inotek Pharmaceuticals Corporation	63
Radiation Monitoring Devices, Inc.	56
Lynntech, Inc.	51
Inflexxion, Inc.	44
Oregon Center For Applied Science	44
New England Research Institutes, Inc.	42
Creare, Inc.	40
Insightful Corporation	40
Hawaii Biotech, Inc.	38
Physical Optics Corporation	33
Biomec, Inc.	30
Surmodics, Inc.	30
Biotek, Inc.	29
Spire Corporation	28
One Cell Systems, Inc.	27
Compact Membrane Systems, Inc.	26
Osi Pharmaceuticals, Inc.	26
Personal Improvement Computer Systems	25
Physical Sciences, Inc.	25
Total	766

SOURCE: National Institutes of Health.

3.2.6 Phase I Awards—Woman- and Minority-owned Firms

In 2005, the NRC identified some significant problems in the collection of data for woman- and minority-owned business firms participating in the NIH SBIR program. Following an NRC request for clarification of the data, NIH investigated the problem and found that it was rooted in the data entry software.[10]

3.2.6.1 Award Shares

The share of awards held by woman- and minority-owned firms has remained relatively constant at about 12 percent, with the exception of 2001-2002 data (see note to Figure 3-9).

---

[10]See National Institutes of Health, letter to the National Research Council "National Academy of Sciences Study of the Small Business Innovation Research (SBIR) Program-NIH SBIR Program Data," February 14, 2005. Since then, NIH has made significant efforts to correct this problem, re-entering the data covering several years of program activity. However, apparent anomalies in the NIH data on the participation of women and minorities in 2001-2002 could not be resolved by the time of publication of this report.

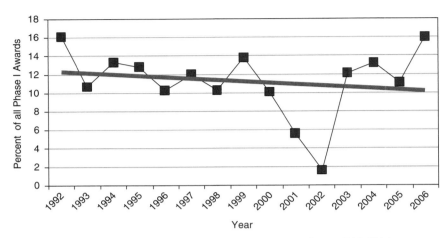

FIGURE 3-9 Award shares of woman- and minority-owned firms, 1992-2006.

NOTE: Following discussions with the NRC staff, the NIH made an effort to recalculate the data for woman- and minority-owners' participation in the SBIR program. In September 2007, the NIH provided corrected data, which is shown in Appendix A and in several figures in this report. However, apparent anomalies in the NIH data on the participation of women and minorities in 2001-2002 could not be resolved by the time of publication of this report.

SOURCE: National Institutes of Health.

However, these data partly conceal a long-term decline in the share of awards going to minority-owned firms, which have fallen from an average of 6.9 percent in 1992-1994 to 4.3 percent in 2003-2006.

3.2.6.2 Share of Applications

One possible explanation for the declining minority share may be in the number of applications received. These are captured in Figure 3-10.

In fact, the data show that applications from minority-owned firms have declined, from more than 10 percent of all applications in 1996, to under 4 percent 2006. Applications share for woman-owned firms has remained approximately constant.

3.2.6.3 Success Rates

A different explanation for declining award shares may lie in relative success rates. These are described in Figure 3-11.

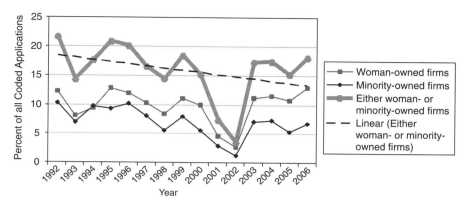

FIGURE 3-10 Woman- and minority-owned firms—Phase I percentage share of all coded applications, by demographic, 1992-2006.

NOTE: Following discussions with the NRC staff, the NIH made an effort to recalculate the data for woman and minority owners' participation in the SBIR program. In September 2007, the NIH provided corrected data, which is shown in Appendix A and in several figures in this report. However, apparent anomalies in the NIH data on the participation of women and minorities in 2001-2002 could not be resolved by the time of publication of this report.

SOURCE: National Institutes of Health.

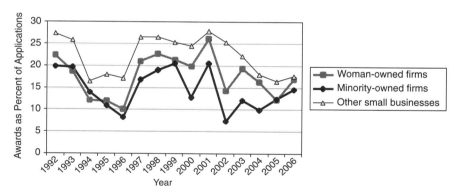

FIGURE 3-11 Success rates for Phase I awards by demographic, 1992-2006.

NOTE: Following discussions with the NRC staff, the NIH made an effort to recalculate the data for woman and minority owners' participation in the SBIR program. In September 2007, the NIH provided corrected data, which is shown in Appendix A and in several figures in this report. However, apparent anomalies in the NIH data on the participation of women and minorities in 2001-2002 could not be resolved by the time of publication of this report.

SOURCE: National Institutes of Health.

The data show that woman- and minority-owned firms are consistently less successful in the selection process—that lower percentages of their applications generate awards. Minority applicants saw a particularly steep decline in success rates from 1999 to 2004, with some recovery in 2005-2006. However, they remain about five percentage points lower than the rates for firms that are neither woman- or minority-owned. Note that it is not possible, using the data provided by NIH, to test the impact of other factors such as whether these woman- and minority-owned firms are disproportionately new applicants. Follow-on research could address this possibility.

The data provide no immediate answer as to why woman- and minority-owned firms have lower success rates. One promising hypothesis is that these firms may tend to be formed more recently, and have both a shorter track record and less experienced principal investigators, both of which may mitigate against success in the NIH selection process.

3.2.6.4 Woman- and Minority-owned Firms in Phase I

While we can conclude that woman- and minority-owned firms receive a significant share of awards, the data suggest that further analysis is needed, and that NIH may in particular need to take more aggressive steps to encourage high-quality applications from these firms. A review of the selection process from this perspective is also warranted.

Finally, it is important to note that while woman-owned firms have maintained and even slightly increased their share of SBIR Phase I awards at NIH, they remain at about 10 percent of the total (over 2003-2006). At the same time, the percentage of women among recent life sciences doctorates has increased dramatically.[11] According to NSF, in 2005 women accounted for more than 48 percent of all biological sciences doctorates awarded.[12] This growth is also reflected in employment data: Women account for 36.5 percent of employed life scientists in 2005.[13] These are the likely founders of firms that might be seeking seed funding from the NIH SBIR program.

In that context, maintaining a ten percent share of awards is much less impressive, and NIH might well wish to undertake further analysis to determine why so few of these new doctorates appear to be applying for NIH SBIR funding (note

---

[11]It is also worth noting that the pool of woman-owned businesses has grown rapidly. For the past two decades, majority woman-owned firms have continued to grow at around two times the rate of all firms (42 percent vs. 24 percent). Woman-owned firms, 50 percent or more owned by women, account for 41 percent of all privately held firms. Source: Center for Women's Business Research, *Key Facts About Women-owned Businesses*, Washington, DC: Center for Women's Business Research, 2006.

[12]National Science Board, *Science and Engineering Indicators*, Arlington, VA: National Science Foundation, 2007, Table F-2.

[13]Ibid, Table H-7.

that there is no requirement that a company exist in order to apply for an award, although a company must be formed in order to accept one.)

3.2.7 Phase I Awards—By IC

The substantial size differences among the various institute centers (IC) at NIH are reflected both in Phase I and Phase II, as can be seen in part in Table 3-5.

Together, the three largest ICs (National Cancer Institute, National Heart Lung and Blood Institute, and NIAID) accounted for 44.4 percent of all NIH SBIR Phase I awards. Only two other ICs account for more than 5 percent of the total. Conversely, the 10 smallest ICs together accounted for 9.7 percent of the total, with none providing more than 2 percent individually.

3.2.8 Phase I—Extended Awards: Year Two of Support

NIH is unique among the granting agencies in providing extended support for Phase I awards. Interviews with NIH personnel and awardees suggest that, to a considerable extent, the normal timeframe for completing a Phase I project has become one year, rather than six months. "No-cost extensions"—extensions of time without additional funding—up to one year total are relatively easy to get.

NIH is also unique in providing additional funding beyond the first year for some Phase I awards. This practice is relatively rare but growing in importance at NIH.

While the median size of second year Phase I awards appears to have stabilized at around $200,000, the number of second year awards continues to climb sharply. From 1992 to 1999, there were always less than five such awards per year. By FY2003, more than 10 percent (80) of all Phase I awards went to second-year support.

3.2.9 Phase I—Supplementary Funding

NIH offers one further form of funding flexibility. Program officers can add limited additional funds to an award in order to help a recipient pay for unexpected costs. While practices vary among individual ICs, it appears that awards of up to 25 percent of the annual funding awarded can be made by a program manager without further IC or NIH review. More substantial supplements must be more extensively reviewed.

For Phase I, supplements remain relatively rare, recently averaging less than 20 per year. They are also not especially large; in no fiscal year have they totaled more then $1 million for all awards combined. They are more important for Phase II, as will be explained below.

TABLE 3-5 Percentage Phase I Awards—By IC, 1992-2002

Funding IC	Percentage Phase I Awards											
	1992	1993	1994	1995	1996	1997	1998	1999	2000	2001	2002	Total
NIAA	0.7	1.2	0.6	1.1	0.8	1.6	0.6	1.5	1.5	1.0	1.2	1.1
NIA	2.2	2.4	5.3	2.6	1.0	4.6	3.6	3.2	3.5	4.1	2.8	3.3
NIAID	11.7	12.3	9.2	11.9	13.4	13.5	11.6	10.0	10.2	11.4	12.9	11.6
NIAMS	0.9	2.7	3.2	4.2	5.5	3.4	2.8	3.0	2.3	3.0	1.5	2.9
NCCAM	0.0	0.0	0.0	0.0	0.0	0.0	0.0	0.2	0.6	0.8	0.7	0.3
NCI	29.6	26.5	21.5	15.3	15.3	20.8	19.3	21.2	18.7	20.5	22.4	20.9
NIDA	2.4	1.5	4.2	3.9	3.6	1.8	2.0	3.5	3.1	3.0	2.1	2.8
NIDCD	1.1	1.7	2.3	1.3	1.0	1.6	1.1	2.2	2.2	2.0	1.0	1.6
NIDCR	1.1	1.3	1.1	2.3	1.3	0.8	2.0	0.9	3.2	3.7	1.3	1.8
NIDDK	3.0	3.7	6.8	7.9	7.5	6.7	5.6	7.7	8.5	7.1	7.1	6.7
NIBIB	0.0	0.0	0.0	0.0	0.0	0.0	0.0	0.0	0.0	0.0	1.8	0.2
NIEHS	1.5	1.3	1.5	1.4	1.9	1.1	1.3	1.2	2.4	3.1	3.2	1.9
NEE	3.0	2.4	4.7	4.5	5.2	2.4	2.5	2.0	2.0	2.1	3.1	2.9
NIGMS	7.8	10.6	9.1	6.8	10.1	9.2	10.3	9.7	8.2	8.8	8.0	8.9
NICHD	5.7	8.8	4.5	4.3	6.7	5.7	4.5	4.1	5.3	3.3	4.3	5.0
NHGRI	1.1	0.8	0.8	0.8	1.0	1.1	2.4	1.9	0.9	0.9	2.3	1.3
NHLBI	12.6	8.6	10.6	16.6	15.1	11.2	10.1	12.0	13.7	12.3	9.0	11.9
NLM	0.4	0.5	0.4	0.0	0.0	0.4	0.8	0.4	0.4	0.1	0.2	0.3
NIMH	3.1	4.0	3.8	5.3	4.0	4.9	4.7	4.5	3.2	4.0	4.4	4.2
NINR	0.7	1.2	0.2	1.0	0.2	0.4	0.8	0.6	0.5	0.2	0.2	0.5
NINDS	4.6	3.9	6.6	6.4	3.6	4.2	6.3	5.6	5.2	4.6	5.3	5.1
NCRR	6.7	4.6	3.8	2.1	2.7	4.4	5.6	4.3	4.2	4.0	3.5	4.2
Shared	0.0	0.0	0.0	0.3	0.2	0.4	2.2	0.1	0.3	0.2	1.7	0.6
Total	100	100	100	100	100	100	100	100	100	100	100	100

SOURCE: National Institutes of Health.

3.3 PHASE II AWARDS

As funding for NIH has increased, with a 5-year doubling of the NIH budget started in 1999, both the number and the average size of Phase II awards have grown in nominal terms, but to different degrees.

After increasing from fewer than 300 new Phase II awards in 1992, the number of new awards at NIH grew to almost 800 in 2002. Since then, the number has remained almost flat, with a slight decline to 774 awards in 2005.

However, this period of stable award numbers coincides with substantial growth in SBIR funding at NIH. Consequently, this additional funding is being distributed in other ways—one of which is increased average award size.

NIH maintains a different record for each year of an award. The average size of the first year of the NIH Phase II award has increased in nominal terms considerably, from around $230,000 in 1994 to more than $500,000 in 2005, with a jump of $55,000 (or 11.9 percent) in 2005 alone. Increased size of year one awards is matched by growth in the size of year two awards as well.

3.3.1 Phase II—Extended Awards

Beyond the size of awards, NIH is also distinctive in the way it provides additional support beyond the second year of Phase II. For example, in FY2003, 39 companies were in their third year of Phase II support, and two more were in their fourth year.

Figure 3-12 shows that there has been a substantial increase in the number of companies receiving third-year Phase II support. In FY2003, 39 companies

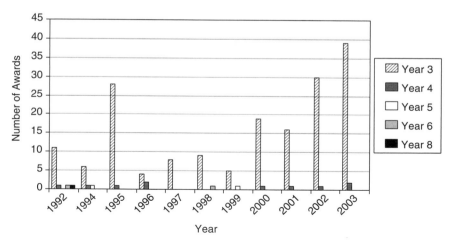

FIGURE 3-12 Phase II—Extended support, 1992-2003.

SOURCE: National Institutes of Health.

received such support; this is equivalent to about 10 percent of the companies that received their initial Phase II awards in FY2001 (the first year of Phase II support for this cohort of awards).

These additional years of support have also provided a growing amount of funding per award.

These data both document the ongoing growth in support beyond year two (see Figure 3-13); they indicate that the average amount of support has continued to grow, and is now at approximately $725,000. Some projects receive more than one year of additional support, as documented in Figure 3-12.

3.3.2 Competing Continuation Awards

NIH has recently initiated a new program aimed at supporting companies during the difficult period of clinical trials, where outsider funding can be especially scarce. Competing Continuation Awards (CCAs) are a competitive program that provides three years of additional support at $1 million per year for companies needed support during trials.

As the data in Table 3-6 indicate, CCAs are now ramping up. It is still much too early to tell whether CCA's will be successful. At least one large IC—NCI— has withdrawn from the program partly on the grounds that it does not require matching funds. This criticism—drawn from the experience at NSF with their Phase IIB program—may however not be easily addressed at NIH, where matching funds are difficult to find for projects before the end of clinical trials.

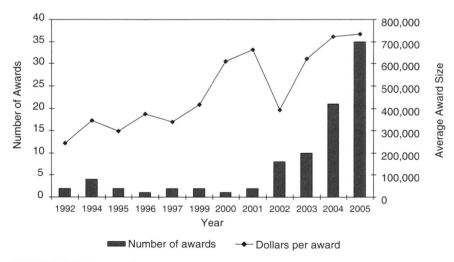

FIGURE 3-13 Support for Phase II beyond Year Two, 1992-2005.

SOURCE: National Institutes of Health.

TABLE 3-6 Competing Continuation Awards at NIH, 2003-2005

Fiscal Year	Total Awards		Competing Awards		Non-competing Awards (Yr2/3)	
	Number	Amount ($)	Number	Amount ($)	Number	Amount ($)
2003	2	975,649	1	799,709	1	175,940
2004	3	2,298,561	2	1,498,562	1	799,999
2005	21	15,494,168	19	13,890,358	2	1,603,810

SOURCE: National Institutes of Health.

It is also worth noting that if CCAs continue to be funded at the current rate, they would account for more than $40 million annually, about 6.5 percent of the NIH SBIR program, and equivalent to 400 standard Phase I awards. That trade-off does not appear to have been systematically addressed by NIH.

3.3.3 Phase II Awards—By Company

As with Phase I, some companies have received numerous Phase II awards, although at levels far lower than at DoD where questions about the role of frequent award winners in the SBIR program have been focused. The companies receiving many Phase I awards are often also successful in applying for multiple Phase II awards.

The correlation is not perfect, however, as shown by Table 3-7. For example, the top Phase I winner—Panorama Research—is only seventh on the list of Phase II winners. Some big winners in Phase I, such as Abiomed, Individual Monitoring, and Sociometrics, have conversion rates from Phase I to Phase II of over 80 percent. Others, by contrast, such as Panorama Research, only convert 20 percent of their Phase I awards into Phase II awards.

In general, NIH awards do not appear to be overly concentrated in a few firms. Only three companies have received 30 or more Phase II awards between 1992 and 2003. Overall, the top 20 winners account for 11.1 percent of all Phase II awards during this period, while more than 1,500 companies received at least one Phase II during this period.[14]

3.3.4 Phase II Awards—By State

The geographical distribution of Phase II awards approximates but does not equal the distribution for Phase I awards. As can be seen by comparing Table 3-1 and Table 3-8, the states with many Phase I award winners tended to get the

---

[14]Note, however, that as NIH has declined to provide employer identification numbers on privacy grounds, the number of awards above was generated by matching company names in awards. This is likely to be less accurate—and to possibly understate the degree of award concentration.

TABLE 3-7 Conversion Rates of Top 20 Phase II Award Winners, 1992-2003

Organization	Number of Phase II Awards	Number of Phase I Awards	Conversion Rate (%)
NEW ENGLAND RESEARCH INSTITUTES, INC.	34	42	81.0
INFLEXXION, INC.	32	44	72.7
RADIATION MONITORING DEVICES, INC.	30	56	53.6
OREGON CENTER FOR APPLIED SCIENCE	27	44	61.4
INSIGHTFUL CORPORATION	22	40	55.0
LYNNTECH, INC.	17	51	33.3
PANORAMA RESEARCH, INC.	15	69	21.7
INOTEK PHARMACEUTICALS CORPORATION	14	63	22.2
SOCIOMETRICS CORPORATION	13	16	81.3
ABIOMED, INC.	13	13	100.0
PERSONAL IMPROVEMENT COMPUTER SYSTEMS	13	25	52.0
CREARE, INC.	13	40	32.5
CLEVELAND MEDICAL DEVICES, INC.	13	23	56.5
INDIVIDUAL MONITORING SYS, INC. (IM SYS)	12	14	85.7
BIOTEK, INC.	12	29	41.4
SURMODICS, INC.	11	30	36.7
ADVANCED MEDICAL ELECTRONICS CORPORATION	11	18	61.1
ELECTRICAL GEODESICS, INC.	11	19	57.9
WESTERN RESEARCH COMPANY, INC.	11	20	55.0
PHYSICAL SCIENCES, INC.	11	25	44.0
Total (top 20 award winners)	335	681	49.2
All Awards	3,027		
Top 20 as percent of all Phase II awards	11.1		

SOURCE: National Institutes of Health.

most Phase II awards. Not surprisingly, states with few Phase I awards had few Phase II awards.

States vary substantially in the degree to which their companies successfully convert Phase I awards into Phase II. Table 3-9 also shows the percentage share of Phase II awards between 1992 and 2003, by state, expressed as a percentage of the Phase I awards between 1992 and 2003, by state. This metric indicates states whose firms appear to be particularly successful at converting Phase I awards into Phase II awards.

More research is needed to understand why companies in some states are so much more likely to be successful in moving from Phase I to Phase II. While part of this phenomenon is due, in part, to the geographic location of particular companies that have in the past been successful, more information is needed to

TABLE 3-8 Phase II, Year One Awards, 1992-2002, By State, 1992-2002

Number of Phase II, Year One Awards

State	1992	1993	1994	1995	1996	1997	1998	1999	2000	2001	2002	1992-2002	Percent of Total	Percent of Phase I
AL	1	1	2	3	1	2	1	2	2	2	4	21	0.8	92.1
AR			1								1	2	0.1	77.9
AZ	1	5	1	1	1	4	3	9	4	2	12	43	1.6	131.3
CA	24	36	21	32	34	58	54	46	40	69	48	462	17.3	85.8
CO		5		5	5	5	4	10	8	5	12	59	2.2	95.7
CT	4	5	4	6	3	5	2	3	6	3	2	43	1.6	91.1
DC		2	1	1	1	4		6	5	4	5	29	1.1	158.5
DE					2	1		3	1	2		9	0.3	77.9
FL	4	8	2	2	1	9		5	7	3	5	46	1.7	94.3
GA		1	2		1	4	3	4	3	1	2	21	0.8	89.6
HI	1	1	1		1	1		1	1	1		8	0.3	60.8
IA	1		1			1	1	1	1	1	2	9	0.3	140.2
ID		1						1				2	0.1	56.6
IL	3	9	1	3	3	8	6	11	7	14	5	70	2.6	101.0
IN	2			3	2	3		4	1	4	3	22	0.8	112.4
KS				2		1	1		1	2	1	8	0.3	118.7
KY	1	1			1			1	1	3	1	9	0.3	96.7
LA		1							1	1	2	5	0.2	91.6
MA	33	33	18	41	39	49	34	51	51	67	51	467	17.5	108.7
MD	9	23	10	15	9	17	20	11	17	27	21	179	6.7	97.7
ME	1		2			2	1	1		1	2	10	0.4	124.6
MI	5	4	3	5	6	8	9	4	6	9	4	63	2.4	118.2
MN	3	6	3	3	7	8	4	3	4	4	9	54	2.0	110.7
MO		1	1			1	1	2	1	2	5	14	0.5	96.9
MS											1	1	0.0	38.9
MT					1		1	1	1	1		5	0.2	82.0

continued

TABLE 3-8 Continued

Number of Phase II, Year One Awards

State	1992	1993	1994	1995	1996	1997	1998	1999	2000	2001	2002	1992-2002	Percent of Total	Percent of Phase I
NC	2	7	5	2	4	8	6	8	2	11	12	67	2.5	100.8
NE	1			2	1	1		1	2	1		8	0.3	178.0
NH	1	1	1	5		3	3	4	2	4		25	0.9	90.6
NJ	4	3	3	4	3	3	6	12	8	12	13	71	2.7	99.2
NM		1	1		2	1			3	2	1	11	0.4	74.5
NV	1		2		1				1	2	2	9	0.3	164.9
NY	6	16	8	13	11	13	9	12	11	18	22	139	5.2	99.5
OH	3	5	4	6	3	10	7	8	7	9	16	78	2.9	104.7
OK	1	1	1				1			1	1	6	0.2	116.8
OR	3	3	5	5	6	8	9	9	11	11	8	78	2.9	133.5
PA	4	4	2	10	6	7	8	11	7	20	16	95	3.6	99.0
PR									1			1	0.0	155.8
RI	1							2	1	2	2	8	0.3	69.2
SC									1		5	6	0.2	69.2
SD								2		1	1	4	0.1	207.7
TN		2	2	2		1	4	1	1	1	2	16	0.6	95.9
TX	5	7	5	10	4	11	8	9	12	11	14	96	3.6	93.5
UT	5	7	3	4	1	2	2	5	3	7	6	45	1.7	96.0
VA	2	7	9	11	5	6	8	5	8	11	10	82	3.1	97.1
VT	1		2	2	1	1			1		1	9	0.3	116.8
WA	4	7	5	9	4	11	18	16	7	22	20	123	4.6	131.7
WI		1		2	3	4	3	4	9	5	3	34	1.3	80.2
WV									1		1	2	0.1	311.5

SOURCE: National Institutes of Health.

TABLE 3-9 Phase II Awards—Over- and Underachieving States

Top Ten Overachievers		Top Ten Underachievers	
State	Percent[a]	State	Percent[a]
WV	311.5	MS	38.9
SD	207.7	ID	56.6
NE	178.0	HI	60.8
NV	164.9	RI	69.2
DC	158.5	SC	69.2
PR	155.8	NM	74.5
IA	140.2	AR	77.9
OR	133.5	DE	77.9
WA	131.7	WI	80.2
AZ	131.3	MT	82.0

[a]Percent of total Phase II awards as percent of Phase I awards (shows successful transitions from Phase I to Phase II)

SOURCE: NRC calculations based on National Institutes of Health data.

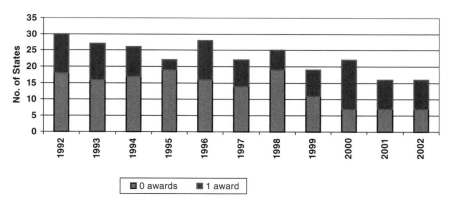

FIGURE 3-14 Phase II—Number of low-award states, 1992-2002.

SOURCE: National Institutes of Health.

explain the award pattern. Understanding this process better would likely be of value to state economic development agencies.[15]

NIH has, as discussed earlier, made considerable outreach efforts toward low-award states. These efforts appear to be generating positive results. Figure 3-14 shows that the number of states receiving zero or one Phase II awards

[15]A significant part of the variation—particularly differences in conversion rates among low-grant states such as West Virginia, Mississippi, and Idaho—could be random. Sample sizes for these states are extremely small.

TABLE 3-10 Phase II—Top Award Winning States (Percent of all new Phase II awards), 2006

State	Percent
CA	18.5
MA	13.1
PA	5.9
MD	4.6
OR	4.4
OH	4.1
IL	3.8
TX	3.6
NY	3.3
WA	3.3

SOURCE: U.S. Small Business Administration, Tech-Net Database.

in a particular year has declined steadily from 30 in 1992 to 16 in 2002. At the other end of the spectrum, California is the top award-winning state, followed by Massachusetts (see Table 3-10).

3.3.5 Phase II Women and Minorities

As with Phase I, several key factors affect the participation of woman- and minority-owned firms in the NIH SBIR program.

The data show that the participation of woman- and minority-owned firms in the NIH SBIR program have diverged over the past ten years.

While participation of woman-owned firms has trended up since 1998, participation of minority-owned firms is both very low and declining. The consistent minority-owned participation at less than 4 percent of awards since 2003 is a matter for considerable concern.

As noted in Section 3.2.6.4, women account for a large and growing percentage of recent Ph.Ds in the life sciences. In light of those figures, a participation level of 12 percent for woman-owned firms is still a matter that merits further analysis by NIH.

One obvious question is whether these award levels are primarily the result of application patterns, or of success rates. Obviously, application patterns in part stem from Phase I patterns overall, as success at Phase I is a requirement before applying for Phase II.

Still, success rates do provide useful information. The data show that minority-owned firms have, as in Phase I, consistently generated lower success rates at Phase II than either woman-owned firms or firms that are neither woman- or minority-owned.

Over the past four years (2003-2006), this gap has averaged 9.3 percentage

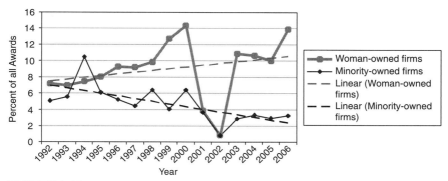

FIGURE 3-15 Share of Phase II awards to woman- and minority-owned firms, 1992-2006.

NOTE: Following discussions with the NRC staff, the NIH made an effort to recalculate the data for woman and minority owners' participation in the SBIR program. In September 2007, the NIH provided corrected data, which is shown in Appendix A and in several figures in this report. However, apparent anomalies in the NIH data on the participation of women and minorities in 2001-2002 could not be resolved by the time of publication of this report.

SOURCE: National Institutes of Health.

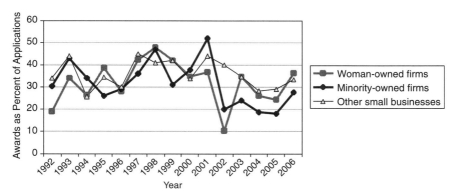

FIGURE 3-16 Success rates for Phase II applications by woman- and minority-owned firms, 1992-2006.

NOTE: Following discussions with the NRC staff, the NIH made an effort to recalculate the data for woman and minority owners' participation in the SBIR program. In September 2007, the NIH provided corrected data, which is shown in Appendix A and in several figures in this report. However, apparent anomalies in the NIH data on the participation of women and minorities in 2001-2002 could not be resolved by the time of publication of this report.

SOURCE: National Institutes of Health.

points (31.3 percent vs. 22.0 percent) as minority-owned firms have succeeded about one third less often than firms that are not minority owned.

This is troubling, and warrants immediate attention from NIH.

3.3.6 Phase II—Awards by IC

Phase II award distribution by IC follows a pattern similar to Phase I awards. The two largest ICs—NCI and NHLBI—account for 30.8 percent of all awards FY1992-2003, while the five largest IC's account for 54.9 percent.

3.4 PHASE I APPLICATIONS

3.4.1 Phase I Applications—By IC

The number of awards made does not closely track success rates at the level of the IC. Success rates vary very widely by IC, from a high of 29 percent at NS to a low of 5 percent at the Library of Medicine (in 2003). Success rates at the three largest ICs average about 24 percent.

It should be noted that the number of applications has declined quite sharply at NIH in 2005 and 2006—down 11 percent and 16 percent respectively.

3.4.2 Resubmissions

It is normal practice at NIH to allow companies to resubmit Phase I proposals. The resubmissions include responses to questions and criticisms from the initial reviewers. This process usually requires a delay of eight or more months, as responses are not usually returned to applicants in time to resubmit during the next funding cycle.[16]

Figure 3-17 and Figure 3-18 show the percentage of resubmissions in total submissions and the relative success rates for resubmissions and original applications, respectively.

Resubmission rates fluctuate somewhat, but the trend since 1992 has been relatively stable; about 20-25 percent of all submissions are resubmissions. Though the success rates of submissions and resubmissions vary by year, resubmissions are overall slightly more likely to be successful than original submissions, even though all resubmissions have been rejected at least once and hence are—one might assume—less convincing applications.

This suggests that there is often room in proposals for improvements and clarifications that would then permit funding. Perhaps the NIH SBIR program should test mechanisms for improving original proposals, thus saving both the applicant and SBIR staff the time and effort of going through the application process twice.

---

[16]See Chapter 5, Program Management, for more details on funding cycles at NIH.

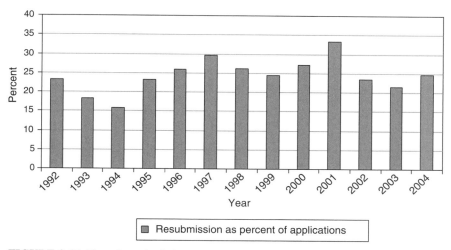

FIGURE 3-17 Phase I resubmission rates at NIH, 1992-2004.

SOURCE: National Institutes of Health.

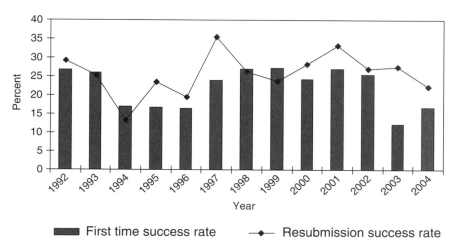

FIGURE 3-18 Phase I success rates for resubmissions and initial proposals, 1992-2004.

SOURCE: National Institutes of Health.

3.5 PHASE II APPLICATIONS

3.5.1 Success Rates

The average success rate over all years is approximately 55 percent, and there is no apparent trend in success rates.

3.5.2 Phase II—Resubmissions

Resubmissions are still an important part of the application and selection process during Phase II. As shown in Figure 3-19, approximately 30 percent of all Phase II applications are resubmissions. This percentage has remained steady in recent years.

These resubmissions can cause a substantial delay in the research of affected companies—in most cases of at least 8 months. Mechanisms for improving the initial selection process to reduce the number of applications where resubmission is needed to clear up minor difficulties should therefore be considered in order to reduce unnecessary delays.

Figure 3-20 shows that resubmission success rates are consistently lower than those for initial submissions. This is not surprising, as initial applications will include all applications, while resubmissions will not include the better applications because they were funded initially. It is not clear why this should be true for Phase II and not Phase I, however.

Success rates do fluctuate, ranging from a low of 15 percent of 1994 to a

TABLE 3-11 Phase II Success Rates, 1992-2005

Fiscal Year	All Applications (#)	Total Funded (#)	Success Rate (%)
1992	551	278	50.5
1993	637	360	56.5
1994	744	351	47.2
1995	780	370	47.4
1996	798	390	48.9
1997	800	468	58.5
1998	827	541	65.4
1999	897	539	60.1
2000	1,023	587	57.4
2001	1,074	683	63.6
2002	1,248	797	63.9
2003	1,299	788	60.7
2004	1,410	792	56.2
2005	1,451	774	53.3

SOURCE: National Institutes of Health.

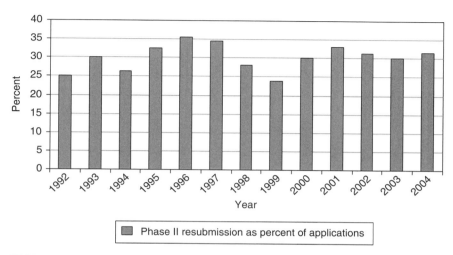

FIGURE 3-19 Phase II—Resubmission rates, 1992-2004.

SOURCE: National Institutes of Health.

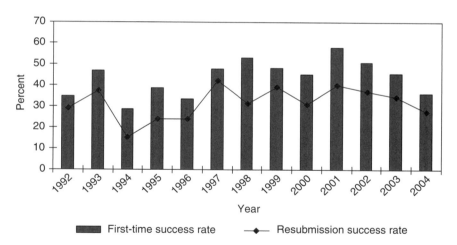

FIGURE 3-20 Phase II—Success rates for resubmitted and initial applications, 1992-2004.

SOURCE: National Institutes of Health.

high of 42 percent in 1997. Since 1997, resubmission success rates have trended downward to 27 percent in 2003. In recent years, changes in resubmission success rates have tracked quite closely with first-time success rates, though at a lower level.

3.6 CONTRACTS AT NIH

As noted earlier, more than 95 percent of all NIH SBIR awards are grants, not contracts. For Phase I in FY2000, there were 28 contracted Phase I awards, for a total of $2.4 million, in comparison to 969 Phase I grants totaling 114.1 million. Phase I contracts were 2.8 percent of all NIH Phase I SBIR awards. Contracts for Phase II are more prominent, accounting for 28 awards worth $9.1 million, while there were 267 grants totaling $119.7 million. Contracts were 9.75 percent of Phase II awards.

Total funding committed for Phase I contracts appears to have increased over time (mostly in line with the general increase in the size of Phase I from $50,000 to $100,000 in FY1998); the number of contracted awards has not increased. Updated to 2005, 42 contracts were awarded for a total value of about $5 million.

It is worth noting that Phase II contracts are relatively more important, accounting for 7.7 percent of all contracts and 5.5 percent of first-year commitments in FY2005. Also, it appears that companies winning Phase I contracts are much more likely to be selected for Phase II contracts than Phase I grantees are to be awarded a Phase II award. Updated to 2005, there were 28 contracts (out of 391 awards), for a total of $21.2 million (out of $409 million total).

3.7 PROGRAM ANNOUNCEMENTS AND REQUESTS FOR APPLICATIONS

NIH uses four different funding avenues to support extramural research.

- **Investigator-Initiated Research. Unsolicited:** The investigator initiates the research and submits a award application within an area that is relevant to the NIH. Most applications for NIH support are unsolicited.
- **Program Announcement (PA). Solicited:** NIH announces funding opportunities through award applications or cooperative agreements in a given research area representing a new, ongoing or expanded interest and/or high-priority program; Generally, no set-aside of funds, and applications submitted in response are often considered investigator-initiated in that the applicant has responsibility for the planning, direction, and execution of the proposed project.
- **Request for Applications (RFA). Solicited:** NIH solicits research grant applications for a one-time competition on a specific topic. They describe an IC initiative in a well-defined scientific area to stimulate research in a priority area; SBIR funds are set aside to cover a certain number of awards.
- **Request for Proposals (RFP). Solicited:** NIH solicits submissions of research proposals for a one-time competition on a specific IC topic. SBIR funds are set aside to cover a certain number of awards.

RFAs (grants/cooperative agreements) and RFPs (contracts) tend to be used more in problem-oriented research efforts, such as disease-specific programs, especially in their beginning stages (for example, in the early years of the War on Cancer and of research on AIDS and Alzheimer's disease).

There has been an important procedural change with electronic submission of grant applications in that all applications must be submitted in response to a Funding Opportunity Announcement (FOA). FOA is Grants.gov's terminology for what NIH refers to as Program Announcement (PA), Request for Application (RFA), Program Announcement with special receipt, referral and/or review consideration (PAR) and Program Announcement with a set aside (PAS).

NIH has issued Parent announcements for the SBIR and STTR program (and developed Omnibus Parent Announcements in November, 2005 for the December 2005 receipt date), for use by applicants who wish to submit, what were formerly termed, "unsolicited" applications. Responding to such an omnibus or umbrella Parent FOA ensures that the correct application package is used and enables NIH to receive the application from Grants.gov. This process in no way diminishes the interest of NIH Institutes and Centers in investigator-initiated, unsolicited research grant applications.

Thus an increased use in RFAs could indicate an ICs shift toward identifying key areas as "high priority." However, PAs are, according to NIH staff, better

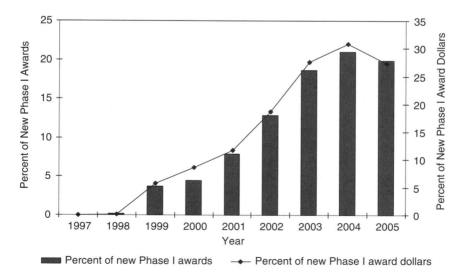

FIGURE 3-21 Phase I—Program announcements, 1997-2005.

SOURCE: National Institutes of Health.

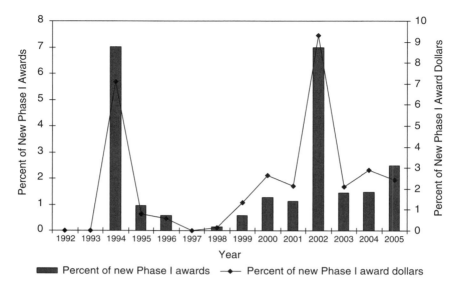

FIGURE 3-22 Phase I RFAs, 1992-2005.

SOURCE: National Institutes of Health.

understood as ways to articulate scientific areas NIH supports, but where all ideas are still investigator-initiated/investigator-driven.[17]

In fact, the data show that SBIR awards have increasingly come from PAs and RFAs. For Phase IIs, RFAs have been used for a minimal share of awards, reaching a maximum of less than 2 percent in 2004 and 2005. PAs have been more important. (See Figures 3-21, 3-22, and 3-23.)

The data show that overall, PAs have become an increasingly important component of the award flow within the NIH SBIR program, and that the trend suggests that this importance will continue to increase. This may be of particular importance as PAs can have extra funding or extra years of support attached to them.[18]

However, it is also apparent RFAs have remained of much lower importance. This is a significant difference, in that RFAs are much more heavily directed by

---

[17]Jo Anne Goodnight, NIH SBIR Program Coordinator, Personal Communication, November 1, 2006.

[18]See for example the recent announcement seeking applications for "New Technology for Proteomics and Glycomics (SBIR [R43/R44])" (*<http://grants.nih.gov/grants/guide/pa-files/PA-06-128. html>*). This provides for up to two years, with $200,000 in support for each year for Phase I, and 4 years and up to $400,000 per year for Phase II. This award was simply the first announcement identified by Google. It was not selected because it was especially large or long term.

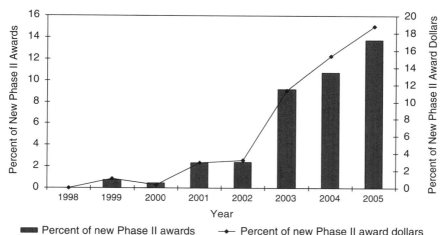

FIGURE 3-23 Phase II use of PAs at NIH, 1998-2005.

SOURCE: National Institutes of Health.

BOX 3-1
A Note on Data

NIH data collection approaches presents some challenges for the purposes of this study. In principle, NIH collects data by award year—which means that each year of an award is separately identified within the database.

There are occasionally difficulties in connecting every award year of each award, as grant ID numbers can change, as do other possible connecting fields such as company name and award title.

As a result, we have for the purposes on this study developed approximations for award numbers and sizes. Numbers of Phase I awards are estimated by using the first year of SBIR support. Numbers of Phase II awards are estimated by using the second year of support where the award ID indicates that this is a Phase II award. Year two of Phase II support is estimated using year three of support where the award is Phase II, and year three of Phase II support is estimated using year four of support where the award is Phase II.

These estimates are undoubtedly not completely accurate. Where companies have received a second year of support during Phase I, the third year of support could be only the first year of Phase II support.

In a similar vein, we have generated estimates for average award size by adding the average for different years of Phase II support. It would facilitate future assessments if NIH would find ways to address these data difficulties as the agency refines its ongoing evaluation and assessment program.

the agency—proposals made in response to PAs remain, in the words of the SBIR Program Coordinator, "viewed as investigator-initiated."

The data suggest that NIH is moving quite cautiously toward a model where some funding is specifically allocated for agency-directed research, and a much larger amount of funding is distributed so as to encourage research in particular areas.

4

NIH SBIR Program—Outcomes

4.1 INTRODUCTION

The Congress has tasked the National Academies to assess whether and to what extent the SBIR program at NIH has met the congressionally mandated objectives for the program, and to suggest possible areas for improvement in program operations. Although Congress has over the years identified a number of objectives for the program, these mandated objectives are usually summarized as follows:

- Supporting the commercialization of federally funded research.
- Supporting the agency's mission.[1]
- Supporting small business and in particular woman- and minority-owned businesses.
- Expanding the knowledge base.

These four areas define the structure and content of this chapter. A subsequent chapter reviews program management in more detail, and provides a basis for possible improvements to the program.

Such an assessment raises difficult methodological challenges, which are discussed and to the maximum extent possible resolved in the NRC's Methodology

---

[1]The mission of NIH is ". . . science in pursuit of fundamental knowledge about the nature and behavior of living systems and the application of that knowledge to extend healthy life and reduce the burdens of illness and disability." Accessed at: *<http://www.nih.gov/about/>*.

Report.[2] One issue however should be briefly discussed here too—the question of comparators.

Assessment is usually done by comparison—comparing programs and activities, in this case. Three kinds of comparison seem possible: with other NIH programs, with SBIR programs at other agencies, and with early stage technology development funding in the private sector, such as venture capital activities.

Yet none of these comparisons is valid.

Other award programs at NIH have fundamentally different objectives, such as promoting basic research (e.g., RO1 awards), developing medical capacity (awards for medical centers), or training. No other NIH award programs have as a primary goal the commercial exploitation of research. This fundamental difference in objectives must be taken into account in evaluating the SBIR program at NIH.

SBIR programs at other agencies are organized very differently and—at DoD and NASA at least—have quite different objectives.

NIH SBIR might be compared with venture capital activities, but these are typically focused closer to market, and include much larger investment (an average investment round of $7 million in 2005 as against less than $1 million for SBIR). VC investments are also focused on companies, not projects, further invalidating comparisons.[3]

Finally, while the question of commercialization is the most readily subject to measurement—through accessible data on sales and licensing revenues and other metrics—Congress has not prioritized among the four mandated objectives and each is equally important to NIH.

4.2 COMMERCIALIZATION

How well has the NIH SBIR program fostered commercialization of funded research? The following sections examine a variety of relevant indicators.

4.2.1 Proposed Commercialization Indicators and Benchmarks

Three sets of indicators are used to evaluate the extent to which SBIR grantees have commercialized their funded research:

1. **Sales and licensing revenues** ("sales" hereafter unless otherwise noted). Revenues flowing into the company from the commercial marketplace

---

[2]National Research Council, *An Assessment of the Small Business Innovation Research Program: Project Methodology*, Washington, DC: The National Academies Press, 2004.

[3]See National Venture Capital Association, *Money Tree Report*, November, 2006. The mean venture capital deal size for the first three quarters of 2006 was $8.03 million. This trend has been accelerated by the growth of larger venture firms. See P. Gompers and J. Lerner, *The Venture Capital Cycle*, Cambridge: The MIT Press, 1999, Ch. 1.

constitute an important measure of commercial success, as sales are an indicator of realized market demand for the output from a project.

There is however no single agreed benchmark against which to measure whether agencies have met the legislative objectives for commercialization. It seems, therefore, reasonable to assess commercialization against a range of benchmarks:

    (a) Reaching the market—any sales.
    (b) Reaching $1 million in cumulative sales—which has been approximately the median size of a Phase I plus a Phase II award at NIH.
    (c) Reaching $5 million in cumulative sales—which could be viewed as a modest commercial success.
    (d) Reaching $50 million in cumulative sales—which could be viewed as full commercial success.

2. **R&D investments and research contracts.** Beyond sales, further R&D investments and contracts are also good evidence that results from the project are moving toward commercialization. These investments and contracts may include partnerships, further grants and awards, or government contracts. The benchmarks for success at each of these levels should be the same as those above, namely:

    (a) Any R&D additional funding.
    (b) Funding of $1 million or more.
    (c) Funding of $5 million or more.
    (d) Funding of $50 million or more.

3. **Sale of equity** constitutes a less clear-cut indicator of commercial activity. A company which is sold because its acquirer is seeking a successful product has generated returns. Key metrics include:

    (a) Equity investment in the company by independent third party.
    (b) Sale or merger of the entire company.

Using these metrics, to what extent have NIH SBIR companies commercialized?

4.2.2 Sales and Licensing Revenues from NIH SBIR Awards

Data from three sources indicate that 30-40 percent of NIH projects funded between 1992 and 2002 have reached the marketplace. (These three data sources all refer here only to NIH projects. Note however that subsequent NIH resurveys suggest that this may substantially understate the eventual commercialization rate.)

The projects underlying the percentages in Figure 4-1 have generated posi-

---

BOX 4-1
A Note on Data Sources

Research on the NIH SBIR program has benefited from the existence for three independent data sources on outcomes from the program.[a] These are:

- The **NRC Phase II Survey** (2005), which sent at least one questionnaire to every Phase II recipient at NIH, 1992-2005. Firms with multiple awards received more questionnaires, but normally not for each award.
- The NIH's **"National Survey to Evaluate the NIH SBIR Program: Final Report"** (hereafter the NIH Survey) (2003), which sent one questionnaire to each firm with a Phase II award 1992-2002. This survey has subsequently been updated.
- The **DoD Commercialization Reports** (CCRs), through which firms applying for future awards at DoD must report on commercialization outcomes for awards at all agencies, including NIH. Data on about 12 percent of NIH awards can be found in the DoD database.

---

[a]For details on the NRC Phase II Survey, see Appendix B. For details on the NIH Survey, see National Institutes of Health, *National Survey to Evaluate the NIH SBIR Program: Final Report*, July 2003. Available online at: *<http://grants.nih.gov/grants/funding/sbir_report_2003_07.pdf>*. The DoD Index is not publicly available.

---

tive revenue from sales or licensing. Follow-on surveys at NIH indicate that this figure could eventually grow to about 60 percent of projects. However, determining that projects have generated some revenues is insufficient, in three respects: First, the distribution of sales by size of revenue is important: Projects generating $50 million in sales have substantially greater commercial returns than those generating $100,000. Second, data on sales to date are insufficient: Accurate analysis requires the adjustment of this raw data set to take account projections of future sales. Third, it is useful to distinguish between sales and licensing revenues.

4.2.2.1 Sales Ranges

Figure 4-2 shows the number of grantees achieving each of the specified sales benchmarks. There are general similarities between the three data sources. The majority of sales (at least 68 percent for all three sources) are concentrated in the $0-$1 million range. None of the sources indicate that more than 10 percent of projects generated $5 million in cumulative revenues. Each data source recorded one (different) project with more than $50 million in revenues.

The DoD database indicates lower commercialization results than the two surveys. Entries in the DoD database constitute a formal part of the SBIR application process, capturing updated data at that time about commercialization from all previous SBIR Phase II awards, and companies may therefore be more

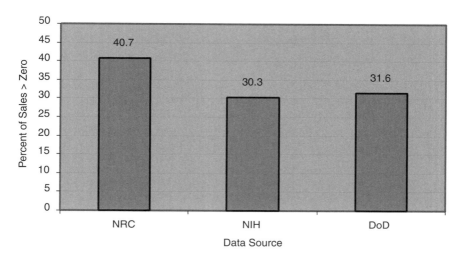

FIGURE 4-1 Percentage of NIH SBIR projects reaching the market from 1992-2002.

SOURCE: NRC Phase II Survey, DoD Commercialization database, and National Institutes of Health, *National Survey to Evaluate the NIH SBIR Program: Final Report*, July 2003.

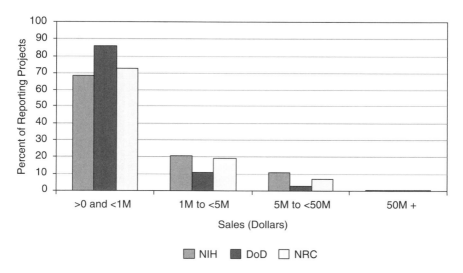

FIGURE 4-2 Sales by sales range, total for 1992-2002.

SOURCE: National Institutes of Health, NRC Phase II Survey, and DoD Commercialization database.

BOX 4-2
Multiple Sources of Bias in Survey Response

Large innovation surveys involve multiple sources of bias that can skew the results in both directions. Some common survey biases are noted below. These biases were tested for and responded to in the NRC surveys.[a]

- **Successful and more recently funded firms are more likely to respond.** Research by Link and Scott demonstrates that the probability of obtaining research project information by survey decreases for less recently funded projects and increases with the award amount.[b] Nearly 40 percent of respondents in the NRC Phase II Survey began Phase I efforts after 1998, partly because the number of Phase I awards increased, starting in the mid 1990s, and partly because winners from more distant years are harder to reach. They are harder to reach as time goes on because small businesses regularly cease operations, are acquired, merge, or lose staff with knowledge of SBIR awards.

- **Success is self-reported.** Self-reporting can be a source of bias, although the dimensions and direction of that bias are not necessarily clear. In any case, policy analysis has a long history of relying on self-reported performance measures to represent market-based performance measures. Participants in such retrospective analyses are believed to be able to consider a broader set of allocation options, thus making the evaluation more realistic than data based on third party observation.[c] In short, company founders and/or principal investigators are in many cases simply the best source of information available.

- **Survey sampled projects at firms with multiple awards.** Projects from firms with multiple awards were underrepresented in the sample, because they could not be expected to complete a questionnaire for each of dozens or even hundreds of awards.

- **Failed firms are difficult to contact.** Survey experts point to an "asymmetry" in their ability to include failed firms for follow-up surveys in cases where the firms no longer exist.[d] It is worth noting that one cannot necessarily infer that the SBIR project failed; what is known is only that the firm no longer exists.

- **Not all successful projects are captured.** For similar reasons, the NRC Phase II Survey could not include ongoing results from successful projects in firms that merged or were acquired before and/or after commercialization of the project's technology. The survey also did not capture projects of firms that did not respond to the NRC invitation to participate in the assessment.

- **Some firms may not want to fully acknowledge SBIR contribution to project success.** Some firms may be unwilling to acknowledge that they received important benefits from participating in public programs for a variety of reasons. For example, some may understandably attribute success exclusively to their own efforts.

- **Commercialization lag.** While the NRC Phase II Survey broke new ground in data collection, the amount of sales made—and indeed the number of projects that generate sales—are inevitably undercounted in a snapshot survey taken at a single point in time. Based on successive data sets collected from NIH SBIR award recipients, it is estimated that total sales from all responding projects will likely be on the order of 50 percent greater than can be captured in a single survey.[e] This underscores the importance of follow-on research based on the now-established survey methodology.

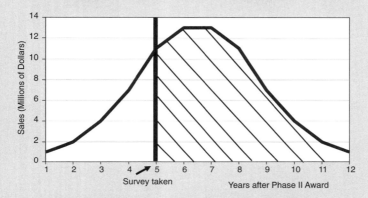

FIGURE B-4-1 Survey bias due to commercialization lag.

These sources of bias provide a context for understanding the response rates to the NRC Phase I and Phase II Surveys conducted for this study. For the NRC Phase II Survey for NIH, of the 1,127 firms that could be contacted out of a sample size of 1,680, 496 responded, representing a 44 percent response rate. The NRC Phase I Survey captured 10 percent of the 7,049 awards made by NIH between 1992 and 2001. See Appendixes B and C for additional information on the surveys.

---

[a]For a technical explanation of the sample approaches and issues related to the NRC surveys, see Appendix B.

[b]Albert N. Link and John T. Scott, *Evaluating Public Research Institutions: The U.S. Advanced Technology Program's Intramural Research Initiative*, London: Routledge, 2005.

[c]While economic theory is formulated on what is called 'revealed preferences,' meaning individuals and firms reveal how they value scarce resources by how they allocate those resources within a market framework, quite often expressed preferences are a better source of information especially from an evaluation perspective. Strict adherence to a revealed preference paradigm could lead to misguided policy conclusions because the paradigm assumes that all policy choices are known and understood at the time that an individual or firm reveals its preferences and that all relevant markets for such preferences are operational. See {1} Gregory G. Dess and Donald W. Beard, "Dimensions of Organizational Task Environments." *Administrative Science Quarterly*, 1984, 29: 52-73. {2} Albert N. Link and John T. Scott, *Public Accountability: Evaluating Technology-Based Institutions*, Norwell, Mass.: Kluwer Academic Publishers, 1998.

[d]Albert N. Link and John T. Scott, *Evaluating Public Research Institutions: The U.S. Advanced Technology Program's Intramural Research Initiative*, op. cit.

[e]Data from NIH indicates that a subsequent survey taken two years later would reveal very substantial increases in both the percentage of firms reaching the market, and in the amount of sales per project.

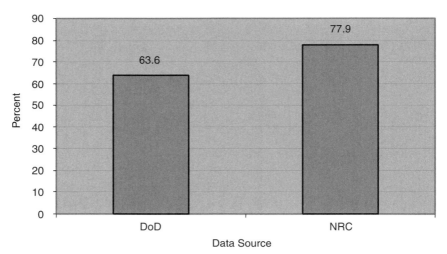

FIGURE 4-3 Degree of all sales concentrated in companies reporting $5 million+ in sales.

SOURCES: NRC Phase II Survey, DoD Commercialization database.

likely to ensure that their responses are conservative. The DoD responses are also from fewer companies, as they include a number of companies with numerous responses: Only 108 companies accounted for all the DoD responses, compared with 495 companies for the NIH Survey.[4]

Sales are highly concentrated. Figure 4-3 shows that the few projects generating at least $5 million per year in revenues account for most of the revenues reported for all projects, ranging from slightly over 60 percent for DoD respondents to more than 75 percent for NRC respondents.

This degree of sales concentration confirms the view that from the perspective of sales, the SBIR program at NIH generates a considerable number of projects that reach the market, no more than 10 percent of which generate sales greater than $5 million in total from the surveyed projects. Two of these larger winners, Optiva[5] and Martek, are discussed in Box 4-3 and Box 4-4.

---

[4]National Institutes of Health, *National Survey to Evaluate the NIH SBIR Program: Final Report*, July 2003 [NIH Survey]. The NIH Survey addressed one questionnaire to every firm winning a Phase II award during the selected period; the DoD data derives from firms applying at DoD who had also won previous NIH Phase II awards, and were thus required to answer commercialization questions about those awards.

[5]Interview with David Guiliani, Optiva founder, July 2006. See also *Puget Sound Business Journal*, "Philips to Acquire Optiva Corp." August 22, 2000.

BOX 4-3
Optiva Corporation

Medicine and dentifrice dispensing sonic brush, sonic toothbrush

Optiva, formed as Tech in 1987 by an entrepreneur and two University of Washington professors, controls more than 26 percent of the U.S. power-toothbrush market, generating a $300 million business and 500 jobs mostly in Snoqualmie, Washington.

By 1997, Optiva was named the fastest-growing company in the U.S. by Inc. magazine, and its CEO was selected as the Small Business Person of the Year.

In August 2000, Philips agreed to acquire Optiva for an undisclosed price (reputed to be approximately $1 billion). At the time, Optiva had more than 600 employees and more than $175 million in annual sales from the Sonicare line. By 2001, Optiva had sold its 10 millionth power toothbrush, and had become the #1 producer of power toothbrushes in the U.S. market.

BOX 4-4
Martek Biosciences Corporation

Products from microalgae

Martek Biosciences Corporation develops and commercializes products from microalgae. Martek's products include fatty acids (omega-3 docosahexaenoic acid and omega-6 arachidonic acid) which are used as ingredients in infant formula and animal feeds. Martek's DHA-rich oil can also be used in nutritional supplements and functional foods for older children and adults. Martek also produces fluorescent algal pigments used for diagnostic and pharmaceutical research purposes.

Martek has become an important player in three markets:

- **Infant formula.** Martek has developed and patented two fermentable strains of microalgae which produce oils rich in docosahexaenoic acid, DHA. In like manner, another patented process was developed for a fungus that produces an oil rich in arachidonic acid, ARA. Both DHA and ARA are found in breast milk and are important nutrients in infant development. Thus the two oils are used in infant formulas.
- **Nutritional supplements.** The DHA-rich oil can also be used in supplements and functional foods for older children and adults.
- **Life sciences and research.** Martek also makes and sells a series of proprietary and nonproprietary fluorescent markers. These products have applications in drug discovery (high-throughput screening), DNA microarray detection and flow cytometry.

Martek developed the technology underlying these products directly as a result of SBIR funding, according to Henry Linsert, founder and CEO. The result has been explosive growth for the company, rising from about $5 million in 2000 to more than $185 million in 2004.

4.2.2.2 Sales Expectations

Nineteen percent of NRC Phase II Survey respondents did not yet report sales but expected sales in the future (see Figure 4-4). Table 4-1 shows that these expectations are strongly concentrated in the immediate out years.

These expectations may, however, be overly optimistic. Table 4-2 shows the elapsed time between the end of the Phase II award and the date of first sales. In some cases, possibly where the award is for improvements to existing technologies, first sales may occur before the Phase II award is even completed.

The data set in Table 4-2 shows that the median elapsed time to sale is less than two years—more than half of all projects reporting sales claim a date of first sale within two years of the start of the Phase II award. This number can be negative in cases where companies were using SBIR to improve products already in the market.

Further, NRC Phase II Survey responses indicate that more than 85 percent of first sales occurred before the end of the 4th year after the date of the award.

About 19 percent of all NRC Phase II Survey respondents claimed that they anticipated sales in the future. However, if the survey data accurately predicts the distribution of first sales across elapsed time since award, these respondents appear to be overly optimistic.

The likelihood of a project generating initial sales diminishes with time elapsed since the award. Table 4-3 focuses on the projects from the NRC Phase II Survey that still anticipate sales. It identifies the award year, and assigns a percentage likelihood of first sales, based on the distribution in Table 4-2. The NRC Phase II Survey data indicate that a vast majority (86.2 percent) of first sales are made within 4 years after the date of award. Consequently, projects that have *not*

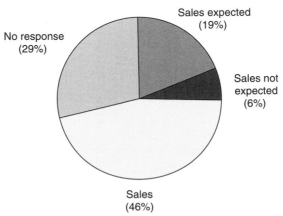

FIGURE 4-4 Sales expectations.

SOURCE: NRC Phase II Survey.

TABLE 4-1 Year of Expected Sales

Year of Expected Sales	Number of Projects
2005	22
2006	20
2007	20
2008	11
2009	1
2010	6
2011	2
2012	1

SOURCE: NRC Phase II Survey.

generated a first sale within four years have a 13.8 percent likelihood that they will do so—historically, 86.2 percent of projects will have reported sales by then if they are going to have sales at all.

These percentages can be used to adjust the claims of respondents, in Table 4-3. They indicate that while 95 projects report that they still expect sales, we estimate that in the end only five will actually reach the market.

It is important to note that this analysis refers only to first sales. The bulk of sales in almost all cases occur at different and unknown periods after the first sale. This is an important point: The sales data from the survey are effectively a snapshot of sales taken at a specific point in the lifetime of a product. Most product revenue returns are bell-shaped—ramping up from initial sales to a maximum

TABLE 4-2 Years Elapsed Between Start of Phase II Award and Year of First Sale

Elapsed Years	Number of Projects	Percentage of Responding Projects
−11	1	0.4
−7	1	0.4
−4	1	0.4
−3	4	1.8
−2	6	2.7
−1	9	4.0
0	18	8.0
1	29	12.9
2	50	22.3
3	48	21.4
4	26	11.6
5	17	7.6
6	8	3.6
7	3	1.3
8	1	0.4
9	2	0.9

SOURCE: NRC Phase II Survey.

TABLE 4-3 Frequency by Award Year for Companies Still Expecting Sales, 1992-2001.

Award Year	Number of Projects	Elapsed Years between Award and Survey	Historical Success (%)
1992	1	13	0.0
1993	3	12	0.0
1994	3	11	0.0
1995	3	10	0.0
1996	3	9	0.9
1997	13	8	0.4
1998	12	7	1.3
1999	20	6	3.6
2000	19	5	7.6
2001	18	4	11.6

NOTE: The results are calculated as follows: Y= time elapsed between date of award and date of NRC Survey); D = 100-sum of percentages from Table 4-2 column three for that number of elapsed years (e.g., for four elapsed years, the sum = sum (all years up to and including 4) = 100; 86.2 percent = 13.8 percent.

SOURCE: NRC Phase II Survey.

and then declining as the product is overtaken in the marketplace. As the bulk of responses to all surveys tend to be concentrated among more recent awards, the "snapshot" in aggregate may therefore be focused on the early ramp up stage.

This hypothesis is supported by recent data from NIH, where the 2002 survey was followed up in 2005. During this period, the percentage of firms with sales increased from 47 percent to 63 percent, and the estimated aggregate sales doubled, to approximately $1.6 billion.[6] None of this subsequent sales growth could be captured during the initial 2002 survey, and we would expect to see a similar trajectory for the NRC survey completed in 2005.

4.2.2.3 Imminent Sales

While the analysis above shows that claims of future sales can be regarded with some caution, focusing attention on imminent sales—those expected to be made within the next 18 months—may provide a more reliable metric, and case studies indicate that company managers have a better understanding of the near future markets for their products.

The NRC Phase II Survey asked firms winning SBIR Phase II awards to estimate the approximate amount of total sales resulting from the technology

---

[6]Jo Anne Goodnight, NIH SBIR/STTR Program Coordinator. Personal communication, April 4, 2007.

developed during the project expected over the next 18 months, by the end of FY2006.

Of the 496 survey recipients, 225 (45.4 percent) anticipated sales within the next 18 months. The overall mean amount of anticipated sales was $559,622. On average, companies without sales to date that anticipated any sales over the 18-month period estimated an average of $1,233,656 in anticipated sales. However, this figure may be optimistic: Less than 30 percent of the projects that reported existing sales claimed that they have had sales of at least this magnitude.

4.2.2.4 Sales by Industry

Do the data show differences in commercialization by industry sector? Based on the NIH Survey (which asked for the primary customer base) Table 4-4 shows that four industry groups (biotechnology, information and research, instrumentation, and medical devices) account for 77 percent of the 205 projects reporting sales.

However, this data set needs to be adjusted to account for the number of respondents in each industry group. Figure 4-5 provides average sales per respondent, by industry. It shows that information and research, and health care provide average sales about twice the amount of other leading sectors. NIH might wish to consider further what makes projects in some sectors more commercially successful than others—and might even consider whether shifting SBIR resources toward those more successful sectors might be warranted.

4.2.2.5 Sales by Size of Company (Employees)

Do commercialization results vary with the number of employees at the time of the award? Although none of the agencies currently gather data about company size during the application process itself, size may be an important predictor of commercial success. Data in Table 4-5 show that there are differences by size of company.

Firms with 10 employees or less account for 41.5 percent of respondents, and 50 percent of projects with some sales but less than $1 million. Firms with 11-25 employees account for 28.3 percent of respondents, 27.2 percent of sub-$1 million returns, and 35.4 percent of respondents with sales of more than $1 million and less than $50 million.

The comparison above shows that companies with no more than 75 employees consistently outperform companies with more than 75 employees in terms of the percentage of projects that generate sales. The former group of companies account for 76.6 percent of respondents to the NRC survey, but 86.1 percent of all projects with sales.

In fact, the "sweet spot" by size is concentrated around 20 employees: com-

TABLE 4-4 Sales—By Industry Sector

Industry	Number of Sales by Industry, by Size								Grand Total
	MISSING	<$50,000	>$50K-<$100K	$100K-<$500K	$500K-<$1M	$1M-<$5M	$5M-<$50M	$50M+	
Biotechnology	3	8	2	11	5	7	6		42
Chemical technology	1	1	2	3	2	1	1		11
Computer hardware, software	2	7	4	8	2	11			34
Diagnostics		1	1	4		1	1		8
Environment, ergonomics	1	1		4					6
Health care	1		2	1	1		3		8
Information & research			2		1		4		7
Instrumentation	1		2	17	3	7	3		33
Medical devices	6	6	2	4	9	11	4	1	43
Medical education, health promotion	4	11	2	6	3	1			27
Other				1	1				2
Pharmaceuticals						3			3
Grand Total	19	35	19	59	27	42	22	1	224
	19	35	19	59	27	159	22	1	

SOURCE: National Institutes of Health, *National Survey to Evaluate the NIH SBIR Program: Final Report*, July 2003.

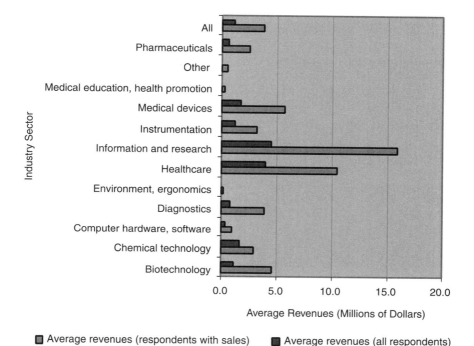

FIGURE 4-5 Per respondent sales, by industry sector (millions of dollars).

SOURCE: National Institutes of Health, *National Survey to Evaluate the NIH SBIR Program: Final Report*, July 2003, NRC estimates.

panies with 11-30 employees accounted for 20.8 percent of respondents, but 31.7 percent of projects with sales.

4.2.2.6 SBIR-only Focus

One question about the role of the SBIR program concerns the extent to which simply acquiring SBIR awards can substitute for further commercial activity. As shown in Table 4-6, some companies' revenues are made up largely of SBIR awards, but the percentage reliance on SBIR awards tend to decline as the size of the company grows.

The data show responding firms' current SBIR focus and current revenue, which may of course be quite different from that during the time period of the SBIR. Very small companies that won SBIRs in the past may not now have one (hence the 31 companies with zero revenues and zero SBIR focus).

Despite these caveats, the data confirm that as companies get larger, their

TABLE 4-5 Sales by Company Size

Employees	Sales ($)				Total	Percent
	<1M	>1M to <5M	>5M to <50M	>50M		
0-5	39	3	1		43	21.3
6-10	35	6			41	20.3
11-15	15	7	6		25	12.4
16-20	16	1	2		18	8.9
21-25	9	4	2		14	6.9
26-30	6	1			7	3.5
31-40	2	5	5		10	5.0
40-50	6	1	1		8	4.0
51-75	3	3	2		7	3.5
76-100		1			1	0.5
100-200	6	2	2	1	11	5.4
201-300	1		1		2	1.0
301-500	2	1			3	1.5
500+	1	2			3	1.5
Missing	6	2	1		9	4.5
Total	147	39	23	1	202	100.0
Percentage	72.8	19.3	11.4	0.5	100.0	

SOURCE: NRC Phase II Survey.

reliance on SBIR funds tends to decline. Of the 38 companies with at least $5 million in revenues, 30 (78.9 percent) reported no more than a 10 percent focus on SBIR. Conversely, of the 102 firms reporting at least 76 percent focus on SBIR, 100 reported annual firm revenues of no more than $1 million.

4.2.2.7 Licensing Revenues

Up to this point, we have focused on sales and licensing revenues accruing to the respondent. However, it is possible that licensing has some kind of multiplier effect by providing the licensee with a critical piece of technology. This could potentially create a substantially larger commercial impact than is captured in the direct sales data of the licensor, and this larger impact would be based on technologies developed with SBIR funding.

Licensing revenues constitute a fairly small fraction of overall sales: The $32,664,380 in licensing revenues reported by NRC respondents constitutes 8.8 percent of all reported revenues. Only a small fraction of SBIR grantees generate substantial revenues from licensing.

This suggests that few companies can rely on licensing alone as a means of generating significant revenues, even though case studies indicate that some companies—and possibly many smaller SBIR recipients without manufacturing capabilities—have business plans that depend on licensing revenues.

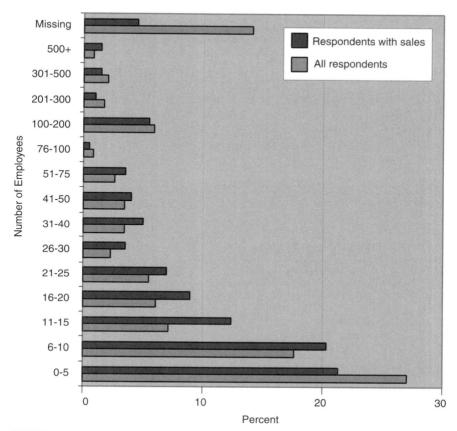

FIGURE 4-6 Distribution of companies by number of employees.

SOURCE: NRC Phase II Survey.

Beyond the firm receiving the award, licensing creates opportunities for the licensee, and the NRC Phase II Survey attempts to quantify how some companies capitalize on this opportunity. It should be borne in mind that these responses are from the licensor company, and may not be an accurate picture of licensee activity.

As with direct sales data, the responses shown in Table 4-8 suggest that a large majority of licensee sales are less than $1 million, and that there are only a few very large responses.

Total sales reported for licensees as $336,677,403. Of this, $324,588,050 (96.4 percent) came from the eight responses (2 percent of all projects responding) reporting at least $5 million in licensee sales. These data indicate that licensing revenues are much more concentrated in a handful of respondent companies

TABLE 4-6 Firm Revenues by Percentage Dependence on SBIR

Firm Revenues ($)	Percent of Firm Revenues that Come from SBIR (Number of responses in each percent range)						Total
	0	1-10	11-25	26-50	51-75	76-100	
0	31	41	2	2	2	14	61
<100K	12	15	7	16	14	29	81
100K-<500K	10	16	7	9	18	25	75
500K-<1M	28	33	17	28	21	32	131
1M-<5M	13	26	3	10	20	2	61
5M-<20M	2	21	5	2	2	0	30
20M-<100M	0	3	0	1	0	0	4
100M+	2	2	0	0	0	0	2

SOURCE: NRC Firm Survey.

TABLE 4-7 Revenues from Licensing

$5M+ (Number of Respondents)	2
$1M-<$5M (Number of Respondents)	3
$1-<$1M (Number of Respondents)	22
Total Dollars	29,184,380
Average Dollars	1,080,903
Average Dollars—All Respondents	58,839

SOURCE: NRC Phase II Survey.

TABLE 4-8 Sales by Licensees, as Reported by Licensor Respondents

Revenues Reported for Licensee ($)	Number of Responses
<1M	39
1M-<5M	5
5M-<50M	5
50M+	3

SOURCE: NRC Phase II Survey.

than direct company sales revenues (discussed earlier in this chapter). Note that respondents indicate three licensees with revenues of more than $50 million. This compares with only one such claim for the responding projects themselves.

Despite the apparent difficulties in generating substantial revenues from licensing, the latter may be the most realistic method of commercializing a product for some companies because, as noted above, small companies may not have the manufacturing, marketing, or distribution resources to effectively sell their own innovations. (See Box 4-5).

BOX 4-5
Applied Health Science and the Wound and
Skin Intelligence System™ or WSIS™

The purpose of the Applied Health Science's (AHS) early SBIR grant work was to validate and automate a standardized assessment instrument (the Pressure Sore Status Tool, originally authored by Dr. McNees' (Dr. McNees is CEO/Chief Scientists for Applied Health Sciences) colleague, Dr. Barbara Bates-Jensen) for use in field settings for describing and tracking status changes in chronic wounds (e.g., pressure ulcers).

The WSIS (Wound and Skin Intelligence System (tm) or WSIS(tm)) provides clinicians with the ability to assess risk and request a "case specific" prevention plan for reducing the probability that a wound will develop. The system tracks prevention and treatment outcomes over time and relates these outcomes to individual risk and wound profiles and interventions employed. Thus, the system has the capacity to "learn" from its own experience.

The product was commercialized through the sale of rights to ConvaTec, a wholly owned unit of Bristol Myers-Squibb and the largest wound products company in the world. ConvaTec provided Phase III funding leading to commercialization. In exchange, it received a right-of-first-refusal for licensing the system, which it subsequently executed. This merged AHS technology and research capabilities with ConvaTec's marketing power—reflected in its presence in about 80 countries world-wide. ConvaTec subsequently bought all rights to the software. AHS retained the worldwide data "pipelines", and analytical functions. AHS also has a right-to-first-review for any elaborations of or changes to the system.

AHS has announced current projections of $30 million in annual sales from the U.S. market, and expects to add one employee for each 75 users of the system. AHS and ConvaTec are also forming a series of strategic alliances with companies prepared to supply or develop add-on capabilities (e.g., a telemedicine home health company in Chicago and a long-term care claims processing company in Nashville).

The sale of technology rights to ConvaTec funded further development, situated AHS strategically where it wanted to be—focused on research and data analysis, not marketing—and took advantage of each partner's strategic strengths.

4.2.2.8 Additional Investment Funding

Further investment in an SBIR project may be further—though by no means sufficient—evidence that the work is of value, at least to the funding party. About 37 percent of NIH Survey companies received some funding other than further SBIR awards, although the NIH Survey did not ask about amounts of investment.

The NRC data differ from the DoD and NIH data in that its respondents reported a higher likelihood of their projects attracting third-party funding other than SBIR.

According to Table 4-9, a substantial number—23-58 percent—of NIH SBIR projects have been able to attract additional funding. A much smaller number—4-9 percent—have been able to attract at least $1 million in additional funding.

A more detailed comparison of the NRC and DoD data is contained in Table 4-10.

The NRC survey reports much stronger further investment than does the DoD database of NIH awards—the average investment per respondent was about $1 million, compared with about $250,000 for the DoD-reporting companies. Fourteen companies—about 7 percent of those reporting investments—received at least $5 million.

Once again, though to a somewhat lesser degree than for sales, investments are heavily concentrated in the few companies receiving substantial investments. The 14 companies with more than $5 million in investments accounted for a total of $383.5 million (76.3 percent) of all investments.

4.2.2.9 Sources of Investment Funding

The NRC Phase II Survey also sought information about the source of third-party funding. Table 4-11 contains the first detailed data on sources of additional

TABLE 4-9 Additional Investment/Funding other than SBIR

Any Investment			Investment >$1M		
	Number of Responses	Percentage		Number of Responses	Percentage
NRC Survey			NRC Survey		
No	193	42.2	No	416	91.0
Yes	262	57.8	Yes	41	9.0
Total	457	100.0	Total	457	
DoD Data			DoD Data		
No	721	76.6	No	901	95.7
Yes	220	23.4	Yes	40	4.3
Total	941		Total	941	
NIH Survey			NIH Survey		
No	487	63.4	Not available.*		
Yes	281	36.6			
Total	768				

NOTE: (*) The NIH Survey did not ask respondents how much funding had been provided, only whether there had been some amount of further non-SBIR funding, as well as the sources of the funding.

SOURCE: NRC Phase II Survey, DoD commercialization database, and National Institutes of Health, *National Survey to Evaluate the NIH SBIR Program: Final Report*, July 2003.

TABLE 4-10 Further Investments in SBIR Projects

	Number of Investments by Size of Investment		Total Investment by Size of Investment	
	DoD Data	NRC Survey	DoD Data	NRC Survey
$50M+	1	3	77,000,000	203,600,000
$5M-<$50M	3	11	32,329,122	179,979,409
$1M-<$5M	36	37	80,492,819	77,691,224
<$1M	180	202	38,637,715	40,699,881
None	721	243	0	0
Total investments	220	253	228,459,656	501,970,514
Percent of all respondents	24.4	51.0		
Average (all)			253,562	1,012,037
Average (with investment)			1,038,453	1,984,073

SOURCE: NRC Phase II Survey, DoD Commercialization Database.

funds for NIH SBIR-funded projects. As expected, venture funding provided both the largest total amount of additional support ($155 million), and also the largest average support per project funded ($10.3 million). However, venture funding supported only 15 projects—less than 4 percent of all responses.

Conversely, internal funding was by far the most widespread form of support, being reported by almost 50 percent of all respondents. Average funding was much lower, at $437,000 per project.

Investments from government and academic sources were relatively few in number (less than 8 percent of the total) and relatively small in amount on a per project basis.

TABLE 4-11 Sources of Investment Funding

Source of Investment	Total Investment ($)	Percent	Number of Investments	Percent	Average Investment ($)
Private Investment from U.S. Venture Capital	154,617,045	33.9	15	3.9	10,307,803
Private Investment from other Private Equity	141,992,212	31.1	40	10.4	3,549,805
Private Investment from Foreign Investment	39,616,075	8.7	12	3.1	3,301,340
Private Investment from other Domestic Private Company	21,624,866	4.7	31	8.1	697,576
Your Own Company	82,118,851	18.0	188	49.1	436,802
State or Local Government	6,290,000	1.4	23	6.0	273,478
Personal Funds	9,850,408	2.2	67	17.5	147,021
College or Universities	236,500	0.1	7	1.8	33,786
Total	**456,345,957**	**100.0**	**383**	**100.0**	**1,191,504**

SOURCE: NRC Phase II Survey.

TABLE 4-12 Most Important Source of Non-SBIR Funding

	Number of Responses	Percent
None	487	
Non-SBIR federal funds	19	6.8
Your own company	85	30.6
Other private company	61	21.9
U.S. venture capital	22	7.9
Foreign venture capital	3	1.1
Private individual investor	37	13.3
Personal funds	22	7.9
State or local government funds	15	5.4
College or university	5	1.8
Other	5	1.8
Foundations	4	1.4
	278	100.0

SOURCE: National Institutes of Health, *National Survey to Evaluate the NIH SBIR Program: Final Report*, July 2003.

The NIH Survey generated responses approximately in line with those from the NRC Phase II Survey. Personal and in-house corporate funds accounted for 20.2 percent of the total funding reported, with other private companies providing another 4.7 percent. It appears that VC funding is underreported: About 40 percent of companies identified by NRC as having received VC funding responded to the NIH Survey as having done so.[7]

However, neither the NIH Survey nor the NRC Phase II Survey align well with a third source of information on further investment—data from venture capital databases.[8] Though reported in more detail in that section our analysis indicates that of the top 200 Phase II award winners at NIH, 50 received venture funding (see Figure 4-7).

We have identified a total VC investment of approximately $1.59BN in these 50 companies, a total that dwarfs the $272 million investment in these companies via the NIH SBIR program.

There are four particularly striking findings regarding the data on external funding:

- Sixty-five percent of all respondents reported no additional funding for their project. Thus, in terms of the external funding indicator only, about two-thirds of all projects did not commercialize.
- Venture capital funding was of mixed importance, accounting for only

---

[7]This illustrates one limitation of the NIH data, namely that it undercounts results from multiple winners, which would presumably include a significant number of the VC-funded companies.

[8]See National Research Council, *Venture Funding and the NIH SBIR Program*, Charles W. Wessner, ed., Washington, DC: The National Academies Press, Forthcoming.

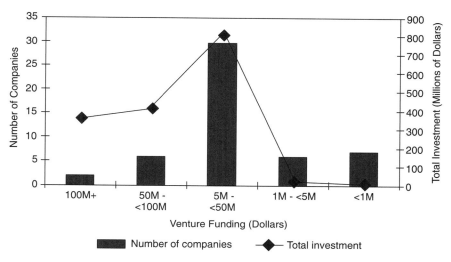

FIGURE 4-7 Venture funding for NIH Phase II winners.

SOURCES: VentureSource and other VC databases; NIH awards database.

3.5 percent of all investments, but almost 30 percent of investments by value.

- The amount of state and local funding provided was small, providing funding for 5.4 percent of projects with funding, or no more than 2.5 percent of all NIH respondents. By contrast, more than half of all respondents received additional SBIR funding related to the project (see below).
- *In fact, on the basis of additional funding alone, it is fair to conclude that SBIR provided additional funding to more projects than did all other sources of additional funding combined.*

4.2.2.10 SBIR Impact on Further Investment

Both the NIH and NRC surveys sought additional information about the impact of the SBIR program on company efforts to attract third-party funding. This "halo effect" was mentioned by some case study interviewees who suggested that an SBIR award acted as a form of validation for external inventors.

Case study interviews provided mixed views on this. Some interviewees strongly supported the view that SBIR helps to attract investment; others claimed that the effect was not that important. This is to be expected insofar as two-thirds of SBIR respondents did not attract outside funding, and only 3.5 percent received venture funding. This suggests that SBIR awards do not in themselves guarantee further external funding.

Survey responses, however, painted a more positive picture of these effects: 69 percent of NIH survey respondents said that the SBIR award had helped them in their efforts to raise additional capital (although only 29 percent reported actually received additional capital). Of the NIH grant recipient respondents that did receive additional funding other than SBIR, 78 percent agreed that this "resulted from" their SBIR participation.

4.2.2.11 Additional SBIR Funding

Aside from third-party investment, the federal government in many cases makes further investments via the SBIR program itself. Both the NIH Survey and NRC Phase II Survey attempted to determine how many additional SBIR awards followed each initial award (see Table 4-13).

Both surveys suggest that over one-third of grant recipients receive at least one additional related Phase II award. Approximately 14 percent of respondents reported receiving at least two additional awards, but as one might expect given the skew in results, and the competition for awards, two-thirds of respondents report no additional related SBIR awards at all.

4.2.2.12 Employment Effects

Employment resulting from the Phase II project is another indicator of commercialization. It is also an indirect indicator of the SBIR program's support for small businesses.

TABLE 4-13 Related Phase II SBIR Awards

	NIH Survey	NRC Survey
Number of Additional Awards	Number of Respondents	Number of Respondents
1	152	92
2	65	30
3	19	11
4	8	8
5	4	12
6	2	0
7	1	1
8	1	0
10	3	2
12	1	0
11		1
> 27		5

NOTE: Overall percentages use total responses + missing responses as denominator.

SOURCES: NRC Phase II Survey; National Institutes of Health, *National Survey to Evaluate the NIH SBIR Program: Final Report*, July 2003.

As shown by Figure 4-8, the median size of companies receiving SBIR awards is relatively small—far lower than the 500 employee limit imposed by the SBA. The median size of grant recipient companies is 10 employees, and 60 percent of respondent companies had 15 employees or fewer at the time of the survey.

However, while the median size of grant winners is small, and most award-ees have 20 or fewer employees, employment is skewed across company size. Total reported employment at the 319 companies is 15,467.5 full-time equivalent employees (FTEs), but 8,090 (52 percent) of those FTEs work for the top ten companies—and three of those companies are no longer eligible for SBIR awards because they employ more than 500 persons.

These results broadly match the data from the NIH Survey, which also shows that most employment is concentrated in the larger companies (Figure 4-9).

4.2.2.12.1 Employment Gains

The NRC Phase II Survey sought detailed information from respondents about the number of employees they had at the time of the award, the number of employees they had at the time of the survey, and the direct impact of the award on their employment levels. Overall, it showed that the mean employment gain at each responding firm since the date of its SBIR award was 29.9 FTEs. In addi-tion, respondents estimated that as a result of their SBIR projects their companies were, on average, able to hire 2.7 FTE employees, and to retain 2.2 FTE existing

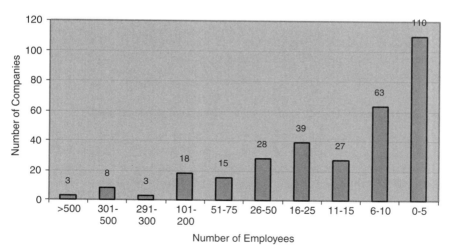

FIGURE 4-8 Distribution of companies, by employees.

SOURCE: NRC Phase II Survey.

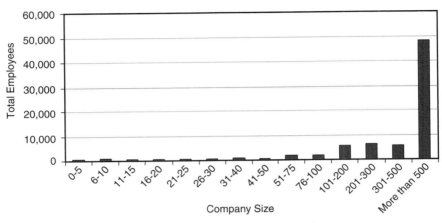

FIGURE 4-9 Employment at SBIR companies, by company size.

SOURCE: National Institutes of Health, *National Survey to Evaluate the NIH SBIR Program: Final Report*, July 2003.

employees that might not otherwise have been retained.[9] Case study interviewees noted that a Phase II award typically provides direct funding equivalent to the addition of slightly more than one full-time researcher plus overhead for two years.

The NRC Phase II Survey results show that the median post-award change in employment was 27.5 FTE employees. Companies that expanded their workforce rapidly pulled the mean employment change up much higher than the median. One company grew by 3,700 employees after receiving the surveyed SBIR award.

In the NIH Survey, 94 percent of respondents claimed that they had increased staff as a direct result of the SBIR award, although the survey did not ask about the size of employment gain.

4.2.2.13 Sales of Equity and other Corporate-level Activities

The NRC Phase II Survey explored several ways in which equity-related activities might be finalized or underway at surveyed projects (see Table 4-14). The data show that marketing-related activities were most widespread, with marketing/distribution agreements related to 33.9 percent of projects, and licensing agreements to 38.1 percent. Agreements likely to involve the direct transfer of equity—mergers (3.2 percent), partial sales of the company (6.5 percent), and complete sales of the company (5.0 percent)—were much less widespread. Note,

---

[9]NRC Phase II Survey, Question 16.

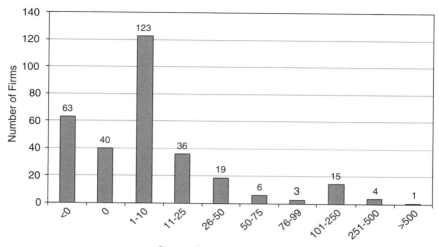

FIGURE 4-10 Employment change at firms since SBIR Phase II.

SOURCE: NRC Phase II Survey.

TABLE 4-14 Equity- and Marketing-related Activities Stemming from the Surveyed SBIR Project

Activities	U.S. Companies/Investors			Foreign Companies/Investors		
	Done (%)	Under way (%)	Total (%)	Done (%)	Under way (%)	Total (%)
Licensing agreement(s)	19	16	35	9	6	15
Sale of company	1	4	5	0	1	1
Partial sale of company	2	4	6	0	1	1
Sale of technology rights	6	7	13	1	1	2
Company merger	0	3	3	0	1	1
Joint venture agreement	3	9	12	1	3	4
Marketing/distribution agreement(s)	21	10	31	12	6	18
Manufacturing agreement(s)	7	4	11	2	2	4
R&D agreement(s)	15	11	26	4	3	7
Customer alliance(s)	8	10	18	3	1	4
Other	2	2	4	0	1	1

SOURCE: NRC Phase II Survey.

however, that the question asked specifically for outcomes that were the "result of the technology developed during this project"[10]—a very tight, and limiting, description for activities that occur at the level of the company, not the project.

---

[10]NRC Phase II Survey, Question 12.

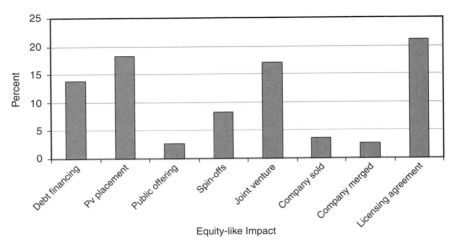

FIGURE 4-11 Equity-like impacts.

SOURCE: National Institutes of Health, *National Survey to Evaluate the NIH SBIR Program: Final Report*, July 2003.

Activities with foreign partners were substantially less common than similar activities with U.S. partners. Once again, marketing-related activities were the most widespread.

Similar results were found from the NIH Survey. Figure 4-11 shows the percentage of NIH respondents who agreed that the specific outcome in question had occurred "because of the product, process, or service developed during this project"[11]

In addition, the NRC Firm Survey determined that three firms with NIH SBIR awards had had initial public offerings, and that a further three planned such offerings for 2005/2006. Seventy-five out of 445 companies (16.9 percent) had established one or more spin-off companies. This percentage is slightly higher than that for all SBIR companies at all agencies during the study time-frame. NIH-related firms accounted for 126 spin-offs, approximately 52 percent of all spin-offs reported.

The impact of these activities on commercialization, on the spread of bio-medical knowledge, and on small businesses is hard to gauge using quantitative assessment tools only. The case study in Box 4-6 illustrates how research conducted using SBIR funding seeded an entire generation of spin-off companies and joint ventures based on a technology of potentially critical significance for homeland security.

---

[11]National Institutes of Health, "National Survey to Evaluate the NIH SBIR Program: Final Report," July 2003, Question 29.

BOX 4-6
Intelligent Optical Systems

Distributed, sensitive chemical and biochemical sensors and sensor networks

Intelligent Optical Systems (IOS) has developed a system for using the entire length of a specially-designed fiber-optic cable as a senor for the detection of toxins and other agents. This bridges the gap between point detection and standoff detection, making it ideal for the protection of fixed assets.

SBIR-supported research has been followed by a focus on the development of subsidiaries and spin-offs at IOS. This activity has generated private investments of $23 million in support of activities oriented toward the rapid transition to commercially viable products.

Since January 2000, IOS has formed two joint ventures, has spun out five companies to commercialize various IOS proprietary technologies, and has finalized licensing/technology transfer agreements with companies in several major industries.

Optimetrics manufactures and markets active and passive integrated optic components based on IOS-developed technology for the telecommunication industry. Maven Technologies was formed to enhance and market the Biomapper technologies developed by IOS. Optisense manufactures and distributes gas sensors for the automotive, aerospace, and industrial safety markets, and will be providing H2 and O2 optical sensor suites designed to enhance the safety of NASA launch operations. OSS, which is IOS's newest spin-off company, was formed to commercialize chemical sensors for security and industrial applications.

The company currently employs 40 scientists, and its current sales mix is almost 80 percent non-SBIR business. IOS currently holds 13 patents, with an additional 13 applications pending.

4.2.2.14 Commercialization and FDA Approval

One final metric is relevant in considering commercialization at NIH: the number of projects that seek and receive FDA approval.

Of the projects surveyed, 20 percent reported that the product they were developing would require FDA approval. Table 4-15 shows the stages of FDA approval that the projects had reached. This data set is comparable to that from the NIH survey, which asked similar questions (see Figure 4-12). NIH data also allow us to review FDA approval stage by industry. (See Table 4-16.)

NIH has recently provided additional data on FDA approval, tracking the same population of projects 3 years later. These data indicate that there has been some increase in meeting FDA milestones. (See Table 4-17.)

These data show that the number of approvals had increased to 60 or 7.8 percent of the projects originally selected for survey. A further 25 (3.1 percent) have reached the intermediate milestone of approval for clinical trials. No data

TABLE 4-15 FDA Approval

Approval Stage	Percent of Responding Projects which Require FDA Approval
Applied for approval	5.0
Review ongoing	3.0
Approved	38.5
Not approved	6.5
IND: Clinical trials	16.0
Other	32.0

SOURCE: NRC Phase II Survey.

are available on the number of projects that would have required FDA approval before they can reach the market.

Further analysis is required to determine whether projects focused on products that will require FDA approval consistently commercialize more or less successfully than others. These data also have implications for the recent NIH Competing Continuation Awards (CCA) SBIR initiative, described in Chapter 3. The CCA aims to support companies through the FDA approval process.

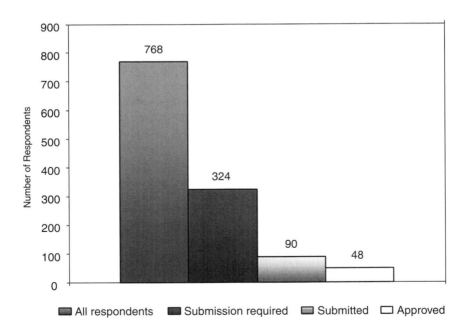

FIGURE 4-12 NIH data on FDA approval stage.

SOURCE: National Institutes of Health, *National Survey to Evaluate the NIH SBIR Program: Final Report*, July 2003.

TABLE 4-16 FDA Approval Requirements by Industry

Business Type	FDA Approval Required	All Respondents	Percent Requiring FDA Approval
Pharmaceuticals	47	58	81.0
Medical devices	102	145	70.3
Biotechnology	87	175	49.7
Diagnostics	21	43	48.8
All Respondents	323	767	42.1
Other	5	12	41.7
Instrumentation	31	88	35.2
Chemical technology	6	20	30.0
Health care	5	21	23.8
Computer hardware, software	14	85	16.5
Engineering, fabrication	2	17	11.8
Environment, ergonomics	1	13	7.7
Information and research	1	25	4.0
Medical education, health promotion	1	65	1.5

SOURCE: National Institutes of Health, *National Survey to Evaluate the NIH SBIR Program: Final Report*, July 2003.

4.2.2.15 Commercialization: Conclusions

The data described above support the view that there has been an effort to bring projects to market, with some measurable success. Even though the number of large (e.g., > \$5M) commercial successes has been few, the overall commercialization effort is substantial. Of the 40 percent of surveyed projects that had already reached the market, more than half did so within two years of the project start date. More than one-third of projects received additional outside funding, and 32.5 percent received additional related SBIR awards. These summary statistics support a conclusion that many award recipients are commercializing their products, services, and processes.

TABLE 4-17 FDA Milestones Updated to 2007

	Number of Projects					
	2002 Survey	2004 Update	January 2005 Update	August 2005 Update	March 2007 Update	Total Unique Projects Approved
FDA approval received	48	9	2	0		60
FDA approval for clinical trials, IND	11	0	7	7	1	25
TOTAL	59	9	9	7	1	85

SOURCE: National Institutes of Health, *National Survey to Evaluate the NIH SBIR Program: Final Report*, July 2003.

BOX 4-7
The NIH Mission

The NIH mission is science in pursuit of fundamental knowledge about the nature and behavior of living systems and the application of that knowledge to extend healthy life and reduce the burdens of illness and disability.

4.3 AGENCY MISSION

NIH's primary mission is improving public health through the development and application of knowledge.

However, measuring the impact of the NIH SBIR program on public health is extremely difficult. By the time the results of SBIR research become part of the health care system, they are deeply intertwined with other inputs, making measurement difficult. And as with commercial outcomes, data collection is a serious problem.

The data provided below, and the cases used to explicate the data, are therefore to be understood as an effort to answer a question for which no conclusive data exist. Instead, we offer a series of efforts to provide indirect evidence about support for agency mission in the NIH SBIR program.

4.3.1 Targeted Populations

One way to evaluate the support provided by SBIR to the agency mission is to assess the populations targeted by SBIR projects, and the NIH Survey seeks to do so. Figure 4-13 shows the distribution of projects by size of affected population for (a) projects reported to have reached the market and be in use, and (b) those projects still in commercialization. Projects still in earlier stages of development or discontinued have been filtered out. Note that percentages do not add up to 100 percent, as respondents were permitted to select more than one affected population.

Quantifying the impact that the products in use have on the affected populations is however problematic for at least two reasons.

First, the distribution of products across user groups does not measure the *intensity* of the benefit received from use. A product that reduces the incidence of hangnails in a potential population of 150,000,000 has a different impact than a product which saves the lives of 1 percent of heart attack victims annually—4,944 people.[12]

---

[12]Heart attack data for 2004 from American Heart Association <*http://www.americanheart.org/ presenter.jhtml?identifier=4591*>.

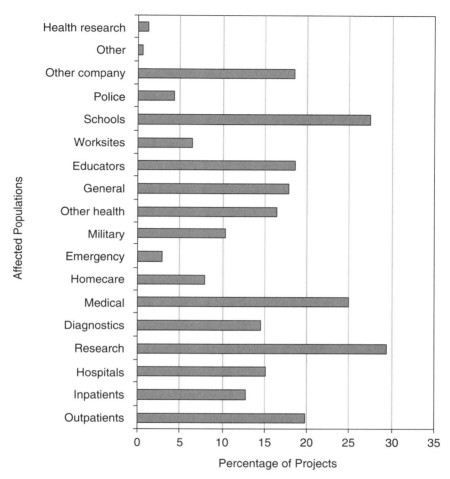

FIGURE 4-13 Distribution of projects, by type of affected population.

SOURCE: National Institutes of Health, *National Survey to Evaluate the NIH SBIR Program: Final Report*, July 2003.

Second, many impacts from products are *indirect*. Medical technology improvements often affect final populations only through a long causal chain, sometimes through indirect effects such as improvements in the efficiency with which the user operates. Chatten Associates, for example, successfully used SBIR to fund technology that automated the review of videotapes used to monitor epileptic patients for seizures. Previously, videos were reviewed manually by nurses, which took many hours of work for each 24-hour tape. By linking the monitoring system to an EEG, and automatically picking up anomalies, the Chatten technology reduced the amount of time spent by nurses reviewing a 24-hour tape from hours to minutes. This dramatic reduction had no obvious impact on epileptic patients—but released nurses for hours of other work. It is also worth noting that there are no data from Chatten or elsewhere on the numbers of nurses affected by the product.

Thus, while we acknowledge the NIH effort described below to quantify the impact of SBIR projects on public health, for both these reasons, it is probably misleading at best to draw solid conclusions from statistics of affected populations derived from untested company estimates.

The data categories themselves are somewhat general and difficult to distinguish from each other, and they provide only limited insight into the markets targeted by each project. The figure contains one surprise—the 27 percent of projects are targeted at schools. However, this result may partly reflect projects targeted at pediatric populations, which are not otherwise identifiable by respondents in the context of this survey.

These data cover projects with products both in use and in the commercialization stage. Figure 4-14 disaggregates the data, and shows that there are significant differences between the two groups. Companies with products in use are much less optimistic about the size of their affected population: Only 27.3 percent of respondents expected to affect at least 500,000 people, while 41.3 percent of respondents with projects still in development felt that they had such a large market.

Table 4-18 focuses on products in use, and distinguishes between "high-impact" projects affecting more than 500,000 users, and other projects.

There are some substantial differences between the distribution of projects among "all projects" and among the "high-impact projects. "All projects" are much more heavily focused on research labs (26.9 percent), and much less focused on medical practitioners and the general public. The table shows that there was some clustering of "high-impact projects" around services to medical practitioners and the general public.

4.3.2 Agency-identified Requirements and SBIR Contracts

At the agencies where the results of SBIR-funded research are purchased for in-agency use (primarily at DoD and NASA), the agency's mission is closely identified with the procurement process. In general, these agencies' SBIR programs support agency goals if the outputs produced by funded projects—weapons or spacecraft, for example—are eventually procured by the agency.

At NIH, in-house use is rare, as contracts account for only about 5 percent of all SBIR awards and the agency directly utilizes very few of its funded projects' outputs.

Still, it is important to recognize that in some cases, the SBIR program has

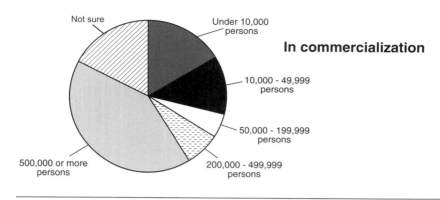

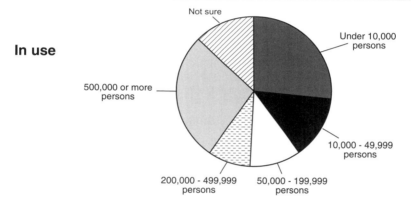

FIGURE 4-14 Distribution of projects by size of most important affected user population.

SOURCE: National Institutes of Health, *National Survey to Evaluate the NIH SBIR Program: Final Report*, July 2003.

generated outcomes that have been of direct use to the agency in fulfilling its mission. A case describing such an outcome is briefly described in Box 4-8.

4.3.3 Identifying Mechanisms for Supporting Public Health Through Qualitative Approaches

The cases completed by the research team, descriptions of successful projects collected by NIH, and interviews with NIH staff paint a complex picture of how SBIR activities can support the agency's mission.

Table 4-19 identifies a number of ways in which SBIR has successfully supported the mission of NIH. SBIR companies have had significant beneficial

TABLE 4-18 High-impact Projects—By Target Sector

Population	High-Impact Respondents	All Respondents	Percent of High-impact	Percent of All
Outpatients	4	28	9.1	11.2
Inpatients	1	22	2.3	8.8
Hospital personnel	4	10	9.1	4.0
Research labs	2	67	4.5	26.9
Diagnostic labs	4	15	9.1	6.0
Medical practitioners	7	24	15.9	9.6
Homecare providers	1	1	2.3	0.4
Other	2	3	4.5	1.2
Other health services	4	12	9.1	4.8
General public	8	26	18.2	10.4
Educators	2	7	4.5	2.8
Worksites	0	1	0.0	0.4
Schools, universities	1	11	2.3	4.4
Other companies, other technologies	2	7	4.5	2.8
Health researchers	1	5	2.3	2.0
MISSING	1	10	2.3	4.0
Total	44	249	100.0	100.0

NOTE: Hi-impact respondents are those with products in use, who expect to affect more than 500,000 people.

SOURCE: National Institutes of Health, *National Survey to Evaluate the NIH SBIR Program: Final Report*, July 2003.

BOX 4-8
Celadon Laboratories, Inc.

Multi-Method Software Platform for Primer and Probe Design

At the National Cancer Institute's (NCI) Core Genotyping Facility (*<http://cgf. nci.nih.gov/home.cfm>*), one critical bottleneck to high-throughput genotyping has been slow, tedious assay design that requires highly-trained personnel, which results in an unacceptably high assay failure rate.

The ProbITy expert system developed by Celadon through SBIR has nearly eliminated that substantial bottleneck. As a result, the NCI expects to recoup the cost of the project within a year.

TABLE 4-19 Mechanisms for Supporting Agency Mission (public health)

Educational impacts	Standards
Cost savings	Knowledge pipeline
Visionary research	Technology platform development
Niche products	Geographical spread
Deployment of public goods	Collaborative technologies
Agency technology needs	Contracting and manufacturing
Diversification and R&D	

effects in all of these areas, though these effects may not directly and obviously contribute to a substantial commercial success. This session discusses some of these areas, and the SBIR activities within them.

4.3.4 Education

NIH has long since recognized that education is a critical component of public health.[13] About 10 percent of SBIR projects are targeted at the general public, and others are focused more tightly on health educators.[14]

Many education-developing companies work on a short product cycle, which allows SBIR project products to reach the market quickly and efficiently. Sociometrics, for example, has claimed that every one of its more than 20 SBIR awards has been directly translated into a product. Similarly, Morphonix has used SBIR funding to develop the award winning children's video game described in Box 4-9.

4.3.5 Cost Savings

Given that health care expenditures have increased at more than twice the general rate of inflation for the past five years,[15] and given that the subsequent competition for scarce health care dollars, projects that generate substantial cost savings are extremely important. However, the fragmented nature of health care markets, and the disconnect between health care patients and health care funding, mean that incentives in this sector are sometimes perverse and the value of cost savings is not always reflected simply in sales data.

One powerful example of cost savings which are reflected only partly in official sales is provided by the case of Chatten Associates, outlined in Box 4-10.

4.3.6 Visionary and Long-term Research

Much policy attention has been focused on the need for measurable outputs from the SBIR program. Yet it is also important to see that the program has been used to support very high quality projects that have large long-term potential pay-offs but a high chance of technical failure.

It is sometimes difficult to distinguish such visionary research from simple failures (projects that have not yet and will never generate any useful commercial outcomes or other important effects). Yet, by looking at individual cases, this

---

[13]All the larger ICs and most of the smaller one's have specific components dedicated to health care education. E.g., the Health Education Programs at the National Institute of Diabetes and Digestive and Kidney Diseases (NIDDK) at *<http://www.niddk.nih.gov/health/edu.htm>*.

[14]Data in this section are derived, unless otherwise noted, from the NIH Survey.

[15]Statistical Abstract of the United States 2007, Table 118, *<http://www.census.gov/compendia/ statab/health_nutrition/health.pdf>*.

BOX 4-9
Morphonix, Inc.

Journey into the Brain

Journey into the Brain is CD-ROM adventure game for children. It is marketed as both a consumer product for 11-14 year olds and a supplemental learning program for middle schools. Morphonix notes that the purpose of the game "is not just to teach about the brain, but to find the fun inherent in the subject."

This kind of product may not generate huge commercial returns, but it may reach a large audience and have a substantial and perhaps long term impact. The evidence gathered by Morphonix suggests that the product:

- Generates increased interest and knowledge in neuroscience among children, ages 7-11, by making science exciting and accessible to them through the use of multimedia.
- Communicates complex concepts so young children can follow their interests in a way that allows for differences in modes of learning. Key concepts of brain structure and function are woven into game play.
- Increases the level of safety awareness among this age group of children regarding issues such as the importance of wearing bike helmets.
- Gives children a sense of awe for their own rapidly developing brains while helping them develop a stronger, more powerful brain.

Journey into the Brain has won many awards including: Best of Show, 1999 Best of the Northbay Awards; the 2000 National Parenting Publications Gold Award; All-Star Rating from Children's Software Review; Finalist, 1999 Educational Title of the Year (The Academy of Interactive Arts and Sciences); and Finalist, Independent Games Festival at the 1999 Computer Game Developers Conference.

Journey into the Brian was released in 1999, and has sold more than 36,000 copies. Many copies were sold to school systems and libraries, meaning that the product has reached a much greater number of final users.

---

kind of project can be identified. One such case is SAM Technologies, of San Francisco.

4.3.7 Niche Products

Many companies working with the SBIR program are focused on small markets, where niche products can make a large difference to the lives of a small client group. Analogous to orphan drug research, projects like these are, according to economists, classic cases for government subsidy or support. One such case is the SmartWheels product created by a small company in Arizona (see Box 4-12).

Another example is the cancer informatics suite developed by Humanitas.

BOX 4-10
Chatten Associates/ Telefactor

Long Term Epilepsy Monitoring

Facilities for long-term monitoring of serous epileptics in specialized facilities and hospitals were traditionally highly labor intensive. Patients would be recorded on synchronized EEG-video 24 hours per day, and nurses would then review the tapes visually by fast forwarding to find epileptic events. This process could take up to 6 hours per patient per day—a huge and expensive burden on highly trained nursing staff.

Chatten worked to automate this process by processing the EEG as it was recorded, and creating a file which highlighted possible epileptic events. Because this occurred in real time, staff in the area could be alerted while an episode was in progress.

The new approach reduced the 6 hours per day spent monitoring an epileptic patient down to a few minutes, providing significant cost savings. However, these technologies were typically embedded into larger systems, so independent sales did not capture their commercial impact. According to Dr. Chatten, the new technology provided the critical edge in the sale of larger integrated systems.

There is no sales data available to suggest the number of nurses affected by this technology, the total amount of time saved, or even the amount of time saved per nurse. The evidence does suggest that this technology—which the company says was developed only because SBIR funding was available—must have released substantial resources for use elsewhere in the hospitals and facilities where it was used. The product's estimated total sales of approximately $30 million (at $5-7,000 per unit) also shows that its use was widespread.

The technology developed includes software for grading toxicity using a hand-held computer, distance learning applications featuring searchable transcripts and audiovisual slide presentations, and an online document/proposal management system (<*http://www.epanel.cc*>). The suite is now distributed free over the Web by Humanitas—by definition generating zero revenues, but delivering value nonetheless in that the project has users.

4.4 SUPPORT FOR SMALL, WOMAN-OWNED, AND MINORITY BUSINESS

SBIR is funding entirely devoted to small business. It is therefore by definition support for small businesses. However, this is not the entire story. Beyond the share of funding going to small business, the quality of that impact is important. We have seen that a variety of commercial and other impacts are associated with the SBIR program. SBIR recipients themselves offer a range of positive testimony about the impact of SBIR on their companies.

In addition, the NRC Firm Survey and NRC Phase II Survey and the NIH

BOX 4-11
SAM Technologies

The Mental Meter Project

SAM Technologies (SAM) was founded by Dr. Alan Gevins in 1986 to pursue a project he had conceived many years earlier as an undergraduate at MIT: to build a Mind Meter (MM) that could directly measure the intensity of mental work in the brain.

The benefits of such a project are likely to be very substantial and to extend beyond the medical applications envisaged for the first product. SAM has high expectations for the Online Mental Meter, a computer peripheral that will provide continuous information about the user's state of alertness and mental overload or under-load. As Gevins notes, "This neuroadaptive capability will enable a system to adapt itself to the user, as contrasted with the current situation in which the user must adapt to the computer."

SAM has been funded by the Air Force, the Navy, DARPA, NASA, NSF and 7 NIH institutes through SBIR and other contracts. It has turned down opportunities with a number of VC firms in order to maintain focus on the long-term objective.

This is a highly focused project, using the same core staff over a long period. The 8 most senior scientists and engineers (out of 13 in total) have been with SAM an average of 11 years.

SAM is now reaching the marketplace. In 2005, SAM will release the first commercial product in the MM line—the world's first medical test that directly measures brain signals regulating attention and memory.

In addition, SAM has generated a substantial flow of knowledge: more than 50 peer reviewed papers, and 18 patents.

In the end, even though there have been peripheral benefits along the way, what is striking is the extent to which the SBIR program has facilitated such an extended research project. According to Gevins, more than 94 percent of annual funding comes from the SBIR program, from multiple agencies. This is a testimony to the flexibility of the program.

Survey all seek to address the question of what would have happened to companies had they not received SBIR awards.

4.4.1 Small Business Shares of NIH Funding

SBIR provides support for small business in that it provides funding only to businesses with no more than 500 employees—the SBA definition of a small business. At NIH, that support is now over $500 million annually (see Figure 4-15). Moreover, SBIR grants and contracts are spread out across a lot of companies. At NIH, few companies receive very large numbers of awards, and many receive one or two.

The very rapid and sustained increase in SBIR funding from 1999 to 2004

BOX 4-12
Three Rivers Holdings, Inc.

The SmartWheel: Development of Wheelchair
Pushrim Force and Measurement Device

SmartWheel is a product designed to measure accurately all the key parameters involved in the propulsion of wheelchairs by their occupants. These include stroke frequency, propulsion angle, acceleration, forces applied to the handrim, velocity, and distance traveled. According to the company, "The SmartWheel is the only commercial product in the world that measures propulsion biomechanics in the natural environment of the wheelchair user."

SmartWheel has by now been in use as a research tool for more than ten years as a means of measuring and analyzing pain and injury among wheelchair users and also as a means of assessing interventions to address problems. Currently, SmartWheels are in use at leading research institutions including the Rehabilitation Institute of Chicago, the University of Michigan, the Rehabilitation Institute of Montreal, the University of Washington, the Kessler Medical Rehabilitation Research and Education Corporation, the University of Pittsburgh, and the University of Alberta.

SmartWheels is now being adapted for use as a clinical product. It has four main uses:

- Justification of equipment decisions for insurance reimbursement, using precise data to identify users who cannot provide the force need to propel a manual chair effectively
- Selection of the appropriate manual wheelchair, once again by the application of precise data to the selection process
- Training that allows wheelchair users to improve propulsion efficiency by reducing the stress on their arms through use of a longer stroke, reducing stroke frequency, and minimizing wasted forces (e.g., pushing directly down on the handrim).
- Creation of an individualized patient database, showing the effect of adjustments and creating a longitudinal record for selected metrics

The company notes that SBIR awards were used to facilitate its transformation from a hard-wired noncommercial research tool to a wireless, user-friendly commercial clinical and research tool. Leading experts were hired as consultants, and speed to market was accelerated.

All the evidence suggests that use of SmartWheel will continue to expand clinically, and that increasing numbers of wheelchair users will benefit from the technology. Yet commercially, this will never be a major success: The company expects that if sales double in 2005 and continue to grow thereafter, revenues will still only be $1 million in 2006. Still, the social benefits for the specific niche of SmartWheel users greatly exceed any commercial return.

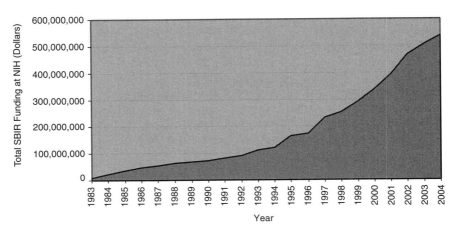

FIGURE 4-15 Total SBIR funding for small business at NIH, 1983-2004.

SOURCE: NIH awards database.

has been driven by the doubling of the overall NIH extramural research budget over that period, with a proportion of that funding allocated for small business.

However, this data set does not answer a related question: To what extent has the SBIR program replaced other funding for small businesses at NIH. This question can be addressed by comparing the level of SBIR funding with that available through all other small business funding mechanisms at NIH (see Figure 4-16).

The awards data show quite clearly that the share of small business funding being disbursed through the SBIR program has fallen steadily since soon after

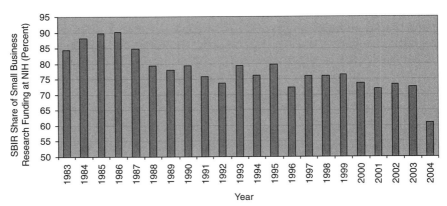

FIGURE 4-16 SBIR share of small business research funding at NIH, 1983-2004.

SOURCE: National Institutes of Health.

the inception of the program at NIH in 1983. After peaking at 90 percent of all small business research funding in the mid-1980's, the SBIR program's share fell steadily to about 72 percent in 2003, before falling further in 2004 (which may be an outlier).

These data clearly invalidate the hypothesis that SBIR has replaced other forms of small business funding at NIH.

4.4.2 The Decision to Begin the Project

Figure 4-17 shows that almost half of NRC Phase II Survey respondents were sure that their projects would not have occurred at all without SBIR funding. Altogether, almost 75 percent thought that would have been the case. NIH Survey data are comparable, with 64 percent of respondents anticipating that the projects would have been a "no go" in the absence of SBIR funding. These

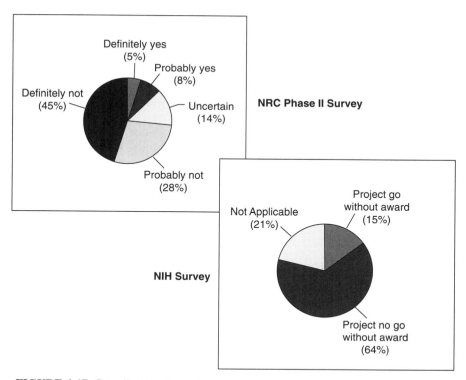

FIGURE 4-17 Greenlighting the project.

SOURCE: NRC Phase II Survey and National Institutes of Health, *National Survey to Evaluate the NIH SBIR Program: Final Report*, July 2003.

figures suggest that SBIR often makes the difference between a research project being pursued or not.

Even for projects that would have continued in the absence of SBIR funding, delays and other changes would have been caused by the resulting paucity of funds. 51 percent of these respondents noted that the scope of their projects would have been narrower; 19 of the 43 firms that would have continued anyway expected their project would have been delayed. Fifteen firms expected this delay would have been at least 12 months, and 13 expected a delay of at least 24 months, generating an average expected delay in project start of 8 months. Sixty-three percent expected that project completion would also have been delayed.

4.4.3 Company Foundation

Responses to the NRC Firm Survey indicate that almost 25 percent of NIH firms that received SBIR Phase II awards were founded entirely or in part as a result of SBIR awards (see Table 4-20).

4.4.4 Company Foundation and Academia

Case study interviews suggest that SBIR has facilitated of the movement of technologies and researchers from university labs to the commercial environment. Data from the NRC Firm Survey strongly support this hypothesis. More than 80 percent of NIH respondent companies had at least one founder from academia (see Table 4-21). The same survey found that about a third of founders were most recently employed in an academic environment before founding the new company. This data set, thus, strongly suggests that SBIR has indeed encouraged academic scientists to work in a more commercial environment.

4.4.5 Growth Effects

While there are no data about the effect of SBIR awards on company growth, except for the employment data discussed above (which do not seek to explain

TABLE 4-20 SBIR Awards and Firm Foundation: Was the Firm Founded Wholly or Partly Because of the Referenced SBIR Award?

	Number of Responses	Percent of Responses
No	342	74.8
Yes	49	10.7
Yes, in part	66	14.4
	457	100.0

SOURCE: NRC Firm Survey.

TABLE 4-21 Academics as Founders

	Number of Responses	Percent
None	86	18.9
At least one	369	81.1
All	455	100.0

SOURCE: NRC Firm Survey.

the cause of growth,) the NRC Firm Survey did ask respondents to provide their own estimates of SBIR impacts on growth (see Table 4-22).

Almost half of respondents indicated that more than half of the growth experienced by their firm was directly attributable to SBIR. This too is evidence of the powerful impact winning an NIH SBIR award can have on the development of a small business.

4.4.6 Support for Woman- and Minority-owned Businesses

One of the congressional mandates for the SBIR program is to support the work of women and minorities in science. The primary metric for this support is the extent to which SBIR programs fund woman- and minority-owned businesses.

There is an extensive analysis of awards to woman- and minority-owned firms in Chapter 3 of this report.

A review of the available data in Chapter 3 draws the following conclusions:

- Together, woman- and minority-owned firms account for an average of about 15 percent of Phase I awards at NIH (2003-2006).
- The trend for minority-owned firms is downward since 1993, with some annual variation, and minority-owned firms have accounted for less than 4 percent of Phase I awards since 2003.

TABLE 4-22 SBIR Impacts on Company Growth
(percentage impact of SBIR on overall company growth)

	Number of Responses	Percent
Less than 25	132	29.5
25 to 50	100	22.4
51 to 75	78	17.4
More than 75	137	30.6
Total	447	100.0

SOURCE: NRC Firm Survey.

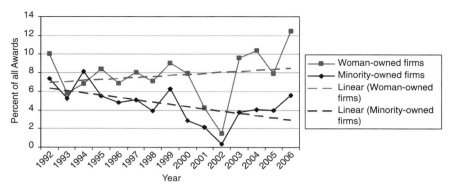

FIGURE 4-18 Phase I Award share of woman- and minority-owned firms, 1992-2006.

SOURCE: National Institutes of Health.

- The share of Phase I applications from woman- and minority-owned firms has declined since early 1992, although absolute numbers have risen.
- This is true in particular of minority-owned firms, whose share of applications has declined from about 10 percent in 1996 to just over 5 percent in 2005.
- Lower levels of awards are partly explained by lower success rates—the rate at which applications are selected to become awards. The data show that woman- and minority-owned firms are consistently less successful in the Phase I selection process—that lower percentages of their applications generate awards. Minority applicants saw a particularly steep decline in success rates from 1999 to 2004, with some recovery in 2005-2006.

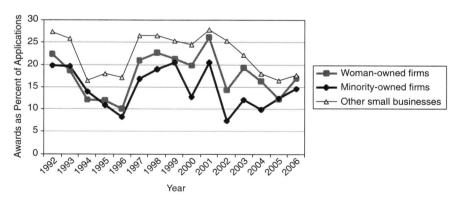

FIGURE 4-19 Success rates for Phase I awards by demographic, 1992-2006.

SOURCE: National Institutes of Health.

However, success rates for minority-owned firms remain about five percentage points lower than the rates for firms that are neither woman- or minority-owned.

The data themselves provide no answer to the question of why woman- and minority-owned firms have lower success rates. One promising hypothesis is that these firms tend to be formed more recently, and have both a shorter track record and less experience principal investigators, both of which may militate against success in the NIH selection process.

Finally, it is important to note that while woman-owned firms have maintained and even slightly increased their share of SBIR Phase I awards at NIH, they remain at an average of about 10 percent of all awards (2003-2006). At the same time, the percentage of women among recent life sciences doctorates has increased dramatically. According to NSF, in 1999 and 2000 women accounted for more than 61 percent of all life sciences doctorates awarded.[16] In that context, maintaining a ten percent share of awards is much less impressive, and NIH might well wish to undertake further analysis to determine why so few of these new doctorates appear to be applying for NIH SBIR funding (note that there is no requirement that a company exist in order to apply for an award, although a company must be formed in order to accept one.)

4.5 SBIR AND THE EXPANSION OF KNOWLEDGE

Metrics for assessing knowledge outputs from research programs are well-known, but far from comprehensive. Patents, peer-reviewed publications, and, to a lesser extent, copyrights and trademarks, are all widely used metrics. They are each discussed in detail below. However, it is also important to understand that these metrics do not capture the entire transfer of knowledge involved in programs such as SBIR.

4.5.1 Patents

The NRC Phase II Survey data indicate that about 34 percent of respondents received patents related to their SBIR-funded project (see Table 4-23). About 41 percent of projects generated at least one patent application, and about 82 percent of those applications were successful.

The NIH Survey generated similar data indicating that 37 percent of respondents received a patent related to their SBIR award (although wording of the question makes it impossible to know whether the patent was awarded for work completed before or after the award). It is possible that a positive response

---

[16]Derived from National Science Foundation, Division of Science Resources Statistics, *Women, Minorities, and Persons with Disabilities in Science and Engineering: 2004*, NSF 04-317, Arlington, VA: National Science Foundation, 2004.

TABLE 4-23 Projects Reporting Patent Applications
and Patent Awards

	Applications		Awarded	
	Number	Percent	Number	Percent
No	249	58.7	280	66.0
Yes	175	41.3	144	34.0
	424	100.0	424	100.0
Total	679		305	

SOURCE: NRC Phase II Survey.

reflects a patent application rather than patent approval. The very small number of "pending items" reported suggests that this may sometimes have been the case.

A negative correlation found between projects with patents and those with marketing activities could indicate differences between projects targeted at products and those focused on knowledge. However, marketing activities are positively strongly correlated with knowledge outputs, indicating that this kind of substitution effect is not detectable.

Once again, relationships between survey results and other variables might provide extremely useful insights. For example, Figure 4-22 shows patenting outputs by size of firm. Analysis of the scientific importance of the patents listed was not possible because the patents themselves were not disclosed in the course of the survey.

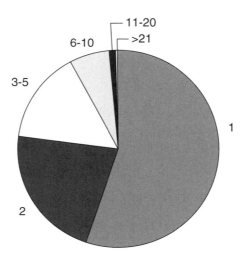

FIGURE 4-20 Number of patents per company reporting patenting activity.

SOURCE: National Institutes of Health, *National Survey to Evaluate the NIH SBIR Program: Final Report*, July 2003.

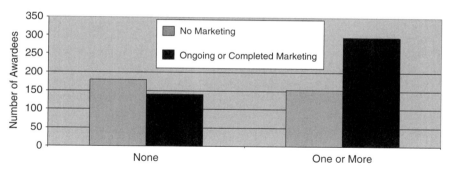

FIGURE 4-21 Awardees with one or more patents, copyrights, or trademarks—by marketing status.

SOURCE: NRC Phase II Survey.

4.5.2 Scientific Publications

The NIH Survey did not distinguish between scientific publications and articles in the trade and popular press. However, the NRC Phase II Survey did so, and it determined that slightly more than half (53.5 percent) of the respondents

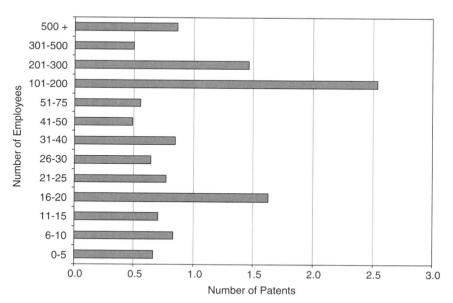

FIGURE 4-22 Number of patents, projects with at least one reported patent—by size of company.

SOURCE: NRC Phase II Survey.

had published at least one scientific paper related to their SBIR grant. About 33 percent of those with publications had published only a single paper, but one company had published 165 papers on the basis of its SBIR project, and several others had published at least 50 (as shown in Table 4-24).

This data set fits well with case studies and interviews, which suggest that SBIR companies are proud of the quality of their research. Publications are featured prominently on many grantee Web sites, and companies like Advanced Brain Monitoring, SAM Technologies, and Polymer Research all made a point of stating during interviews that their work was of the highest technical quality, as measured in the peer-reviewed publications.

Publications therefore fill two important roles in the study of SBIR programs.

First, they provide an indication of the quality of the research being conducted with program funds. More than half of the funded projects appear to be of sufficient value to generate at least one publication.

Second, publications are themselves the primary mechanism through which knowledge is transmitted within the scientific community. The existence of articles based on SBIR projects is therefore direct evidence that the results of these projects are being disseminated widely. This, in turn, implies the NIH SBIR is meeting its congressional mandate to support scientific outcomes. It is useful to note that the non-SBIR portion of the NIH research program does not have any mechanism in place for determining whether similar knowledge effects are being generated at the same rate as in the SBIR program. Note also that comparisons with SBIR programs at other agencies may be less than completely valid, as the publishing culture may be different outside the biomedical scientific world.

4.5.3 SBIR and Universities

SBIR can have further effects on the spread of knowledge through the involvement of university staff and students in SBIR projects. For example,

TABLE 4-24 Publications

Number of Publications	Number of Responses	Total Publications
1	72	72
2	52	104
3	32	96
4	19	76
5	15	75
6-10	15	133
11-30	9	146
30+	7	420
Totals	236	1,122

SOURCE: NRC Phase II Survey.

TABLE 4-25 University Involvement in SBIR Projects

4%	The Principal Investigator (PI) for this Phase II project was a faculty member.
7%	The Principal Investigator (PI) for this Phase II project was an adjunct faculty member.
34%	Faculty or adjunct faculty member (s) work on this Phase II project in a role other than PI, e.g., consultant.
15%	Graduate students worked on this Phase II project.
16%	University/College facilities and/or equipment were used on this Phase II project.
5%	The technology for this project was licensed from a University or College.
6%	The technology for this project was originally developed at a University or College by one of the participants in this Phase II project.
24%	A University or College was a subcontractor on this Phase II project.

SOURCE: NRC Phase II Survey.

Advanced Targeting Systems, in San Diego, has forged an extended and very successful research partnership with a senior scientist at the University of Utah. Other companies have made similar arrangements.

Just over half (54 percent) of all respondents indicated that there had been involvement by university faculty, graduate students, and/or a university itself in developed technologies. This involvement took a number of forms, as shown by Table 4-25.

The wide range of roles played by university staff and students indicates once more the multiple ways in which SBIR projects feed the knowledge base of the nation. Involvement in these projects provides opportunities for university staff different than those available within the academy.

5

Program Management at NIH

5.1 INTRODUCTION

The congressional charge to the National Academies was to assess the SBIR program at NIH, and to suggest possible areas for improvement.

In this chapter, we focus primarily on the latter: areas where NIH might make improvements to its SBIR program. In doing so, we primarily utilize case studies, interviews with NIH staff and other stakeholders, and secondary materials, as well as data from the NRC surveys and other statistical sources.

The focus of the chapter is to provide an objective review of the management of the NIH SBIR program, with a view to providing recommendations for improvement. The latter are described in a separate chapter. The structure on this chapter follows the logic of the awards cycle at NIH starting, with outreach activities to attract the best applicants, through topic development, selection, and funding, and concluding with commercialization support and a discussion of metrics and data.

5.2 BACKGROUND

The NIH SBIR program started soon after the program was launched, in 1983. It has expanded steadily with the growth of extramural research at NIH, and effectively doubled over the past four years as NIH funding doubled. The program is now the second-largest, after DoD, and funded approximately $552 million in SBIR awards in FY2006.

Most of these awards are made in the form of grants; about 5 percent are contracts focused on specific NIH needs. Almost all others are not designed to generate results that are purchased by NIH, unlike the procurement-oriented programs at DoD and NASA.

The NIH program has a number of defining characteristics, some of which are addressed in more detail in the remainder of this chapter.

- **Investigator-initiated research.** NIH is the only agency where the topics areas in the program solicitation (request for applications) are guidelines, not mandatory limitations on research topics.
- **Larger awards.** NIH now consistently exceeds the SBA awards size guidelines for Phase I and Phase II, utilizing a blanket SBA waiver to do so.
- **Peer-driven selection procedures.** NIH appears to depend more than most other SBIR programs on external peer review for advice on award selection, although final decisions remain the responsibility of NIH staff.
- **Regulatory concerns.** NIH is the only agency whose research often requires approval from the FDA before it can reach the market. This creates an important barrier to commercialization.
- **Multiple awarding components.** Twenty-three Institutes and Centers (ICs) at NIH award fund their own SBIR awards, using a range of procedures and with different degrees of integration with other programs.

Together, these characteristics give the NIH program a unique character, and have informed management of the program in a number of important ways.

5.3 OUTREACH

Outreach activities at NIH are extensive, compared to some other agencies, and have received significant attention from the NIH SBIR/STTR Program Office in recent years.

The activities appear in general to have had three primary objectives:

- To ensure that SBIR attracts the most qualified applicants;
- To reach geographical areas often perceived to be underserved; and
- To reach specific demographic groups that are perceived to be underserved (e.g., businesses owned by women and minorities).

Mechanisms for achieving these objectives include:

- **National SBIR conferences,** which twice a year bring together representatives from all of the agencies with SBIR programs, usually at locations far from the biggest R&D hubs (e.g., the spring 2005 national conference was in Omaha, Nebraska).
- **The National NIH SBIR conference** held annually, in Bethesda, MD.

- **The annual Program Administrators' bus tour.** An annual swing through several "under-represented" states, with stops at numerous cities along the way. Participants always include the NIH Program Coordinators.
- **Web sites and listservs.** NIH maintains an extensive Web site[1] containing application information and other support information. A number of explanatory presentations are available online. NIH also allows users to sign up for a news list-serve.
- **Agency publications and presentations.** NIH does not appear to use print publications to any significant degree to publicize SBIR (except as NIH events are reported in other publications, for example at the state level). NIH does use electronic publications, such as the NIH Guide for Grants and Contracts, to publicize Funding Opportunity Announcements as well as the Commercialization Assistance Program and the Niche Assessment Program.
- **Demographic-focused outreach.** NIH regularly participates in several conferences designed to reach specific demographics.

Overall, there are currently no metrics in place to determine whether the above three objectives have been met in the past or are now being met. Interviews at NIH suggest that the staff believes more outreach is required, and that raising the size of awards has been the most important recent NIH outreach initiative. Some staff members suggest that bigger awards attract better applicants.

NIH has strongly supported the SWIFT bus tour, and the NIH SBIR/STTR Program Coordinator has gone on all recent tours personally.[2] Staff members claim to have noticed a spike in applications from visited states and regions, but have no empirical evidence matching bus tours with increased applications.

A review of IC Web sites also indicates that they provide a range of online information from very basic to "fancy bells and whistles." The Institutes and Centers (ICs) vary greatly in their resources and talent to launch attractive and informative Web pages. It could therefore be helpful if the NIH SBIR/STTR Program Office could develop a standard information package that the Institutes could then adapt for their particular programs, e.g., to display their own particular list of initiatives.

5.3.1 Attracting the Best Applicants

The NIH staff notes that average scores for SBIR awards have trended upward (NIH scores range from 100 (best) to 500 (worst), so an upward trend indicates relatively weaker applications.) Some staff members have stated that the

---

[1]Accessed at: *<http://www.nih.gov/grants/funding/sbir.htm>*.

[2]SWIFT is a multistate bus tour periodically undertaken by SBIR Program Administrators from different agencies to fuel technology growth and development across different regions by promoting awareness of the SBIR programs.

rapid expansion of funding in the program, together with the trend in marginally funded scores, means that relatively weak applications are being funded.

This observation raises two questions:

- Is this perception accurate—is the quality of funded SBIR applications low relative to those that receive other NIH funding?
- If so, does this mean that there are other better-qualified companies who are not applying for SBIR?

Low relative scores. From discussions with staff, it appears that the pay-lines[3] for SBIR awards at the different IC's are substantially higher than for RO1 awards,[4] and these gaps have grown recently. This implies that projects funded through SBIR are receiving worse peer-review scores than projects funded through other mechanisms.

NIH management decided not to share scoring data with the research team, so it is difficult to determine whether or to what extent reality matches perceptions in this area. However, it seems likely that these different scores may well be the result of using a selection process that is primarily aimed at selecting academic applications for basic research and adapting it for use with SBIR, which has different objectives and indeed different selection characteristics. For example, commercialization plans are supposed to play an important role in selection for SBIR, but not for other NIH awards. It does not appear that program staff has undertaken research either to substantiate this perception or to investigate possible alternative explanations for differential scores between RO1 and SBIR applications.

New companies are applying. More than 30 percent of winning applications are from companies not previously funded by the NIH SBIR program.[5] New companies participate in the annual conferences, and hits on the Web site continue to increase. The new entrants in the program illustrate the attractiveness of SBIR awards but do not address the qualifications of the applicant companies.

Burden on staff. There are "cultural" issues that may affect perceptions of project or company quality. In interviews and responses to the NRC Program Manager Survey, many NIH staff noted that SBIR applicants and awardees placed a disproportionately high burden on agency staff, compared to similar applicants and awardees in other programs. Michael-David Kerns of NIA may have expressed this issue most clearly, observing that "We spend a disproportionately large amount of time with program administrators interacting with both

---

[3]The payline is defined as the score for the worst-scoring application that is still funded.

[4]RO1 awards are grants made to individual researchers. They constitute the most common form of NIH award, and are also sometimes used as an informal comparison group for SBIR awards. However, as explained below, they are different, and comparisons between these groups are invalid.

[5]See Section 3.2.3.2: New Winners.

potential and actual SBIR-STTR applicants.[6] These potential and actual SBIR-STTR applicants send emails and telephone much more than other categories of applicants (for basic research grant programs at NIH-NIA), making tremendous demands upon the time of program administrators and the grants management specialists. . . . Some of the reluctance and the comparatively low regard for the SBIR-STTR Programs, is the amount of time that would-be applicants attempt to and actually engage program administrators in marketing and selling their project and product idea. Even after having explained, usually more than once, that program administrators at NIH-NIA are not in the position of "buying" any project and/or product, the SBIR-STTR potential applicants persist in marketing and selling their projects and products. NIH-NIA program administrators are not accustomed to and do not welcome attempts by individuals to "sell" anything"[7]

5.3.2 Applications and Awards from Underserved States

Chapter 3 on program awards illustrated the extent to which awards have been concentrated geographically. A single zip code in San Diego has received more than twice as many awards as any other zip code in the country. Massachusetts and California alone account for 36 percent of Phase I awards 1992-2005.

Even though there has been some increase in awards to underserved states, data for FY2005 shows that six states received zero Phase I awards, and a further four states received one or two.[8]

A better approach to the issue of underrepresentation would be to look at applications per scientists and engineer. The distribution of the latter reflects the distribution of scientific and engineering talent, which should tend to predict applications and awards as well.

As Table 5-1 shows, there are wide variation in the number of applications per 1,000 scientists and engineers, indicating that scientists and engineers in some states use the SBIR program much more—in fact up to twenty times more—than those in other states.

This does raise some important practical questions for the NIH program. To begin with, it points to a somewhat different set of "underserved" states. While

TABLE 5-1 NIH SBIR Phase I Applications per 1,000 Scientists and Engineers

MA	50.5	NJ	13.7	MN	10.1	AL	7.4	NV	5.0
MD	33.4	HI	13.5	OH	9.9	TX	7.0	LA	4.8
UT	23.2	CO	13.5	IL	9.7	ND	6.8	NE	4.7
NH	23.1	CT	13.0	MT	9.7	FL	6.5	KS	4.7
CA	19.1	WA	12.7	DC	9.7	KY	6.4	AR	4.4
VT	18.7	NY	12.6	WY	9.3	IA	6.3	OK	3.9
RI	16.2	SD	12.3	NM	9.0	MO	6.1	ID	3.4
DE	16.0	PA	11.5	WI	8.0	GA	6.0	SC	3.1
VA	15.2	NC	11.0	TN	7.9	IN	5.7	MS	2.5
OR	14.5	ME	11.0	AZ	7.4	MI	5.7	WV	2.2
								AK	1.4

SOURCE: U.S. Census; National Institutes of Health.

states with low numbers of applications per scientists and engineer tend to have low numbers of applications overall and hence low numbers of awards, only five of the bottom ten states in Table 5-1 are also among the bottom ten states in overall awards.

Some underserved states have made substantial efforts to win more awards in recent years. This approach has been partly supported by the FAST program.[9] While a comprehensive analysis of the FAST program is not available, interviews with state agency staff and program participants suggest that, despite its limited funding, the program has been successful in helping to generate additional applications.

Additional applications do not, however, always translate into increased awards. For example, the state of Louisiana has made significant outreach efforts that have resulted in an increase in the number of Phase I applications to NIH from six in 1998 to 20 in 2001. However, during that period the number of awards increased only modestly, from 0 to 2. More experience with the application process may generate a more positive outcome over time.

5.3.3 New Applicants

Awards and applications data from NIH (described in detail in Chapter 3) suggest that about 40 percent of applicants for Phase I have not previously won an NIH SBIR award, and that about 30 percent of Phase I awards go to these companies.

[9]The Federal and State Technology Partnership Program (FAST) Program is operated by the SBA, and provides states with a limited amount of matching funds to be used to strengthen the technological competitiveness of small business concerns in states. See *<http://www.sba.gov/sbir/indexfast.html>*.

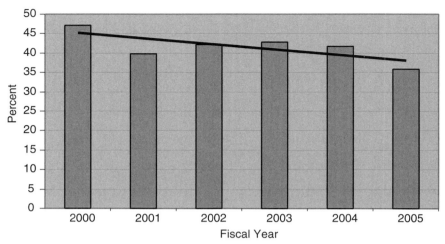

FIGURE 5-1 Percentage of winning companies new to the NIH SBIR program, 2000-2005.

SOURCE: National Institutes of Health.

Figure 5-1 shows that that the number of new winners has fallen slowly but steadily in recent years. However, this is likely explained by the fact that there are many more previous winners in the potential applicant pool each year.

5.3.4 Conclusions

In general, the data above support the hypothesis that the NIH SBIR program is open to new companies, and continues to attract them, and that it is also open to companies from outside the major biomedical research hubs in states such as California, Massachusetts, and Maryland. However, it is also worth noting that some at NIH—including NCI in its institutional response to the NRC Program Manager Survey, suggested that funding for this outreach was severely constrained:

> We need to have annually committed funds to support a reasonable number of HSA and Grants Management staff to travel to the two national meetings as well as the annual NIH SBIR/STTR Conference which is now being held offsite. If the NIH Conference is held in Bethesda, then logistics funds are needed to support the Conference. Either funds should be made available from the SBIR/STTR set-aside for outreach, or Institutes should make a standing commitment to support these activities.[10]

---

[10]NCI response to NRC Program Manager Survey, April 2006.

5.4 TOPICS

Like other agencies, NIH publishes areas in which it is interested in funding research, known as "topics." These topics are published in the annual NIH Omnibus Solicitation. But unlike the other SBIR agencies, where technical topic descriptions tightly limit awards, NIH topics are guidelines, not boundaries. The agency is proud of this "investigator-initiated" approach. Researchers are encouraged to submit applications on any topic that falls within the broad mandate of the IC funding agencies—which covers the entire universe of biomedical research.

This description of the SBIR funding as "investigator-initiated" is broadly accurate. However, in recent years, an increasing percentage of awards have been made through alternative mechanisms. The Program Announcement (PA) mechanism operates through the regular selection procedure, but marks certain areas as being of special interest to NIH; the Request for Applications (RFA) mechanism goes further, and earmarks dollars within the SBIR set-aside specifically for selected topic areas. PAs now account for about 20 percent of Phase Is, and RFAs for a further 5 percent. These are discussed briefly below, and in more detail in Chapter 3.

5.4.1 Standard Procedure at NIH—The Omnibus Annual Solicitation

The Annual Omnibus Solicitation lists all the topics from all of the ICs at NIH (and two other HHS SBIR participating agencies, CDC and FDA who use NIH to manage their SBIR program). The Solicitation describes areas in which research applications are encouraged, but applications outside these topic areas are welcomed. The topics listed in the annual solicitation are broad guides to the current research interests of the ICs.

These topics are developed by individual ICs for inclusion in the annual Omnibus Solicitation. Typically, the NIH SBIR/STTR Program Office sends a request to the individual Program Administrators (PMs), the SBIR points of contact at each IC. These PMs in turn meet with division directors and determine the focus of SBIR topics within the IC.

Division directors review the most recent Omnibus Solicitation (with their staff), and suggest changes and new topics based on recent developments in the areas of particular interest to the IC, or agency-wide initiatives with implications within the IC. The revised topics are then resubmitted for publication by the SBIR office at the Office of Extramural Research (OER), which provides a further review.

5.4.2 Procedures for Program Announcements (PAs) and Requests for Applications (RFAs)

PAs and RFAs are NIH's version of the mission-driven approach to topics used in particular by the procurement agencies—DoD and NASA. Essentially, they are tools through which the Institutes and Centers (ICs) can encourage firms to propose project that meet IC research priorities.

RFAs are announcements of research funding areas that the IC expects to prioritize. The two types of announcement indicate different levels of IC interest. RFAs are high priority areas that have funding from SBIR set aside for them. In effect, they are operated much like the more rigid topics at other agencies.

PAs are simply announcements of interest—applications received in response go through the same SBIR application process as other applications. However, as described in Chapter 3, ICs may announce that awards made under a PA can be for a longer time period (several additional years) and also for more money than the standard guidelines or even than the average award at NIH. While the PA applications go through the same selection process as other SBIR applications, IC's may exercise discretion and decide to fund an application under a PA over other better-scoring applications. Discussions with agency staff suggest that this occurs only at the margin (i.e. a decision between two projects both close to the payline).

RFAs indicate more interest from the IC in two respects. First, applications in response to a RFA compete for a separate pool of SBIR funding that the IC carves out of its general SBIR pool specifically to serve the RFA. Second, these applications are not selected using the normal Center for Scientific Research (CSR)[11] process. Instead, RFA applications go through a separate review process, normally internal to the relevant IC.

Both PA and RFA announcements are published by one or more ICs and reflect the top research priorities at the ICs. NIH tries to ensure that while PAs and RFAs define a particular problem, they are written broadly enough to encompass multiple technical solutions to the defined problem.

PAs and RFAs appear to be the result of efforts to develop a middle ground between topic-driven and investigator-initiated research. Essentially, by layering PA/RFA announcements on top of the broad, standard solicitation, NIH seeks to focus some resources on problems that it believes to be of pressing concern, while retaining the flexible investigator-initiated approach that has served the agency well. In a recent interview, the NIH SBIR/STTR Program Coordinator indicated that NIH plans to increase the percentage of SBIR funds allocated to more targeted research through these mechanisms.

5.5 SELECTION

The peer review process at NIH is by far the most elaborate of all the SBIR agencies. It is operated primarily through the Center for Scientific Research

---

[11]The Center for Scientific Research manages the review process for all NIH awards, except the small number managed in-house by individual ICs (such as the SBIR RFAs).

(CSR). CSR is a separate IC which serves only the other ICs—it has no direct funding responsibilities of its own.

The system has been criticized on a number of fronts, most notably for being inhospitable to innovation,[12] and because in tests of peer review processes elsewhere in biomedical research a significant degree of randomness in results has been identified.[13] Nonetheless, peer review is deeply entrenched at NIH, and the selection of SBIR awards at NIH operates through the peer review that has been implemented agency wide.

5.5.1 Study Sections

Applications for NIH SBIR awards are received at CSR and are assigned to a particular study section (as review panels are known at NIH) based on the technology and science involved in the proposed research. Panels can either be permanent panels legally chartered (established and defined) by Congress, or temporary panels designated for operation by NIH, called Special Emphasis panels (SEPs). Most SBIR applications are assigned to temporary panels, many of which specialize in SBIR applications only.

Specialized panels at NIH are increasingly used because the requirements for assessing SBIR applications—notably the commercialization component—are quite different from the analysis required to assess the basic research conducted under other NIH grant programs. However, several respondents to the NRC Program Manager Survey at NIH noted that some study sections did consider all kinds of applications, and they did not believe this was the optimal way to review SBIR applications. A program manager at NCI observed that "More and more mixing of mechanisms is occurring in study sections once devoted to SBIRs, thus diluting the focus."[14]

CSR is organized into four divisions, each of which is divided into Integrated Review groups (IRGs) by science/technology (e.g., infectious diseases, immunology). Each IRG manages a number of study sections.[15] Neither CSR nor the study sections are organized by either disease or IC—they reflect scientific distinctions only.

Special Emphasis Panels (SEPs) are reconstituted for each funding cycle. Almost all SBIR applications are reviewed by SEPs, which have a broader technology focus than the permanent chartered panels. Members can attend no more than 12 SEP study sections in 6 years. Section membership shifts with scientific trends.

---

[12]D. F. Horrobin, "The philosophical basis of peer review and the suppression of innovation," *Journal of the American Medical Association*, 263:1438-441, 1990.

[13]T. Jefferson, et al., "Measuring the Quality of Editorial Peer Review," *Journal of the American Medical Association*, 287:2786-2790, 2002.

[14]Response to NRC Program Manager Survey, April 2006.

[15]For example, the immunology IRG has seven permanent and two temporary study sections.

The second kind of study section, known as chartered (permanent) study sections, usually has a narrow technical focus (e.g., host defenses, innate immunity). Most sections are chartered, and their members are semi-permanent; sitting for 4-8 years out of every 12.

Most SEPs draw the majority of their applications from a subset of ICs. For example, the immunology IRG covers applications that refer to about 15 ICs, but 50 percent of its work comes from NAIAD, with a further 33 percent from NCI, reflecting the technical specialization of the SEP.

NIH guidelines are that at least one panelist (member of the study section) should have small business background. However, some Scientific Review Administrators (SRAs) appear to be making a greater effort to get panelists with entrepreneurship experience. One recent panel, for example, had 13 small business representatives out of 25 panelists.[16] That constituted a change for that panel: Previous panels in that technical area had been dominated by academics. NIH guidelines mandate 35 percent female and 25 percent minority panelists on each panel.[17]

There were numerous comments from agency staff and awardees about the difficulties of getting study sections with an appropriate mix of expertise. Some respondents to the NRC Program Manager Survey also focused on the need for more training for reviewers. Connie Dresser at NCI, for example, noted that "SBIR training needs to be mandatory for all SBIR reviewers in that they need to know what they should not be focusing on or why they should not be comparing SBIR content with R01 content. Also, we need people with marketing training and experience in review. The university types know text book information about marketing, not real-world marketing."[18] Other comments were more trenchant: "One basic flaw, in addition to the fundamental methodological deficiencies, is the reliance upon academic scientists to conduct reviews of SBIR-STTR applications. To put it simply: They are not qualified."[19]

One additional point on this subject was made by an NIH staff member. She noted that the selection process would be improved by the addition of professional consumers of medical producers, e.g., users of MRI technology, as well as experts in its development.[20]

---

[16]NIH staff interview.

[17]See Center for Scientific Review, "Overview of Peer Review Process" for detailed discussion of the peer review process at NIH. <*http://cms.csr.nih.gov/ResourcesforApplicants/PolicyProcedure Review+Guidelines/OverviewofPeerReviewProcess/*>.

[18]Response to NRC Program Manager Survey, April 2006.

[19]Michael-David Kerns, NIA, Response to the NRC Program Manager Survey, April 2006.

[20]Amy Swain, NCRR. Response to NRC Program Manager Survey, April 2006.

BOX 5-1

"Competitive pressures have pushed researchers to submit more conservative applications, and we must find ways to encourage greater risk-taking and innovation and to ensure that our study sections are more receptive to innovative applications."

Dr. Toni Scarpa, Director, CSR.
"Research Funding: Peer Review at NIH"
Science, 311(5757):41, January 6, 2006.

5.5.2 Selection Procedures

Each application is assigned to a subset of outside reviewers on the relevant panel—two lead reviewers and one discussant.

These three panelists begin by separating out the bottom half of all applications. These applications are not formally scored, though the applicants do receive a written review explaining why they were not selected.

At the review panel meeting, the three reviewers provide their scores on the remaining half of the applications before there is any discussion. Following a panel discussion, the three reviewers make changes to their scores if they wish. The entire review panel then scores the application.

Scoring is based on five core criteria:

- Significance of the proposed research.
- Effectiveness of the proposed approach.
- Degree of innovation.
- Principal Investigator's reputation.
- Environment and facilities.

There is no set point value assigned to each of these. Scores of individual reviewers are averaged (no effort is made to smooth results for example by eliminating highest and lowest scores). This average is multiplied by 100 to generate the reported score, between 100 (best) and 500 (worst). Fundable scores are usually in the 210-230 range or better, although this varies widely by IC and by funding year. Scores are computed and become final immediately.

According to an experienced NIH SBIR program manager, Gregory Milman, "most reviewers feel that NIH funds should be used for research and not for development."[21] This reflects the view that reviewers are generally biased

---

[21]Gregory Milman, "Advice on NIH SBIR and STTR Applications," April 2005, Slide 10. Accessed at: *<http://www.niaid.nih.gov/ncn/sbir/advice/advice.pdf>*.

toward the kind of basic research funded by more standard NIH programs, such as RO1. Currently, there are no data to substantiate this view, but it is held by several senior staff members. For example, the NIDA response to the NRC Program Manager Survey noted that "Grants are currently reviewed mostly from a research perspective (which reflects the characteristics of the review group and NIH priorities) with minimal emphasis on commercialization potential."[22]

Milman further notes that "Academic reviewers are most comfortable with hypothesis-driven research . . . the collection and analyses of data necessary for your product. Research is not developing something, building something, or discovering something. You can use grant funds to develop, build, and discover but only as necessary to collect and analyze data."

Reviewers are instructed not to base their evaluation of applications on the size of the funding requested. They are required to note if the funding requested is appropriate for the work proposed. As a result, reviewers do not consider possible trade-offs between different size applications (i.e. whether one large high scoring project is "worth" giving up for two or three similarly meritorious smaller projects). This is increasingly important as the size of applications varies from the standard SBA and NIH guidelines. These trade-offs are supposed to occur within the IC as it makes funding decisions, but interviews with IC staff suggest that the degree to which it does so is highly variable, and nontransparent.

Reviewers are also instructed not to consider in their evaluation the number of SBIR awards previously given to the applicant. The application form asks proposing companies to note if they have received more than 15 Phase II awards, but this question is for administrative purposes only. Otherwise, the application forms have no place to list previous awards. While companies with strong track records seek to ensure that these previous successes are reflected in the text of their application, there is no formal mechanism for indicating the existence or outcomes of past awards. Reviewers also do not know the minority or gender status of the PI or of the company.

5.5.3 Post-meeting Procedures

Once the study section has completed its meeting, scores are tallied immediately. These scores are then sent to the funding IC, which receives scores for all other SBIR applications that have been assigned to it.

Budget officers then work through procedures designed to establish the payline—the score above which applications will be funded for this funding cycle. These procedures include identifying the overall size of the funding pool for SBIR (2.5 percent of the total budget for extramural research), identifying and tallying all noncompeting SBIR awards (e.g., Phase II, year two awards) to which the NIH is already committed, setting aside funds needed for RFAs, and

---

[22]NRC Program Manager Survey, April 2006.

finally calculating the amount of available funds. These funds are then allocated to applications by the IC (ICs appear to use different procedures for doing this), primarily according to their scores. The payline is established at the point at which all available funds have been expended.

Typically, the payline for each ICs SBIR awards is in the range of 210-230, but it can be considerably higher or lower depending on the specific IC and the specific application cycle.

At this point, IC staff may intervene to make marginal adjustments to the list, perhaps moving one or two nonfunded applications up above the payline, and consequently defunding marginal applications with higher scores. Staff at NIH report that these adjustments are minimal, but there are no available data on this important point.

The funding procedure at NIH does not appear to have changed even though the size of awards has increased substantially. A new element—trade-offs—has been added into the funding equation. Applications asking for relatively large funding amounts can potentially preclude multiple smaller awards of similar merit. It does not appear that any IC staffers are explicitly charged with assessing these possible trade-offs within the SBIR program, nor is there any additional formal layer of review for unusually large SBIR awards. Extra-large RO1 applications, by contrast, must receive special approval.

5.5.4 Positive and Negative Elements of NIH Peer Review Process

On the positive side, outside review results in:

- Strong endorsements within the agency for applications derived from formal peer review;
- Alignment of the program with other programs at NIH, which operate primarily via peer review;
- Perceptions of fairness related to outside review in general;
- Absence of claims that awards are prewired for particular companies; and
- Access to reviewers with specialized expertise.

On the negative side, difficulties with the outside review process expressed by staff, awardees, and other experts in interviews appear to have been exacerbated by recent efforts to infuse commercialization assessment. Problems include:

- Deteriorating quality of reviews as workload increases, and difficulties in recruiting peer reviewers with appropriate expertise; NIH now handles 80,000 applications annually, and recruits more than 15,000 peer reviewers.[23]

- Significant perceptions that scoring has a large random component (a view presented by many case study interviewees, and also by a number of NIH SBIR program officers).
- Conflict of interest problems related to commercialization (an issue raised forcefully by several interviewees and by other stakeholders knowledgeable about the program, but not accepted in the course of NIH agency interviews).
- Substantial delays in processing (accepted by NIH as a problem).
- Questions about the trade-offs between different size awards (see above). These questions are likely to grow as the number and diversity of extra-sized awards continues to expand (see Chapter 3 for details).

Overall, outside review appears to add fairness and legitimacy but also complexity and delay. Companies interviewed and NIH program officers both pointed out that in many ways, the NIH process had not been adjusted to address the needs of companies trying to work fast in an increasingly competitive environment. Delays that might be acceptable at an academic institution focused on basic research with multiyear timeframes may have a more harmful effect on smaller businesses working within a much shorter development cycle. These issues are to a considerable extent understood at NIH, and the agency has started to initiate changes to address these problems. (See Section 5.5.8.)

5.5.5 Confidentiality and IP Issues

Applications are, in theory, strongly protected. They are not made public and reviewers sign confidentiality agreements before seeing the applications. Only the summaries of awards are published.

Nonetheless, confidentiality remains an important issue at NIH. Several case study interviewees (e.g., those at Neurocrine, Advanced Brain Research) were concerned that competitors are able to act as reviewers—in some cases despite written appeals for their removal to the Scientific Review Administrator (SRA), the NIH health scientist administrator in charge of review and advisory groups.

These concerns were reflected in some of the responses to the NRC Program Manager Survey (although others specifically saw no problems with conflict of interest). Connie Dresser of NCI, for example, noted that "Conflict of interest is a major concern in my review sessions. While the SRA is very good about reminding reviewers to excuse themselves from the room, I have had reports from

---

[23]Dr. Toni Scarpa, "Research Funding: Peer Review at NIH," *Science*: 311(5757):41, January 6, 2006.

grant applicants about reviewers who presented similar information or projects to theirs at conferences.[24]

There appears—from interviews—to be some evidence that peer review panels are requiring more detailed data from applicants, especially at Phase II, and that these demands present further difficulties: Neurocrine noted that this raised problems because the data requested were confidential, commercially critical, and not yet legally protected because patenting every advance at the earliest stage was simply not economically feasible. This left an "IP gap" between the initial identification of a promising compound or molecule, and the date at which testing results were sufficiently promising to justify the time and expense of patenting. Conversely, CSR officials noted that review panels had every right to require sufficient data on which to make a reasoned judgment about the viability of a particular technical approach, and that with increasing numbers of applications, more attention was focused on the technical details of each proposal.

These concerns are reflected in the advice from Gregory Milman, SBIR Program Manager for NAIAD, who warns applicants in advance: "I strongly recommend that you protect your intellectual property before you describe it in a grant application. I would not depend upon confidentiality agreements signed by reviewers or the fact that grant applications are not public documents."[25]

5.5.6 Metrics for Assessing Selection Procedures

Assessment of the SBIR selection process is complicated because the program serves many objectives and hence must meet multiple distinct criteria. Discussions with agency staff, award winners, and other stakeholders (such a bio-oriented venture firms, congressional staff) suggest that the following criteria best reflect a "successful" selection process:[26]

- **Fair.** Award programs must be fair and be seen as fair; the selection process is a key component in establishing fairness.
- **Open.** The program should be open and accessible to new applicants.
- **Efficient.** The selection process must be efficient, using the time of applicants, reviewers, and agency staff efficiently.
- **Effective.** The selection process must select the applications that show the most promise for meeting congressionally mandated goals, as interpreted by NIH.
- **Mission-oriented.** The selection process must help the program to meet the agency mission.

---

[24]Response to NRC Program Manager Survey, April 2006.

[25]Gregory Milman, "Advice on NIH SBIR and STTR Applications," op. cit., Slide 16.

[26]While these are the criteria against which all SBIR agencies develop their selection procedures, the criteria are not explicitly recognized or articulated in any agency, and the agencies balance them quite differently.

The last two criteria are best considered in light of outcomes (see Chapter 4). The remaining components of the selection process are discussed below.

5.5.6.1 Fairness

Discussions with case study interviewees and agency staff indicate that the perceived fairness of selection procedures is a function of several factors. These may include:

- Transparency—is the process well known and understood?
- Implementation—are procedures implemented consistently?
- Checks and balances—are outcomes effectively reviewed by staff with the knowledge and the authority to correct mistakes?
- Conflicts of interest—are there procedures in place and effectively implemented to ensure that such conflicts are recognized and eliminated?
- Appeals and resubmissions—are there effective appeals and/or resubmission procedures in place?
- Debriefings—is there a debriefing procedure that increases the perception of fairness among unsuccessful applicants?

Both agency staff and applicants noted that the considerable degree of apparent randomness in the process to some extent undercut perceptions of fairness. Karen Peterson, of NIAAA, noted that "This is the weakest point of all in the program. While scores have been improving for applicants to our institute, the quality of reviews especially in the behavioral sciences is widely variable."[27]

Transparency. At NIH, the selection process is almost the same process that is used for all other NIH awards. The process is explained on the Web, and in written materials sent to applicants. However, NIH staff report that they spend considerable more effort supporting SBIR applicants and awardees than they do applicants from universities, where the NIH application process is often supported by more experienced staff.

Implementation. The NIH review procedures are formalized, and are implemented under the supervision of professional and independent review staff at CSR; procedures appear to be followed consistently and predictably.

Checks and balances. Scores are highly influenced by the three core reviewers of each proposal, and within them, by the lead reviewer. Once the study section has scored and reviewed the panel, IC staff may decide to fund or not fund "across the payline,"[28] essentially reversing decisions by the study section. Interviews with NIH staff suggest that this is rare, though NIH has provided no

---

[27]Response to NRC Program Manager Survey, April 2006.
[28]See discussion of Payline below.

data on this subject. Decisions by IC staff are reviewed and usually approved by the IC's advisory council, which usually meets three times annually.

Appeals. The appeals process is largely moribund. NMIH staff and interviewees agreed that the resubmission process was much faster, simpler, and likely to be more effective. NIH does provide a written response to every application, with detailed information about why awards were not accepted. Applicants indicated in interviews that this debriefing was critical to the resubmission of applications—although some noted that changes in the composition of review sections meant that fixing criticisms was often not enough to ensure selection next time around.

Conflicts of interest. NIH has clear conflict of interest regulations in place for reviewers, and also has procedures in place that would allow applicants to seek to exclude an individual panel member from reviewing their proposal.

However, a number of interviewees among the companies and other stakeholders such as VC firms noted that these regulations largely operate on the context of an honor system: CSR undertakes no systematic or random checks on reviewers. Their experiences had been mixed, and several noted that as NIH seeks to introduce more commercial expertise into the review process for SBIR awards, the potential for conflict of interest problems may increase (although others noted that academics may also have conflicts of interest). The extent to which this works in practice is not clear, and it may depend on individual CSR officers. Interviewed awardees have repeatedly mentioned potential conflicts of interest as a problem with the SBIR review system.

Resubmissions are the standard mechanism for appeal at NIH, and about 33 percent of all awards are eventually made after at least one resubmission. This ability to resubmit enhances perceptions of fairness.

Finally, respondents to the NRC Program Manager Survey from NHLBI noted that there were inequities between the larger and smaller ICs with regard to paylines: "it seems unfair for smaller Institutes to have to forego paying outstanding applications when the larger Institutes fund at much higher (i.e., lower quality) scores."[29]

5.5.6.2 Openness

Some useful metrics for assessing the degree of openness relate to new companies entering the program; others relate to the concentration of awards going to certain companies within the program.

5.5.6.2.1 New Winners

Figure 5-2 shows the annual percentage of previous nonwinners at NIH (who

---

[29]NHLBI composite responses to the NRC Program Manager Survey, April 2006.

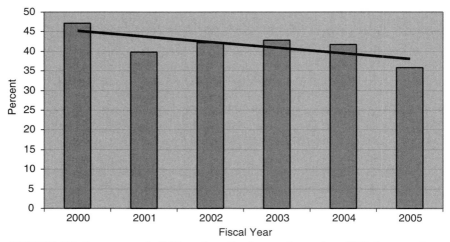

FIGURE 5-2 Percentage of all Phase I applications and awards at NIH from previous non-winners at NIH, 2000-2005.

SOURCE: National Institutes of Health.

may however have received SBIR awards from other agencies) applying for and winning Phase I awards from 2000-2005.

The data show that the Phase I share of previous nonwinners has remained above 35 percent, although it has declined since 2000. The latter is possible due to the increasing number of previous winners in the pool of potential applicants.

The fact that one-third of all applications and awards involve companies who have not previously won an NIH SBIR grant strongly suggests that the program is reasonably open. These levels are comparable to those at other agencies.[30]

5.5.6.2.2 Award Concentration and Multiple-Award Winners

Another view of openness might consider the extent to which awards are concentrated among the top award winners. Table 5-2 shows the distribution of

---

[30]See NRC Reports on the SBIR programs at DoD, NSF, DoE, and NASA: National Research Council, *An Assessment of the Small Business Innovation Research Program at the Department of Defense*, Charles W. Wessner, ed., Washington, DC: The National Academies Press, 2009; National Research Council, *An Assessment of the Small Business Innovation Research Program at the National Science Foundation*, Charles W. Wessner, ed., Washington, DC: The National Academies Press, 2008; National Research Council, *An Assessment of the Small Business Innovation Research Program at the Department of Energy*, Charles W. Wessner, ed., Washington, DC: The National Academies Press, 2008; National Research Council, *An Assessment of the Small Business Innovation Research Program at the National Aeronautics and Space Administration*, Charles W. Wessner, ed., Washington, DC: The National Academies Press, 2009.

TABLE 5-2 Top 20 Companies—Phase II Awards at NIH, 1992-2005

Name of Organization	Number of Phase II Awards
RADIATION MONITORING DEVICES, INC.	45
NEW ENGLAND RESEARCH INSTITUTES, INC.	38
OREGON CENTER FOR APPLIED SCIENCE, INC.	37
INFLEXXION, INC.	37
SURMODICS, INC.	28
INSIGHTFUL CORPORATION	22
LYNNTECH, INC.	21
CREARE, INC.	21
INOTEK PHARMACEUTICALS CORPORATION	17
BIOTEK, INC.	17
CLEVELAND MEDICAL DEVICES, INC.	16
ABIOMED, INC.	16
OSI PHARMACEUTICALS, INC.	15
PHYSICAL SCIENCES, INC.	15
GINER, INC.	15
PANORAMA RESEARCH, INC.	14
SOCIOMETRICS CORPORATION	14
WESTERN RESEARCH COMPANY, INC.	14
CANDELA CORPORATION	13
PERSONAL IMPROVEMENT COMPUTER SYSTEMS	13
Total	428
Percent of all Phase II awards	10.4

SOURCE: National Institutes of Health.

Phase II awards to the "top 20" winners at NIH—the 20 companies receiving the most Phase II awards at NIH.

The data set above shows that Phase II awards at NIH are not highly concentrated. The most frequent recipient of Phase II awards received 45 over 14 years—just over three per year. In all, these top 20 winners account for 428 Phase II awards, 10.4 percent of the total awarded.

5.5.6.3 Efficiency

Efficiency can be defined in many ways. Box 5-2 includes several possible external and internal efficiency goals towards which the NIH SBIR program should strive.

5.5.6.3.1 *Efficiency for Applicants*

There are a number of positive components of the current system from the perspective of applicants. These include:

- The possibility of resubmission;
- Broad topic design, which ensures that highly promising research applications are not arbitrarily excluded;

BOX 5-2
Possible Efficiency Indicators for SBIR Selection Process

External: Efficiency for the Applicant
- Shorten time from application to award.
- Reduce effort involved in application.
- Reduce red tape involved in applying.
- Output from application (not including award).
- Re-use of applications.

Internal: Efficiency for the Agency

- Move the grant money quickly to right recipients.
- Minimize use of staff resources.
- Maximize agency staff buy-in.
- Reduce appeals and bad feelings.

- Widespread support for the notion of peer review; and
- The existence of multiple annual application windows, which effectively shorten the time from idea to funding.

At the same time, interviews with NIH staff and SBIR awardees indicate considerable areas for possible improvement; these include:

- **Random outcomes.** Many interviews and NIH staff asserted that that there is a substantial element of randomness in the selection process. While this clearly impacts fairness, it also impacts efficiency: Firms contribute time and resources in the form of applications, without a belief that these will generate a return commensurate with their quality.
- **Reliance on resubmissions.** While the availability of resubmissions does promote fairness, its extensive use within the NIH SBIR application process is inefficient: It imposes significant additional costs and substantial delays on applicants, the latter almost always amounting to at least 8 months between applications. From a small business's perspective, this delay could be disastrous. A second resubmission—which is not uncommon—results in a further 8-month delay.
- **Application procedures at NIH are still largely nonelectronic.** NIH has now moved to all-electronic submission of applications. However, the study section process remains based on in-person meetings and written documentation, and there appears to be room for considerable improve-

ment and experimentation, as noted by Dr. Scarpa, Director of CSR, in a recent article in *Science*.[31]

- **Delays.** The delays imposed by the current process, again as accepted by Dr. Scarpa, are substantial and could clearly be reduced. Eighty percent of NRC Phase II Survey respondents reported a gap between Phase I and Phase II. The median length of the gap was 13 months, and 11 percent of respondents reported a gap of 2 years or more. NIH is now beginning to experiment with a number of pilot changes to the selection process, focused on this issue.

5.5.6.3.2 *Efficiency for NIH*

Program efficiency can be measured in a number of ways, and—based on interviews with staff and awardees—these provide a mixed picture for NIH:

- **Moving the money.** The process is 100 percent successful in moving SBIR funds from NIH to awardees.
- **Low overhead.** Program costs appear to be low; NIH has simply imposed additional work on existing staff as grant applications have increased.
- **Return on Investment (ROI).** NIH has only a limited knowledge of the ROI from its SBIR investment, partly because efforts to minimize overhead have led to insufficient investment in monitoring and evaluation.
- **Staff buy-in.** The SBIR process is not designed to encourage staff buy-in (see staffing issues section). Nevertheless, some SBIR Program Administrators are enthusiastic and effective.
- **Minimizing appeals.** Resubmission effectively replaces appeals within the NIH framework. Appeals are unusual.

Overall, it is fair to say that NIH has little idea whether the SBIR program is efficient for the institution, or whether efficiency varies by IC. SBIR has generated more data on outcomes than other NIH research funding programs, but not enough to make those kinds of determination. It is however true that some NIH staff strongly believe that SBIR programs place a significant additional burden on NIH administrators, compared to other programs, largely because the applicants are working in an environment that they are not familiar with: "Grants management specialists also report hugely disproportionate (vis-à-vis other principal investigators, organizations, & research-grant mechanisms) demands from SBIR-STTR potential & actual applicants (& applicant organizations) (vis-à-vis other research-grant program applicants). The vast majority of problems, including violations worthy of formal investigation, encountered by our grants management specialists within NIA's entire research portfolio, derive from SBIR-STTR research grants & the small-business organizations. The grants management

---

[31]Dr. Toni Scarpa, "Research Funding: Peer Review at NIH," op. cit.

specialists have indicated that they spend anywhere from 40-60 percent more time and effort working up and administrating SBIR-STTR grant projects. The frequency and persistence of problems with SBIR-STTR projects are such that within NIA's GCMO (contracts office), there is a trenchant lack of enthusiasm for the SBIR-STTR programs."[32]

5.5.7 Funding Cycles and Timelines: The NIH Gap-reduction Model

Many SBIR awardees rely heavily on SBIR funding to pay for their operations. Gaps in funding can be deadly to small businesses without other stable sources of revenues.

NIH has recognized this issue, and several characteristics of the NIH SBIR program fall within what the Summary Report describes as the "gap reduction model" for managing funding cycles and timelines.[33] This model is distinguished by its emphasis on supporting applicants using a range of features designed to reduce gaps in funding and decrease the time from initial conception to final product deliverable. Elements in use at NIH include:

Multiple annual submission dates. NIH provides three annual submission dates for awards, in April, August, and December. This is a substantial improvement on the one annual date in effect at some other agencies because it potentially reduces time lags related to these deadlines by 8 months. Dr. Scarpa has indicated that CSR will experiment in 2007 with open submission—submissions throughout the year with no set deadline.

Topic flexibility. Topics are discussed extensively above, but they have important implications for the gap reduction model. Narrow, topic-bounded application processes can harm small businesses because they have to wait for an appropriate topic to show up in a solicitation before they can apply for an SBIR grant. NIH is in this respect highly flexible, with its investigator-initiated research approach, which in largely preclude "topics-based" delays. This should therefore be seen as an important component of the overall gap-reduction model at NIH.

Phase I - Phase II gap funding. Two mechanisms have been developed at NIH to bridge the funding gasp between the conclusion of Phase I and the start Phase II funding: "work-at-risk" and the NIH Fast Track.

- **"Work at risk."** Companies that anticipate winning an NIH Phase II award can work for up to three months at their own risk, and the cost of that work will be covered if the Phase II award eventually comes through. If it does not, the company must swallow the cost.
- **Fast Track.** Fast Track efforts are designed primarily to reduce the amount

---

[32]Michael-David Kerns, NIA, Response to the NRC Program Manager Survey, April 2006.

[33]Described in more detail in National Research Council, *An Assessment of the Small Business Innovation Research Program*, Charles W. Wessner, ed., Washington, DC: The National Academies Press, 2008.

of time between the end of Phase I and the start of Phase II. At NIH (unlike DoD[34]), applicants must apply for Fast Track status during the Phase I application, as it is in effect an application to do a joint Phase I—Phase II application. The advantage of Fast Track is that acceptance should—at least in theory—mean that funding gap is dramatically reduced. (See Section 5.6 for more details).

Phase II plus programs. Phase II plus programs are designed to help bridge the gap between the end of Phase II and commercialization (sometimes known as "Phase III").

NIH has implemented a new initiative targeted at helping to fund companies through the first stages of the clinical trials process, with funding for up to three years, at up to $1 million per year.

5.5.8 NIH Selection Initiatives

NIH is well aware of complaints about cycle times, and about the burden placed on companies and other grant applicants. As Dr. Toni Scarpa, Director of CSR, notes, "Our system can be particularly frustrating for those who may need to make only minor revisions, because results from our reviews typically come too late for them to reapply for the next review round."[35]

CSR is now working to reduce cycle time. In particular,

- As of October 2005, NIH now posts summary statements of most reviews within 1 month after the study section meeting, instead of 2-3 months after the meeting. This gives important guidance to applicants.
- In February 2006, NIH began a pilot study to cut 1½ months from the review process. Forty CSR study sections will participate in this pilot, which will speed the reviews of R01 applications submitted by new investigators. Resubmission deadlines will be extended to allow these new investigators to resubmit immediately if only minor revisions are necessary. Specifically, CSR will: (i) schedule study section meetings up to a month earlier; (ii) provide scientists their study section scores, critiques, and panel discussion summaries within a week after the section meeting; (iii) shave days from the internal steps involved in assigning applications to study sections; and (iv) extend resubmission deadlines by 3 weeks.

Dr. Scarpa notes that "we are experimenting with new electronic technologies that permit reviewers to have discussions with greater convenience and to spend less of their precious time in traveling. For example, asynchronous Inter-

---

[34]The DoD Fast Track program is completely different from the NIH Fast Track effort; the only operational similarity is the name.

[35]Dr. Toni Scarpa, "Research Funding: Peer Review at NIH," op. cit.

TABLE 5-3 Fast Track Applications and Success Rates, 1997-2004

Fiscal Year	Number of Applications	Number of Awards	Fast Track Success Rate (%)	Phase I Success Rate (%)
1997	41	13	31.7	26.6
1998	63	11	17.5	26.8
1999	129	45	34.9	26.5
2000	120	34	28.3	25.1
2001	129	38	29.5	28.6
2002	183	50	27.3	25.8
2003	273	61	22.3	15.1
2004	329	58	17.6	17.9

SOURCE: National Institutes of Health.

net-assisted discussions—secure chat rooms—allow reviewers to "meet" and to comment independently of time as well as place."[36]

5.6 FAST TRACK AT NIH

Fast Track at NIH is a completely different program than Fast Track at DoD. At NIH, Fast Track offers the promise of accelerated flow of funds by eliminating the reselection process at Phase II. Instead, companies with approved Fast Track awards simply provide an approved final report for Phase I, and Phase II begins automatically.

Fast Track has attracted to a growing number of companies in recent years (as shown by Table 5-3). To be eligible for Fast Track, an applicant must submit complete Phase I and Phase II applications at the same time, along with:

- Clear, measurable milestones for Phase I, used to judge whether Phase I objectives have been met;
- A full Phase II Product Development Plan; and
- Evidence of commitment from a commercial partner.

In theory, Fast Track should reduce funding gaps and application time by up to seven months, as the diagram in Figure 5-3 shows.

Milman notes however that in many cases, Fast Track is not an appropriate route, particularly where the specific milestones are unclear. For example, he contrasts a drug company with a drug candidate selected, now planning small mammal trials in Phase I and primate trials in Phase II, with a drug company whose candidate drug has not yet been identified and which will rely on Phase I

---

[36]Ibid.

Normal application, review, award process

Submit Phase I	Review 7-9 months	Award 6 months	Prepare & Submit Phase II	Review 7-9 months	Award 24 months
Apr 1	~July	~Nov	~Apr	~July	~Nov

Fast Track application, review, award process

Submit Phase I & Phase II	Review 7-9 months	Award 6 months	Phase I Progress Report	Program Review 1 month	Award 24 months

>7 months earlier

FIGURE 5-3 Fast track and normal timelines at NIH.

SOURCE: Gregory Milman, NAIAD.

results in designing its Phase II research plan. The latter case is, according to Milman, better suited to the standard Phase I-Phase II progression.

Karen Peterson of NIAAA also notes that "Fast Track is not very useful in its current incarnation." She goes on to say that "Most reviewers are very reluctant to give these applications good scores because of the time and money commitment they feel they are making."[37] This view is also reflected in comments from NIDA: "For some reason, reviewers do not like fast track and all most always give them worse scores than they would normally receive. We now recommend, even to the best of companies, not to submit using a fast track because it definitely reduces their chances of funding."[38]

Other reasons for avoiding Fast Track include:

- Difficulties in attracting a commercial partner on appropriate terms, which is likely if the product is early in the development cycle.
- The proposal work required, which Milman estimates at four times the work of a standard Phase I.
- The existence of alternative paths across the funding gap which may be less risky and resource-intensive.
- Reluctance, according to other NIH staff, among reviewers to accept Fast Track applications. Study sections can recommend that fast Track appli-

---

[37]Response to NRC Program Manager Survey, April 2006.
[38]Ibid.

cations be approved for Phase I only, returning the application to standard format.

These points and the data above suggest several observations:

- Fast Track is rapidly growing in importance, expanding from 41 applications and 13 awards in 1997 to more than 300 applications and almost 60 awards in 2004, or from 1.4 percent to 5.7 percent of all applications during that period.
- Success rates for Fast Track are on average close to those for Phase I (26.1 percent for Fast Track, 24.0 percent for Phase I).
- Fast Track appears to be working well enough that companies are applying in growing numbers.
- Fast Track is still an uncommon choice for applicants—95 percent of awardees use the standard progression. Milman's analysis suggests that relatively few additional companies will qualify for this approach in the future.
- Projects for which the experimental design is known and accepted are good candidates for Fast Track.
- NIH has undertaken no outcomes analysis to assess whether Fast Track awards generate more positive outcomes than standard awards.

5.7 FUNDING: AWARD SIZE AND BEYOND

NIH's SBIR program gives out awards that are different than those of other agencies in three ways:

- In some cases, NIH has made much larger awards than are given out by other agencies (see Chapter 3).
- NIH has begun to offer additional years of support including a second year of Phase I support in some cases, compared to the 6-month limit imposed by most other agencies.
- NIH provides administrative supplements that boost Phase I awards when additional resources are needed to complete the proposed research.

5.7.1 Larger Awards at NIH

Figure 5-4 shows that, starting in 1999, NIH began to provide an increasing number of Phase I awards of more than $250,000. There have been similar increases in the number of awards between $100,000 and $250,000. NIH has also in a few, but increasing, number of cases provided Phase I funding of more than $1 million.

An extensive discussion of larger awards can be found in Chapter 3. Here,

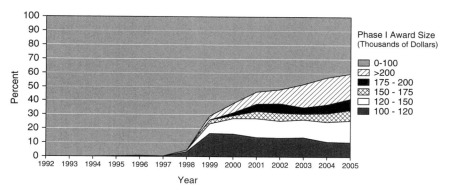

FIGURE 5-4 Extra-large Phase I awards at NIH, 1992-2005.

SOURCE: National Institutes of Health.

we simply note that the trend toward larger awards has continued, and that awards beyond the size of the SBA guidelines are rare except at NIH.

It is also worth noting that views among the Program Administrators responding to the NRC Program Manager Survey varied widely on this issue. Many recommended increased funding and extended time for awards; others indicated that they would prefer to see the limits more strictly enforced. However, this appears to depend on the kind of research being pursued. For example, Melissa Raccioppo of NIDA noted "Since our SBIR/STTR grants tend to involve a clinical trial of some sort, the limits on budget seem too restrictive for our investigators' purposes."[39] These comments applied to the new Competing Continuation Awards as well; Program Administrators with few likely recipients of these awards were concerned that they might take a disproportionate amount of SBIR program funding.

Finally, one of the respondents to the NRC survey noted that "these larger awards further point to a dire need for a solid outcomes tracking and evaluation capability."

5.7.2 Supplementary Funding

NIH officials have observed that the availability of supplementary awards adds further flexibility in helping companies to handle the unexpected costs that can easily arise in high-risk research.

In principle, program officers can add limited additional funds to an award

---

[39]Ibid.

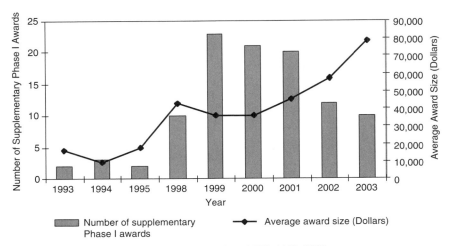

FIGURE 5-5 Supplementary Phase I awards at NIH, 1993-2003.

SOURCE: National Institutes of Health.

in order to help a recipient pay for unexpected costs. While practices vary at individual ICs, it appears that up to 25 percent (or up to $50,000) of current annual funding for an individual grant can be awarded by the program manager without further IC or NIH review (budget permitting). More substantial supplements must be more extensively reviewed, but are not unknown.

All supplemental requests require documentation. Full applications are required for competing supplements, and administrative supplements need at least a budget page and a letter justification.

For Phase I, supplements remain relatively rare, averaging less than 20 annually in recent years. They are also not especially large, and in no cases have NIH Phase I supplements totaled more than $1 million for a given fiscal year, Still, the data indicate that the size of Phase I supplementary awards are growing at NIH (see Figure 5-5).

Supplementary awards are also available for Phase II, where they are more significant. As shown in Figure 5-6, the number of Phase II supplement awards has hovered around 30. Thus about 10 percent of all Phase II awards receive supplementary funding.

5.7.3 Duration of Awards

Just as the size of awards has grown, NIH has extended the period of support as well. In FY2002 and FY2003, more than 5 percent of all Phase I awards received a second year of support, with a median value of about $200,000.

Year one and year two awards cannot be easily aggregated into a single

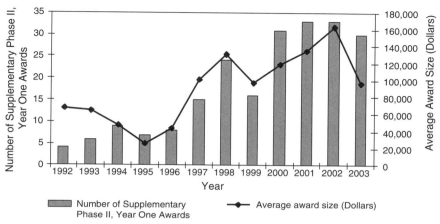

FIGURE 5-6 Supplementary Phase II, Year One awards at NIH, 1992-2003.

SOURCE: National Institutes of Health.

"Phase I award" at NIH owing to characteristics of the NIH awards database. However, the rapidly growing number of year two awards—which in FY2003 were equal to 6.3 percent of all 2002 Phase I, year one awards—as well as the jump in median size in 2000, suggests that this mechanism is of growing importance at NIH.

NIH staff and recipients alike agree that 6 months is too short to complete Phase I work in many biomedical disciplines. NIH usually approves requests for "no-cost" extensions to one year or even longer. No-cost extensions simply ex-

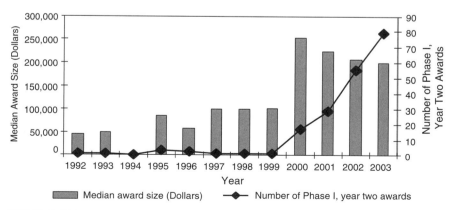

FIGURE 5-7 Phase I, Year Two awards at NIH, 1992-2003.

SOURCE: National Institutes of Health.

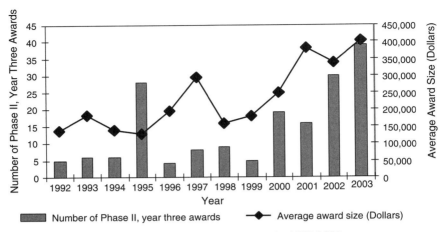

FIGURE 5-8 Third year of support for Phase II awards, 1992-2003.

SOURCE: National Institutes of Health.

tend the term of the award without providing additional funding. No other agency offers such a liberal extension program.

For Phase II, NIH also offers extended funding beyond the standard 24 months of support. Figure 5-8 contains estimates of Phase II, year three support calculated on the basis of NIH data (see Chapter 3 for detailed calculations).

The steadily rising numbers of Phase II, year three grants in recent years suggest that third year support is becoming an important component of NIH SBIR activity. In FY2002 and FY2003, more than 10 percent of awards received a third year of support.

In a few cases, NIH goes further. Ten grantees have received a fifth overall year of SBIR support, a few for even longer period.

5.7.4 Award Size: Conclusions

The data shown in Chapter 3 indicate that the size of awards at NIH is rising, that additional administrative support is of increasing importance, and that the duration of awards (and support) is expanding as well.

One important question might be why NIH is making these large awards. A second question might concern the growing number of extended awards. Both are discussed in Chapter 3, but conclusive answers are not available partly because neither question has been directly addressed by NIH, at least in materials that are publicly available.

One final point should, be noted, drawn from conversations with agency staff and from responses to the NRC Program Manager Survey: NIH has repeatedly sought to convert Phase I STTR's to Phase II SBIR's and vice versa, as the circumstances related to the research change. SBA has denied these appeals, for reasons that are not clear to NIH staff. Unless SBA can find convincing justifications for this position, it would appear that a change of policy here could be warranted.

5.8 COMMERCIALIZATION SUPPORT

5.8.1 Background

Since its inception in 1982, the SBIR program has aimed to increase "commercialization innovations derived from Federal research and development" (Public Law 97-219). After reauthorization in 1992, agencies were required to consider commercialization potential as part of its review process. The reauthorization also included a provision for technical assistance services to help grantees "develop and commercialize new commercial products and processes." SBA then issued a rule stating that assistance efforts focused on bringing products to market could be supported by up to $4,000 per Phase I award and up to $4,000 per year for each Phase II award. Subsequent interpretations of the rule by SBA supported aggregation of these funds for an SBIR technical assistance program.

5.8.2 Overview

NIH has recognized that many SBIR Phase II winners struggled to survive the period between the end of SBIR Phase II and market entry, and in June 2002, the Office of Extramural Programs at NIH (OEP) began to provide commercialization assistance to SBIR winners in June 2002.

This assistance is now rendered through the Technical Assistance Program (TAP). Thus far, OEP has initiated three pilot assistance programs and two follow-on, full-scale assistance program under the TAP:

- The Pilot NCI Commercialization Assistance Program (PCAP) supported 47 SBIR Phase II winners (related to NCI only) and concluded in March 2003.
- The Pilot Niche Assessment Program (PNAP) was made available to a maximum of 100 SBIR Phase I winners on a first-come, first-serve basis. The pilot program had finished assisting 45 projects as of February 16, 2005, and ended in August 2005.
- **Pilot Manufacturing Assistance Program.** In FY2007, NIH plans to pilot an additional assistance program targeting the many manufacturing issues small companies face when trying to commercialize their SBIR-funded products. In partnership with the NIST Manufacturing Extension Partnership (MEP) program, the pilot is aimed at providing transitional support as Phase II awardees move to a manufacturing stage. The goal

is to help companies make better decisions when developing their operational transition strategies (method of scale up, cost estimation, quality control, prototyping, design for manufacturability, facility design, process development/improvement, vendor identification and selection, plant layout, etc.) NIH has engaged Dawnbreaker of Rochester, NY, to operate this program. Twenty-five (25) NIH SBIR Phase II awardees are expected to participate.

- The Commercialization Assistance Program (CAP) was launched in July 2004 as the first full-scale, ongoing commercialization assistance program. Two cohorts of 114 firms each have completed the program as of January 2007.

5.8.3 The Commercialization Assistance Program (CAP)

The perceived success of PCAP prompted OEP to launch the Commercialization Assistance Program (CAP) as its first full fledged, ongoing TAP "menu" item. It is open to companies funded by all NIH ICs.

Larta Institute (Larta) of Los Angeles, CA,[40] was selected by a competitive process to be the contractor for this program.[41] The Larta contract began in July 2004, and will run for five years. During the first three years, three cohorts of SBIR Phase II winners will receive assistance. Years four and five will cover follow-up work, as each cohort is tracked for 18 months after completion of the assistance effort.

CAP Program details. *The assistance process for each group typically includes:*

- Provision of consultant time for business planning and development.
- Business presentation training.
- Development of presentation materials.
- Participation in a public investment event organized by Larta.
- Eighteen months for follow-up and tracking.

Participants. *Based on interviews with NIH staff and Larta, the typical CAP participant is:*

- A small technology-oriented business;
- Founded by an engineer or physician turned entrepreneur;
- In operation for 5 to 10 years; and

---

[40]Larta Web site, accessed at: <*http://www.larta.org*>.
[41]Larta was founded by Rohit Shukla who remains as its Chief Executive Officer. It assists technology oriented companies by bringing together management, technologies, and capital to accelerate the transition of technologies to the marketplace.

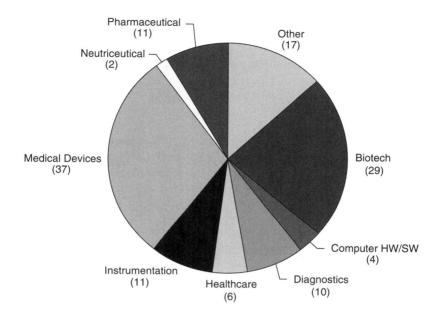

Industry Sector and Number of Participants

FIGURE 5-9 CAP participants, by industry sector.

SOURCE: National Institutes of Health.

• Substantially reliant on government grants because of limited outside funding.

These companies have typically not yet generated meaningful sales, but appear to have significant commercial upside.

As of January 1, 2004, NIH had 634 active SBIR Phase II projects from 455 companies across 23 Institutes and Centers. All of these companies were invited to participate in the CAP program[42] and a total of 114 companies participated. Approximately 75 chose to participate in a series of investment workshops offered in Orange County, CA; San Francisco, CA; Washington, DC; Chicago, IL; and Boston, MA, which allowed participants to present their respective business opportunities to a group of investors, and to receive feedback on the effectiveness of their presentations.

Participation by industry. The two largest industry sectors in CAP are Medical Devices (37 or 29 percent of total participants) and Biotech (29 or 23

---

[42]NIH SBIR Technical Assistance Program, Office of Extramural Programs, Enrollment Criteria.

percent of total participants). The Northeast region accounts for 35 percent of total participants and the West 32 percent of total participants.[43]

Areas of focused assistance. Three primary "Tracks," areas of focused assistance, were added by NIH after the pilot based on participant feedback. The three tracks are:

- The *Regulatory Track*, for participants in need of a strategy for FDA approval.
- The *Licensing Track*, for participants in need of documentation for establishing relationships with potential licensees.
- The *Strategic Alliance Track*, for participants in need of documentation for establishing joint ventures, collaborative agreements, or other similar partnerships.

Each Track is further adapted to the special needs of two industry sectors: Biomedical Devices (includes all medical devices and device-based products) and Biotechnology (includes all drugs and biologic-based products). The distribution of the current CAP participants by "Track" is represented in Figure 5-10.[44]

5.8.4 Niche Assessment Program (NAP) (for Phase I Winners)

Sometimes scientific researchers do not have the entrepreneurial skills to assess other applications or niches for their SBIR-developed technology. As a result, they may underestimate its true market value. This program assesses the market opportunities and needs and concerns of the end-users and helps to discover new markets for possible entry.

The NAP aims to assist SBIR Phase I winners in identifying and evaluating various market opportunities for commercialization (e.g., licensing, sales, partnering). This effort is operated by Foresight Science and Technology, Inc. (Foresight) of New Bedford, MA.[45] It has three phases:

1. Foresight gathers relevant information on the technology from the participant and begins to identify potential commercial applications.
2. Foresight and the participant determine the technology application that warrants detailed analysis. This application is analyzed by Foresight to determine end-user needs, current and emerging competing technologies, market dynamics, socioeconomic trends, market drivers, market size, the

---

[43]NIH CAP Participants by State, March 1, 2005.

[44]Update, SBIR Technical Assistance Program, February 16, 2005.

[45]Foresight is a scientific consulting firm offering market research, technology assessment, and valuation and licensing services to the medical, pharmaceutical and biotechnology industries. They focus on helping move technology from the laboratory to the marketplace and assess approximately 300 new technologies annually.

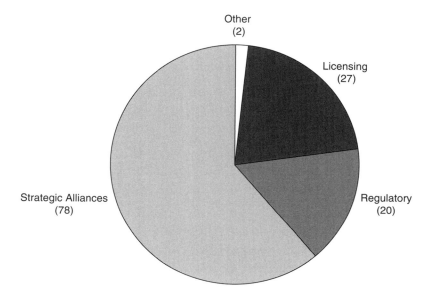

Assistance "Tracks" and Number of Participants

FIGURE 5-10 CAP distribution by track.

SOURCE: National Institutes of Health.

potential technology's possible market share, potential technology's current competitive advantages, and strategies for improving the technology's competitiveness.

3. Foresight develops a market entry strategy including how to market the technology to end-users and attract Phase III partners. The strategy also projects revenues from the sale or licensing of the technology, possible "launch" customers, testing centers, suppliers, manufacturers, and other parties potentially interested in the technology (e.g., beta testers). Foresight may also make introductions to potential partners.

Each step concludes with an electronic report plus follow-on discussions.

5.8.5 Outcomes and Metrics

5.8.5.1 Pilot NCI Commercialization Assistance Program

Evaluations were completed at 6, 12, and 18 months following culmination of this pilot program, when 32 participants presented at an investor/partner Forum in March 2003.

Participants were not obligated to provide feedback. However, 13 (40 percent) of the 32 companies reported that they had received additional private sector investment and/or sales related to the technology opportunity they presented at the Forum. Cumulative private sector funding and sales received within 18 months following completion of this program totaled almost $38 million.[46] Unsurprisingly, these results were highly skewed: A majority of these funds were received by five of the companies—Computer Science Innovations, Focus Surgery, High Throughput Genomics, Phoenix Pharmacologics, and Vaccinex. Approximately $18 million—or about 47 percent of the total—was generated through the sale of one of these companies.[47]

Of course, this minimal assessment does not provide or even suggest grounds for a causal link between the program and these results.

5.8.5.2 Commercialization Assistance Program

Two cohorts (2004/2005 and 2005/2006) have completed the CAP training program, and results have been very encouraging though not yet definitive. Evaluation data are collected from the companies at the conclusion of the program, and at 6, 12, and 18 months afterwards. These data indicate that firms going through the CAPM program are attracting funding, as Table 5-4 illustrates.

NIH has also developed some intermediate metrics that indicate project impact. However, as these metrics are not compared with other groups of companies that have not gone through the CAP program, it is difficult to draw conclusions from them.

Data collected six months after the CAP showed a strong increase in commercialization, and in particular in the conclusion of commercialization agreements, which increased for the 2004/2005 cohort by 87 percent (up from 23 at the baseline to 43 6 months later).

These data are encouraging, and are bolstered by discussions with individual participants that indicate that participants find this program to be of considerable value. Development of a control group of some kind would add considerably to the power of this analysis.

5.9 EVALUATION AND ASSESSMENT

Traditionally, NIH has not conducted outcomes assessment on its SBIR and STTR programs, or indeed on other programs. More recently, the NIH SBIR/STTR Program Office has initiated a number of activities aimed at infusing more data into the operation of the program, Most notably, in 2003 NIH followed on from its agreement to fund the NRC study with a separate NIH Survey of Phase II

---

[46]NIH Office of Extramural Programs.
[47]OER would not disclose the exact details of these outcomes citing confidentiality restrictions.

TABLE 5-4 Funding for CAPM Firms

	Year	
	2004/2005	2005/2006
Number of companies in CAP	114	114
Number receiving investments	24	13
Percent of total	21.1	11.4
Total investment to date	$22,414,078	$45,636,520

SOURCE: National Institutes of Health.

recipients. Using somewhat different methodologies from the NRC Phase II Survey, with concomitantly different strengths and weaknesses, the NIH Survey broke important ground, and provided results that have been used throughout this analysis.

Discussions with agency staff and responses to the NRC Program Manager Survey indicate widespread views that the program does not have the resources needed to develop an evaluation and assessment program sufficient to manage a program of this size and scope. Phil Daschner, from NCI, for example noted that "More resources should be available to program staff that track and evaluate objective benchmarks for past institutional and investigator productivity." In its institutional response to the NRC survey, NCI observed that "we still do not have reliable tools to capture in an ongoing way success stories from our grantees. It is a considerable undertaking to get evaluation funds and go through the OMB process. Methods have been identified to capture outcomes, but funds are not

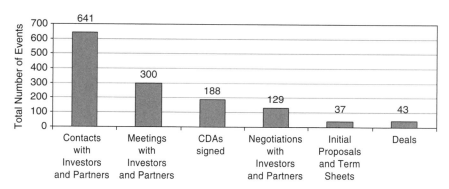

FIGURE 5-11 Aggregate number of partnership- and deal-related activities by category.

SOURCE: National Institutes of Health.

available to support a sustainable effort to track SBIR/STTR outcomes. This is a critical and long-term need."[48]

More specifically, as NIDA noted, "More time should be spent following up on grants near their end and after they no longer received NIH funding. We know little about Phase III and whether or not it actually occurs. Most time is spent funding the grant and administering it, but little or no time is spent on follow-up and evaluation."[49]

Currently, the NIH SBIR/STTR Program Office must seek one-time funding for any significant assessment activity; this largely precludes longitudinal approaches needed for effective use of evaluation and assessment.

---

[48]NRC Program Manager Survey, April 2006.
[49]Ibid.

Appendixes

Appendix A

NIH SBIR Program Data

NOTES:

"Year one" awards: NIH maintains data by fiscal year, and in a number of cases changes award number between years. As a result, it is often hard to track a complete award from first year to last. Our approach has therefore been to focus on "award years," identifying for example the first year of each award for analysis. This focus provides a complete data set for all NIH awards, and also allows analysis of awards in subsequent years of support.

Following discussions with the NRC staff, the NIH made an effort to re-calculate the data for woman and minority owners' participation in the SBIR program. In September, 2007, the NIH provided corrected data, which is shown in Appendix A and in several figures in this report. However, apparent anomalies in the NIH data on the participation of women and minorities in 2001-2002 could not be resolved by the time of publication of this report.

BOX App-A-1
Institute and Center Codes for the National Institutes of Health

In data tables throughout this report, the following codes are used to reference National Institutes of Health institutes and centers.

Code	Institute Name and Acronym
AA	National Institute on Alcohol Abuse and Alcoholism (NIAAA)
AG	National Institute on Aging (NIA)
AI	National Institute of Allergy and Infectious Diseases (NIAID)
AR	National Institute of Arthritis and Musculoskeletal and Skin Diseases (NIAMS)
AT	National Center for Complementary and Alternative Medicine (NCCAM)
CA	National Cancer Institute (NCI)
DA	National Institute on Drug Abuse (NIDA)
DC	National Institute on Deafness and Other Communication Disorders (NIDCD)
DE	National Institute of Dental & Craniofacial Research (NIDCR)
DK	National Institute of Diabetes and Digestive and Kidney Diseases (NIDDK)
EB	National Institute of Biomedical Imaging and Bioengineering (NIBIB)
ES	National Institute of Environmental Health Sciences (NIEHS)
EY	National Eye Institute (NEI)
GM	National Institute of General Medical Sciences (NIGMS)
HD	National Institute of Child Health and Human Development (NICHD)
HG	National Human Genome Research Institute (NHGRI)
HL	National Heart, Lung and Blood Institute (NHLBI)
LM	National Library of Medicine (NLM)
MD	National Center on Minority Health and Health Disparities (NCMHD)
MH	National Institute of Mental Health (NIMH)
NR	National Institute of Nursing Research (NINR)
NS	National Institute of Neurological Disorders and Stroke (NINDS)
RR	National Center for Research Resources (NCRR)
TW	Fogarty International Center (FIC)

TABLE App-A-1 Applications: Phase I and Phase II, 1992-2005

Phase I				Phase II			
Fiscal Year	All Applications (#)	Total Funded (#)	Success Rate (%)	Fiscal Year	All Applications (#)	Total Funded (#)	Success Rate (%)
1992	1,982	541	27.3	1992	551	278	50.5
1993	2,297	594	25.9	1993	637	360	56.5
1994	3,225	530	16.4	1994	744	351	47.2
1995	3,453	624	18.1	1995	780	370	47.4
1996	3,051	525	17.2	1996	798	390	48.9
1997	2,789	743	26.6	1997	800	468	58.5
1998	2,689	717	26.7	1998	827	541	65.4
1999	3,430	908	26.5	1999	897	539	60.1
2000	3,907	986	25.2	2000	1,023	587	57.4
2001	3,203	940	29.3	2001	1,074	683	63.6
2002	3,735	1,001	26.8	2002	1,248	797	63.9
2003	4,812	1,137	23.6	2003	1,299	788	60.7
2004	5,856	1,150	19.6	2004	1,410	792	56.2
2005	5,071	937	18.5	2005	1,451	774	53.3

SOURCE: National Institutes of Health.

TABLE App-A-2 Phase I Applications by IC, 1992-2004

Number of Applications

IC	1992	1993	1994	1995	1996	1997	1998	1999	2000	2001	2002	2003	2004
AA	25	52	38	37	50	53	30	40	39	24	29	65	62
AG	84	94	149	159	146	122	105	103	134	130	95	134	162
AI	154	201	350	413	349	333	290	365	461	356	592	676	844
AR	111	101	165	126	167	126	127	132	173	152	142	169	201
AT								5	36	58	55	41	35
CA	334	349	607	559	507	514	465	663	817	695	701	889	1,173
DA	59	45	74	92	78	56	57	80	74	75	70	90	112
DC	44	47	68	69	51	41	45	65	60	45	38	60	65
DE	71	80	67	64	63	49	61	70	89	84	57	57	83
DK	107	121	164	201	180	150	179	261	267	211	255	335	376
EB											71	247	331
ES	46	48	47	83	62	51	45	44	67	79	115	114	133
EY	93	85	130	135	119	78	75	105	84	52	66	104	124
GM	157	226	204	215	219	203	207	254	248	234	234	313	387
HD	121	198	224	236	184	168	178	212	206	128	171	245	305
HG	24	18	36	35	33	29	42	57	42	27	53	68	67
HL	221	228	336	382	356	347	334	430	478	375	386	472	528
LM	10	27	24	27	40	57	36	53	40	32	38	53	59
MD												1	8
MH	93	90	135	191	152	124	120	124	122	110	120	148	172
NR	39	65	66	61	37	16	22	23	19	17	22	16	37
NS	85	95	129	123	113	105	129	148	187	132	162	221	225
RR	102	123	209	240	141	166	128	179	222	170	190	198	242
Total	1,980	2,293	3,222	3,448	3,047	2,788	2,675	3,413	3,865	3,186	3,662	4,716	5,731

Percentage of All Applications

IC	1992	1993	1994	1995	1996	1997	1998	1999	2000	2001	2002	2003	2004
AA	1.3	2.3	1.2	1.1	1.6	1.9	1.1	1.2	1.0	0.8	0.8	1.4	1.1
AG	4.2	4.1	4.6	4.6	4.8	4.4	3.9	3.0	3.5	4.1	2.6	2.8	2.8
AI	7.8	8.8	10.9	12.0	11.5	11.9	10.8	10.7	11.9	11.2	16.2	14.3	14.7
AR	5.6	4.4	5.1	3.7	5.5	4.5	4.7	3.9	4.5	4.8	3.9	3.6	3.5
AT	0.0	0.0	0.0	0.0	0.0	0.0	0.0	0.1	0.9	1.8	1.5	0.9	0.6
CA	16.9	15.2	18.8	16.2	16.6	18.4	17.4	19.4	21.1	21.8	19.1	18.9	20.5
DA	3.0	2.0	2.3	2.7	2.6	2.0	2.1	2.3	1.9	2.4	1.9	1.9	2.0
DC	2.2	2.0	2.1	2.0	1.7	1.5	1.7	1.9	1.6	1.4	1.0	1.3	1.1
DE	3.6	3.5	2.1	1.9	2.1	1.8	2.3	2.1	2.3	2.6	1.6	1.2	1.4
DK	5.4	5.3	5.1	5.8	5.9	5.4	6.7	7.6	6.9	6.6	7.0	7.1	6.6
EB	0.0	0.0	0.0	0.0	0.0	0.0	0.0	0.0	0.0	0.0	1.9	5.2	5.8
ES	2.3	2.1	1.5	2.4	2.0	1.8	1.7	1.3	1.7	2.5	3.1	2.4	2.3
EY	4.7	3.7	4.0	3.9	3.9	2.8	2.8	3.1	2.2	1.6	1.8	2.2	2.2
GM	7.9	9.9	6.3	6.2	7.2	7.3	7.7	7.4	6.4	7.3	6.4	6.6	6.8
HD	6.1	8.6	7.0	6.8	6.0	6.0	6.7	6.2	5.3	4.0	4.7	5.2	5.3
HG	1.2	0.8	1.1	1.0	1.1	1.0	1.6	1.7	1.1	0.8	1.4	1.4	1.2
HL	11.2	9.9	10.4	11.1	11.7	12.4	12.5	12.6	12.4	11.8	10.5	10.0	9.2
LM	0.5	1.2	0.7	0.8	1.3	2.0	1.3	1.6	1.0	1.0	1.0	1.1	1.0
MD	0.0	0.0	0.0	0.0	0.0	0.0	0.0	0.0	0.0	0.0	0.0	0.0	0.1
MH	4.7	3.9	4.2	5.5	5.0	4.4	4.5	3.6	3.2	3.5	3.3	3.1	3.0
NR	2.0	2.8	2.0	1.8	1.2	0.6	0.8	0.7	0.5	0.5	0.6	0.3	0.6
NS	4.3	4.1	4.0	3.6	3.7	3.8	4.8	4.3	4.8	4.1	4.4	4.7	3.9
RR	5.2	5.4	6.5	7.0	4.6	6.0	4.8	5.2	5.7	5.3	5.2	4.2	4.2
Total	100.0	100.0	100.0	100.0	100.0	100.0	100.0	100.0	100.0	100.0	100.0	100.0	100.0

SOURCE: National Institutes of Health.

TABLE App-A-3 Applications by IC—Phase II, 1992-2004

Number of Applications

	1992	1993	1994	1995	1996	1997	1998	1999	2000	2001	2002	2003	2004
AA	11	9	5	9	6	14	6	9	9	3	6	5	20
AG	38	24	19	27	22	16	21	26	30	27	26	24	32
AI	41	44	47	56	47	57	59	64	58	60	66	68	100
AR	19	17	14	12	16	22	19	24	26	23	35	24	14
AT								1	1	1	4	3	10
CA	43	97	148	166	134	110	72	95	123	117	126	147	159
DA	11	18	10	16	15	29	21	16	18	27	24	27	29
DC	7	15	9	14	11	6	8	11	11	19	26	14	27
DE	16	13	10	11	13	10	7	13	7	14	33	28	20
DK	25	36	21	24	34	48	29	35	30	52	57	59	58
EB											14	11	18
ES	9	15	5	9	8	6	12	10	9	11	17	24	31
EY	17	17	11	26	21	27	24	21	19	16	13	30	27
GM	37	25	44	53	45	47	38	46	56	63	44	69	81
HD	23	23	43	43	36	39	29	48	51	42	42	45	55
HG	3	5	3	7	3	5	4	11	10	8	6	11	12
HL	61	73	62	59	72	91	87	98	66	105	114	130	94
LM	1	1	3	1				2	4	3	2	3	
MD												1	
MH	20	16	23	22	28	28	27	36	39	37	21	33	45
NR	4	7	2	6	6	2	1	4	5	5	5	4	4
NS	16	24	14	23	37	37	32	29	39	54	49	30	39
RR	7	10	29	35	21	18	8	28	42	45	42	41	33
Total	409	489	522	619	575	612	504	627	653	732	772	831	908

Percentage of Total Applications, by IC

	1992	1993	1994	1995	1996	1997	1998	1999	2000	2001	2002	2003	2004	92-03
AA	2.7	1.8	1.0	1.5	1.0	2.3	1.2	1.4	1.4	0.4	0.8	0.6	2.2	1.3
AG	9.3	4.9	3.6	4.4	3.8	2.6	4.2	4.1	4.6	3.7	3.4	2.9	3.5	4.1
AI	10.0	9.0	9.0	9.0	8.2	9.3	11.7	10.2	8.9	8.2	8.5	8.2	11.0	9.1
AR	4.6	3.5	2.7	1.9	2.8	3.6	3.8	3.8	4.0	3.1	4.5	2.9	1.5	3.4
AT	0.0	0.0	0.0	0.0	0.0	0.0	0.0	0.2	0.2	0.1	0.5	0.4	1.1	0.1
CA	10.5	19.8	28.4	26.8	23.3	18.0	14.3	15.2	18.8	16.0	16.3	17.7	17.5	18.8
DA	2.7	3.7	1.9	2.6	2.6	4.7	4.2	2.6	2.8	3.7	3.1	3.2	3.2	3.2
DC	1.7	3.1	1.7	2.3	1.9	1.0	1.6	1.8	1.7	2.6	3.4	1.7	3.0	2.1
DE	3.9	2.7	1.9	1.8	2.3	1.6	1.4	2.1	1.1	1.9	4.3	3.4	2.2	2.4
DK	6.1	7.4	4.0	3.9	5.9	7.8	5.8	5.6	4.6	7.1	7.4	7.1	6.4	6.1
EB	0.0	0.0	0.0	0.0	0.0	0.0	0.0	0.0	0.0	0.0	1.8	1.3	2.0	0.3
ES	2.2	3.1	1.0	1.5	1.4	1.0	2.4	1.6	1.4	1.5	2.2	2.9	3.4	1.8
EY	4.2	3.5	2.1	4.2	3.7	4.4	4.8	3.3	2.9	2.2	1.7	3.6	3.0	3.3
GM	9.0	5.1	8.4	8.6	7.8	7.7	7.5	7.3	8.6	8.6	5.7	8.3	8.9	7.7
HD	5.6	4.7	8.2	6.9	6.3	6.4	5.8	7.7	7.8	5.7	5.4	5.4	6.1	6.3
HG	0.7	1.0	0.6	1.1	0.5	0.8	0.8	1.8	1.5	1.1	0.8	1.3	1.3	1.0
HL	14.9	14.9	11.9	9.5	12.5	14.9	17.3	15.6	10.1	14.3	14.8	15.6	10.4	13.9
LM	0.2	0.2	0.6	0.2	0.0	0.0	0.0	0.3	0.6	0.4	0.3	0.4	0.0	0.3
MD	0.0	0.0	0.0	0.0	0.0	0.0	0.0	0.0	0.0	0.0	0.0	0.1	0.0	0.0
MH	4.9	3.3	4.4	3.6	4.9	4.6	5.4	5.7	6.0	5.1	2.7	4.0	5.0	4.5
NR	1.0	1.4	0.4	1.0	1.0	0.3	0.2	0.6	0.8	0.7	0.6	0.5	0.4	0.7
NS	3.9	4.9	2.7	3.7	6.4	6.0	6.3	4.6	6.0	7.4	6.3	3.6	4.3	5.2
RR	1.7	2.0	5.6	5.7	3.7	2.9	1.6	4.5	6.4	6.1	5.4	4.9	3.6	4.4
Total	100.0	100.0	100.0	100.0	100.0	100.0	100.0	100.0	100.0	100.0	100.0	100.0	100.0	100.0

SOURCE: National Institutes of Health.

TABLE App-A-4 Woman- and Minority-owned Firms—Application Shares, 1992-2006 (percent of all applications)

	Woman-owned Firms				Minority-owned Firms		
Fiscal Year	Phase I (%)	Phase II (%)	All (%)	Fiscal Year	Phase I (%)	Phase II (%)	All (%)
1992	12.2	13.0	12.3	1992	10.3	5.6	9.5
1993	8.0	9.0	8.2	1993	6.9	5.7	6.7
1994	9.3	7.3	9.0	1994	9.6	7.9	9.4
1995	12.6	7.1	11.8	1995	9.1	8.1	9.0
1996	11.8	9.9	11.5	1996	10.0	5.4	9.3
1997	10.2	9.8	10.1	1997	8.0	5.5	7.6
1998	8.3	8.4	8.3	1998	5.5	5.6	5.5
1999	10.8	12.8	11.1	1999	7.8	5.5	7.3
2000	9.8	13.9	10.5	2000	5.4	5.7	5.5
2001	4.5	4.7	4.6	2001	2.9	3.1	2.9
2002	2.6	3.0	2.7	2002	1.2	1.5	1.3
2003	11.1	10.8	11.0	2003	7.0	4.1	6.4
2004	11.4	11.7	11.5	2004	7.3	5.1	6.9
2005	10.6	11.9	10.9	2005	5.3	4.8	5.2
2006	12.9	12.8	12.9	2006	6.7	4.0	6.1

SOURCE: National Institutes of Health.

TABLE App-A-5 Woman- and Minority-owned Firms—Award Shares, 1992-2006 (percent of all awards)

	Woman-owned Firms Award Share (%)				Minority-owned Firms Award Share (%)		
Fiscal Year	Phase I	Phase II	Total	Fiscal Year	Phase I	Phase II	Total
1992	10.0	7.2	9.4	1992	7.4	5.0	6.9
1993	5.8	7.0	6.1	1993	5.2	5.6	5.3
1994	6.8	7.5	7.0	1994	8.2	10.4	8.6
1995	8.4	8.0	8.3	1995	5.5	6.1	5.7
1996	6.9	9.3	7.5	1996	4.8	5.2	4.9
1997	8.1	9.2	8.4	1997	5.1	4.4	4.9
1998	7.1	9.8	7.8	1998	3.9	6.4	4.5
1999	9.0	12.7	10.1	1999	6.3	4.0	5.6
2000	7.9	14.3	9.4	2000	2.9	6.4	3.7
2001	4.3	3.9	4.1	2001	2.1	3.7	2.6
2002	1.5	0.8	1.3	2002	0.3	0.8	0.5
2003	9.6	10.8	9.9	2003	3.8	2.8	3.5
2004	10.4	10.6	10.4	2004	4.0	3.4	3.9
2005	8.0	10.0	8.6	2005	4.0	2.9	3.7
2006	12.5	13.9	13.0	2006	5.6	3.3	4.7

SOURCE: National Institutes of Health.

TABLE App-A-6 Woman- and Minority-owned Firms—Success Rates, 1992-2006 (percent)

| | Woman-owned Firms | | | | Other Small Businesses | | | | Minority-owned Firms | | |
| Fiscal Year | Success Rate (%) | | | Fiscal Year | Success Rate (%) | | | Fiscal Year | Success Rate (%) | | |
	Phase I	Phase II	All		Phase I	Phase II	All		Phase I	Phase II	Total
1992	22.3	18.9	21.7	1992	27.3	34.0	28.4	1992	19.7	30.4	20.8
1993	18.6	34.1	21.6	1993	25.8	44.0	29.0	1993	19.6	42.9	23.1
1994	12.0	26.3	13.6	1994	16.4	25.7	17.7	1994	13.9	34.1	16.2
1995	12.0	38.6	14.4	1995	18.0	34.2	20.4	1995	10.8	26.0	12.9
1996	10.0	28.1	12.5	1996	17.2	29.9	19.2	1996	8.2	29.0	10.1
1997	21.0	42.2	24.9	1997	26.5	45.0	30.1	1997	16.7	36.1	19.5
1998	22.7	47.9	27.3	1998	26.5	40.9	29.1	1998	18.9	46.9	24.0
1999	21.3	41.8	25.8	1999	25.3	42.1	28.5	1999	20.5	31.0	22.0
2000	19.7	34.5	23.2	2000	24.4	33.6	26.0	2000	12.8	37.8	17.3
2001	26.1	36.6	28.5	2001	27.7	44.1	31.4	2001	20.5	51.9	27.8
2002	14.3	10.3	13.3	2002	25.4	39.9	28.6	2002	7.3	20.0	10.7
2003	19.2	34.7	22.3	2003	22.2	34.6	24.7	2003	12.0	23.9	13.6
2004	16.3	25.9	18.2	2004	18.0	28.4	19.9	2004	9.9	18.8	11.2
2005	12.3	24.3	14.9	2005	16.4	28.9	18.9	2005	12.4	17.9	13.4
2006	16.9	36.2	21.4	2006	17.5	33.4	21.2	2006	14.5	27.7	16.5

SOURCE: National Institutes of Health

TABLE App-A-7 Phase I by State—Compared with Scientists and Engineers Employed

State	Applications per 1,000 pop	Applications per 1,000 pop	State	AREA	Engineers	Scientists	Total Scientists and Engineers	All Employees	Scientists/Engineers/1,000 Employees	Applications
MA	90.2	2.4	AK	1	6,290	6,690	12,980	1,830,360	7.1	18
DC	57.2	12.8	AL	2	36,220	10,330	46,550	293,130	158.8	343
MD	54.5	12.0	AR	4	11,560	6,740	18,300	2,313,520	7.9	80
UT	33.1	11.9	AZ	5	56,460	14,790	71,250	1,125,830	63.3	528
NH	31.8	6.2	CA	6	319,420	147,100	466,520	14,534,620	32.1	8,929
VT	27.3	8.5	CO	8	53,000	25,260	78,260	2,110,640	37.1	1,053
CA	26.4	9.5	CT	9	37,290	16,970	54,260	1,633,350	33.2	708
DE	25.5	49.3	DC	10	13,350	20,520	33,870	408,270	83.0	327
CO	24.5	10.6	DE	11	6,090	6,380	12,470	603,130	20.7	200
VA	23.4	8.0	FL	12	113,390	46,750	160,140	7,330,880	21.8	1,034
WA	23.0	7.8	GA	13	55,680	23,280	78,960	3,806,550	20.7	472
OR	21.1	6.8	HI	15	7,920	4,710	12,630	564,850	22.4	170
IL	21.1	9.7	IA	16	15,880	9,780	25,660	574,270	44.7	161
CT	20.8	10.4	ID	17	13,670	9,100	22,770	5,719,070	4.0	77
AZ	19.8	2.6	IL	18	89,150	42,290	131,440	2,866,350	45.9	1,280
RI	19.3	15.0	IN	19	49,320	17,800	67,120	1,423,170	47.2	382
NM	17.3	11.2	KS	20	26,960	9,170	36,130	1,297,710	27.8	170
MN	16.8	8.2	KY	21	23,300	9,300	32,600	1,728,300	18.9	210
NJ	15.8	9.7	LA	22	28,580	13,050	41,630	1,861,000	22.4	201
WY	14.6	6.9	MA	23	73,160	40,340	113,500	595,120	190.7	5,728
MT	14.2	6.3	MD	24	52,070	34,340	86,410	2,458,140	35.2	2,886
HI	14.0	6.3	ME	25	9,800	4,250	14,050	3,125,930	4.5	154
IN	13.1	12.1	MI	26	134,090	30,900	164,990	4,294,310	38.4	939
PA	12.8	9.8	MN	27	55,180	26,780	81,960	2,602,230	31.5	826
IA	12.4	3.9	MO	28	35,600	18,550	54,150	1,095,450	49.4	331
NC	12.2	8.8	MS	29	15,120	6,740	21,860	2,630,780	8.3	55
ME	12.1	23.0	MT	30	5,620	7,530	13,150	402,930	32.6	128
SD	11.9	4.8	NC	31	51,390	38,010	89,400	883,450	101.2	984

NY	11.9	7.2	ND	32	4,210	2,510	6,720	1,116,860	6.0	46
MO	11.6	4.8	NE	33	11,300	8,290	19,590	613,090	32.0	93
OH	11.3	8.7	NH	34	12,820	4,190	17,010	3,881,440	4.4	393
WI	10.9	4.5	NJ	35	57,810	39,250	97,060	755,620	128.5	1,328
MI	9.4	9.7	NM	36	22,060	12,880	34,940	8,248,470	4.2	314
TX	9.0	7.0	NV	37	14,910	6,980	21,890	3,722,700	5.9	110
AL	7.7	3.0	NY	38	105,190	74,280	179,470	318,520	563.4	2,253
ND	7.2	3.1	OH	39	96,320	33,550	129,870	5,308,970	24.5	1,286
TN	6.9	6.0	OK	40	21,360	11,490	32,850	1,421,270	23.1	128
FL	6.5	4.8	OR	41	31,790	17,890	49,680	1,556,470	31.9	721
KS	6.3	10.2	PA	42	88,560	47,440	136,000	5,507,880	24.7	1,570
GA	5.8	7.8	RI	44	8,150	4,310	12,460	480,420	25.9	202
NV	5.5	8.7	SC	45	36,360	10,830	47,190	1,772,760	26.6	144
NE	5.4	6.6	SD	46	3,960	3,370	7,330	368,100	19.9	90
KY	5.2	4.6	TN	47	35,600	13,660	49,260	2,634,450	18.7	390
LA	4.5	6.6	TX	48	198,990	69,090	268,080	9,299,360	28.8	1,871
OK	3.7	5.7	UT	49	19,700	12,150	31,850	1,059,720	30.1	739
SC	3.6	6.0	VA	50	75,070	33,620	108,690	293,650	370.1	1,657
AK	2.9	6.9	VT	51	5,780	3,080	8,860	3,451,890	2.6	166
WV	1.9	9.7	WA	53	66,240	40,040	106,280	2,583,080	41.1	1,355
AR	1.6	8.1	WI	54	47,940	25,000	72,940	685,110	106.5	584
MS	1.0	7.3	WV	55	9,470	5,970	15,440	2,689,160	5.7	34
ID	0.6	9.7	WY	56	3,640	4,100	7,740	245,050	31.6	72

SOURCE: National Institutes of Health.

TABLE App-A-8 Phase I—New Applicants and New Winners, 2000-2005

Number of "New" organizations by Fiscal Year, Based on Funded Competing R43, U43 SBIR Phase 1 Grants
Baseline 1986 (program started in 1983, but recovering data from the mainframe files is difficult)
Calculated by using the year of first SBIR award post baseline
Source: impacii irdb pubfiles (1986 thru 2005)

Fiscal Year	Number of "New" Winners	Number of "New" Applicants	All Applicants (#)	All Winners (#)	"New" Winners (%)	"New" Applicants (%)	Success Rate of New Applicants (%)	Success Rate of Previous Winners (%)
2000	306	1,263	2,006	649	47.1	63.0	24.2	46.2
2001	240	963	1,660	603	39.8	58.0	24.9	52.1
2002	274	1,198	1,937	650	42.2	61.8	22.9	50.9
2003	305	1,499	2,353	712	42.8	63.7	20.3	47.7
2004	294	1,703	2,680	705	41.7	63.5	17.3	42.1
2005	211	1,493	2,449	589	35.8	61.0	14.1	39.5

NOTE New is based on organizations with no previous SBIR support; however, these organizations may have submitted more than one application in previous years.

SOURCE: National Institutes of Health.

TABLE App-A-9 Phase I Awards—By State, 1992-2005

State	Number of Phase I Awards by Fiscal Year														Grand Total	Percent of Total
	1992	1993	1994	1995	1996	1997	1998	1999	2000	2001	2002	2003	2004	2005		
AK												2		1	3	0.0
AL	5	11	6	4	7	5	6	10	8	11	9	8	6	4	100	0.8
AR	2					1		3	1	2	1	3	4	5	22	0.2
AZ	14	11	2	10	8	7	10	11	11	15	26	14	12	8	159	1.2
CA	110	114	133	144	117	153	136	201	226	240	238	275	232	178	2,497	19.5
CO	8	14	8	12	17	22	20	19	36	22	31	36	34	27	306	2.4
CT	17	11	13	13	9	19	12	15	13	13	21	26	19	11	212	1.7
DC	4	6	6	6	2	3	6	7	11	11	16	10	5	4	97	0.8
DE		2	2	2	2	5	1	6	8	7	4	7	6	4	56	0.4
FL	10	11	9	14	11	18	8	11	30	18	23	32	20	16	231	1.8
GA	5	3	3	6	8	5	4	4	16	12	12	14	19	6	117	0.9
HI	8	5	4	3	3	3	3	2	3	6	5	1	1	1	46	0.4
IA		1			1	1	1	3	3	4	5	9	4	2	34	0.3
ID	1	1					1		2	4	1	1	1	1	13	0.1
IL	13	21	15	22	11	25	22	19	36	35	28	31	22	29	329	2.6
IN	5	5	5	6	5	5	3	7	7	9	7	14	11	12	101	0.8
KS	1	2		1	1			4	2	4	5	6	4	1	31	0.2
KY	2	3	2	1	2	4	1	3	9	13	10	2	4	4	60	0.5
LA	3	2	3			2			5	2	3	3	2	1	27	0.2
MA	91	114	105	100	105	111	120	175	212	193	189	170	152	106	1,943	15.2
MD	68	58	52	54	24	60	38	67	63	84	77	114	85	58	902	7.1
ME	2	2	1	6	2		1	3	3	2	5	3	2	2	34	0.3
MI	15	18	13	17	10	9	22	20	19	35	24	31	27	24	284	2.2
MN	11	13	10	21	11	12	16	7	17	21	24	28	18	19	228	1.8
MO	6	8	2	1	3	3	4	5	6	8	12	12	10	8	88	0.7
MS			1	1	1	1		2	1	1	2		1		13	0.1
MT			2		1	1	3	2	3	4	3	1	4	2	25	0.2

continued

TABLE App-A-9 Continued

State	Number of Phase I Awards by Fiscal Year														Grand Total	Percent of Total
	1992	1993	1994	1995	1996	1997	1998	1999	2000	2001	2002	2003	2004	2005		
NC	22	10	15	10	14	21	23	21	32	37	39	35	39	28	346	2.7
ND	1									1	2	1	3		8	0.1
NE	1	3	1	1		4	2	1	3	3	4	3	3	5	34	0.3
NH	4	7	6	3	8	13	4	6	12	18	18	20	7	3	129	1.0
NJ	12	20	14	15	21	18	17	27	34	34	42	33	29	23	339	2.7
NM	7	2	1	3	1	4	2	8	8	10	8	12	11	4	81	0.6
NV						3	1		1	2	3	5		3	18	0.1
NY	47	38	32	31	25	28	42	43	62	54	71	53	52	45	623	4.9
OH	17	19	12	19	12	26	34	26	39	47	47	50	37	35	420	3.3
OK	4	3		2	2		3	2	2	3	7	5	4	5	42	0.3
OR	9	11	13	19	16	16	14	18	32	27	34	29	19	12	269	2.1
PA	22	19	14	24	21	27	28	24	46	57	56	49	47	41	475	3.7
PR							1	1	1						3	0.0
RI							1	3	8	9	10	7	10	8	58	0.5
SC				1	1	1	3	2	3	6	12	8	3	2	41	0.3
SD				1	1	1	1	2		2	3	1	1		13	0.1
TN	3	3	4	3	2	8	3	2	6	9	8	9	4	4	68	0.5
TX	29	25	23	24	22	35	24	39	64	43	46	65	55	37	531	4.2
UT	14	11	10	10	7	14	16	14	23	25	23	11	17	8	203	1.6
VA	20	23	21	21	14	29	23	22	36	50	37	57	30	20	403	3.2
VT	3	6	3	3	1	1	2	2	3	2	2	8	4	2	40	0.3
WA	16	18	17	22	24	30	28	44	48	47	59	48	37	28	466	3.6
WI	4	5	12	9	7	10	15	15	16	29	25	27	18	15	207	1.6
WV									1	1					2	0.0
WY						1	2	3	2		4	2			15	0.1
Grand Total	636	659	596	666	560	764	725	931	1,233	1,293	1,339	1,393	1,135	862	12,792	100.0

SOURCE: U.S. Small Business Administration, Tech-Net Database.

TABLE App-A-10 Resubmission Rates—Phases I and II, 1992-2004

Phase I

Fiscal Year	All Applications (#)	Resubmissions (#)	Resubmission as Percent of Applications	Total Funded (#)	1st-time Winners (#)	Funded on Resubmission (#)	Not Funded (#)	1st-time Success Rate (%)	Resubmission Success Rate (%)
1992	1,980	334	23.2	540	442	98	1,440	26.9	29.3
1993	2,293	312	18.4	593	514	79	1,700	25.9	25.3
1994	3,222	427	15.9	530	473	57	2,692	16.9	13.3
1995	3,448	655	23.2	621	467	154	2,827	16.7	23.5
1996	3,047	657	26.0	523	395	128	2,524	16.5	19.5
1997	2,788	607	29.7	742	526	216	2,046	24.1	35.6
1998	2,675	514	26.2	716	581	135	1,959	26.9	26.3
1999	3,413	612	24.4	906	761	145	2,507	27.2	23.7
2000	3,865	786	27.1	969	746	223	2,896	24.2	28.4
2001	3,186	760	33.4	911	658	253	2,275	27.1	33.3
2002	3,662	636	23.4	945	773	172	2,717	25.5	27.0
2003	4,716	864	21.6	710	472	238	4,006	12.3	27.5
2004	5,731	1,167	24.8	1,028	767	261	4,703	16.8	22.4

continued

TABLE App-A-10 Continued

Phase II

Fiscal Year	All Applications (#)	Resubmissions (#)	Resubmission as Percent of Applications	Total Funded (#)	1st-time Winners (#)	Funded on Resubmission (#)	Not Funded (#)	1st-time Success Rate (%)	Resubmission Success Rate (%)
1992	409	103	25.2	137	107	30	272	35.0	29.1
1993	489	147	30.1	215	160	55	274	46.8	37.4
1994	522	138	26.4	131	110	21	391	28.6	15.2
1995	619	202	32.6	210	162	48	409	38.8	23.8
1996	575	204	35.5	173	124	49	402	33.4	24.0
1997	612	211	34.5	280	191	89	332	47.6	42.2
1998	504	141	28.0	237	193	44	267	53.2	31.2
1999	627	149	23.8	289	231	58	338	48.3	38.9
2000	653	197	30.2	267	206	61	386	45.2	31.0
2001	732	242	33.1	381	284	97	351	58.0	40.1
2002	772	242	31.3	358	269	89	414	50.8	36.8
2003	830	249	30.0	349	264	85	481	45.4	34.1
2004	908	287	31.6	302	223	79	606	35.9	27.5

SOURCE: National Institutes of Health.

TABLE App-A-11 Phase I, Year One Awards at NIH, 1992-2005

Fiscal Year	Number of Awards	Total Amount ($)	Average Amount ($)
1992	541	26,616,441	49,199
1993	594	29,560,122	49,765
1994	530	39,249,711	74,056
1995	624	59,005,464	94,560
1996	525	50,936,972	97,023
1997	743	72,528,667	97,616
1998	717	70,077,801	97,738
1999	908	96,125,835	105,865
2000	986	117,779,337	119,452
2001	940	120,072,266	127,736
2002	1,001	137,504,731	137,367
2003	1,137	168,520,060	148,215
2004	1,150	187,091,805	162,689
2005	937	160,982,684	171,806

SOURCE: National Institutes of Health.

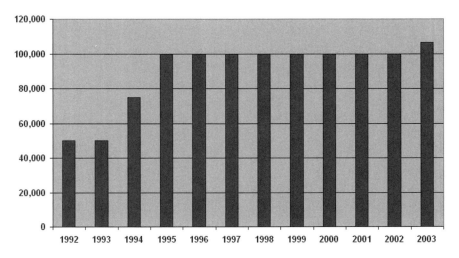

FIGURE App-A-1 Phase I median award size, 1992-2003.

SOURCE: National Institutes of Health, NRC calculation.

TABLE App-A-12 Oversized SBIR Phase I Awards, 1992-2005

Fiscal Year	$0-$100K (%)	$>100K- $120K (%)	$>120K- $150K (%)	$>150K- $175K (%)	$>175K- $200K (%)	>$200K (%)
1992	100.0	0.0	0.0	0.0	0.0	0.0
1993	100.0	0.0	0.0	0.0	0.0	0.0
1994	100.0	0.0	0.0	0.0	0.0	0.0
1995	99.7	0.3	0.0	0.0	0.0	0.0
1996	99.4	0.6	0.0	0.0	0.0	0.0
1997	99.7	0.1	0.1	0.0	0.0	0.0
1998	95.4	3.5	1.2	0.0	0.0	0.0
1999	70.8	16.6	7.0	2.5	0.1	2.9
2000	62.1	16.4	11.1	2.0	2.2	6.3
2001	53.8	13.9	13.5	5.1	5.1	8.6
2002	51.8	13.5	12.2	5.8	6.6	10.1
2003	48.0	13.9	12.7	4.1	4.6	16.7
2004	43.4	10.9	14.2	6.0	6.4	19.1
2005	40.5	10.5	15.6	7.6	7.8	18.0

SOURCE: National Institutes of Health.

TABLE App-A-13 Phase I Awards by States—Per 1,000 Life and Physical
Scientists

	Life & Physical Scientists, 2003	NIH Phase I Awards, 2003	NIH Phase I Awards per 1,000 Life & Physical Scientists, 2003
New Hampshire	1,480	14.0	9.5
Vermont	850	6.0	7.1
Massachusetts	20,380	140.0	6.9
Maryland	17,910	90.0	5.0
Oregon	5,870	23.0	3.9
Connecticut	5,670	22.0	3.9
California	64,390	248.0	3.9
Virginia	13,030	40.0	3.1
Ohio	15,100	45.0	3.0
Iowa	3,130	9.0	2.9
Colorado	11,710	33.0	2.8
Indiana	4,070	11.0	2.7
Rhode Island	1,580	4.0	2.5
Michigan	9,390	23.0	2.4
Arizona	5,580	13.0	2.3
Nevada	2,510	5.0	2.0
Wyoming	1,510	3	2.0
Delaware	2,020	4.0	2.0
Minnesota	11,200	22.0	2.0
Washington	16,940	33.0	1.9
New Mexico	3,200	6.0	1.9
Wisconsin	11,220	21.0	1.9
New Jersey	17,530	32.0	1.8
Utah	5,060	9.0	1.8
South Carolina	4,610	8.0	1.7
Maine	1,830	3.0	1.6
Pennsylvania	25,080	41.0	1.6
District of Columbia	5,210	8.0	1.5
Oklahoma	3,350	5.0	1.5
North Dakota	1,420	2.0	1.4
New York	30,330	41.0	1.4
North Carolina	17,770	24.0	1.4
Missouri	9,240	12.0	1.3
Kansas	3,910	5.0	1.3
Florida	19,440	24.0	1.2
Alabama	5,170	6.0	1.2
Texas	42,440	49.0	1.2
Illinois	18,300	21.0	1.1
Arkansas	2,700	3.0	1.1
Louisiana	5,540	5.0	0.9
Nebraska	3,920	3.0	0.8
Kentucky	2,660	2.0	0.8
Alaska	2,800	2.0	0.7
South Dakota	1,420	1.0	0.7

continued

TABLE App-A-13 Continued

	Life & Physical Scientists, 2003	NIH Phase I Awards, 2003	NIH Phase I Awards per 1,000 Life & Physical Scientists, 2003
Georgia	11,410	8.0	0.7
Tennessee	7,130	4.0	0.6
Hawaii	1,790	1.0	0.6
Montana	2,790	1.0	0.4
Idaho	3,100	1.0	0.3
Mississippi	3,650	1.0	0.3
West Virginia	2,510	0.0	0.0
Average			3.2

SOURCE: National Institutes of Health; National Science Board, *Science and Engineering Indicators 2005*, Arlington, VA: National Science Foundation, 2005.

TABLE App-A-14 Top 20 Zip Codes, 1992-2003

Zip Code	State	Total Number of Grants
92121	CA	311
02139	MA	143
94043	CA	114
02472	MA	99
01801	MA	92
01915	MA	73
20850	MD	73
20877	MD	64
97403	OR	60
84108	UT	57
02138	MA	57
53711	WI	52
98104	WA	50
92037	CA	47
77840	TX	46
94545	CA	45
98109	WA	44
92008	CA	44
27713	NC	41
02464	MA	40
02142	MA	39
		1,591

SOURCE: National Institutes of Health.

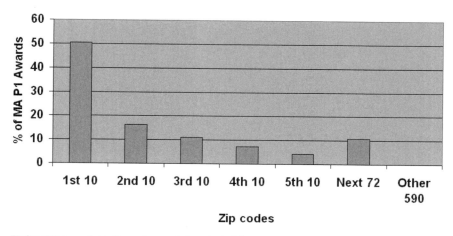

FIGURE App-A-2 Phase I awards by zip in Massacussetts.

Distribution of Phase I awards in Massachusetts 1992-2002 by zip code

SOURCE: National Institutes of Health.

TABLE App-A-15 Phase I Awards by IC, 1992-2002

Funding IC	Number of Awards											
	1992	1993	1994	1995	1996	1997	1998	1999	2000	2001	2002	Total
AA	4	7	3	7	4	12	4	14	15	9	11	90
AG	12	14	28	16	5	34	26	29	34	37	26	261
AI	63	73	49	74	70	100	83	91	99	104	122	928
AR	5	16	17	26	29	25	20	27	22	27	14	228
AT								2	6	7	7	22
CA	160	157	114	95	80	154	138	192	181	187	212	1,670
DA	13	9	22	24	19	13	14	32	30	27	20	223
DC	6	10	12	8	5	12	8	20	21	18	9	129
DE	6	8	6	14	7	6	14	8	31	34	12	146
DK	16	22	36	49	39	50	40	70	82	65	67	536
EB											17	17
ES	8	8	8	9	10	8	9	11	23	28	30	152
EY	16	14	25	28	27	18	18	18	19	19	29	231
GM	42	63	48	42	53	68	74	88	79	80	76	713
HD	31	52	24	27	35	42	32	37	51	30	41	402
HG	6	5	4	5	5	8	17	17	9	8	22	106
HL	68	51	56	103	79	83	72	109	133	112	85	951
LM	2	3	2			3	6	4	4	1	2	27
MH	17	24	20	33	21	36	34	41	31	36	42	335
NR	4	7	1	6	1	3	6	5	5	2	2	42
NS	25	23	35	40	19	31	45	51	50	42	50	411
RR	36	27	20	13	14	33	40	39	41	36	33	332
Shared				2	1	3	16	1	3	2	16	44
Total	540	593	530	621	523	742	716	906	969	911	945	7,996

Percentage of Awards

Funding IC	1992	1993	1994	1995	1996	1997	1998	1999	2000	2001	2002	Total
AA	0.7	1.2	0.6	1.1	0.8	1.6	0.6	1.5	1.5	1.0	1.2	1.1
AG	2.2	2.4	5.3	2.6	1.0	4.6	3.6	3.2	3.5	4.1	2.8	3.3
AI	11.7	12.3	9.2	11.9	13.4	13.5	11.6	10.0	10.2	11.4	12.9	11.6
AR	0.9	2.7	3.2	4.2	5.5	3.4	2.8	3.0	2.3	3.0	1.5	2.9
AT	0.0	0.0	0.0	0.0	0.0	0.0	0.0	0.2	0.6	0.8	0.7	0.3
CA	29.6	26.5	21.5	15.3	15.3	20.8	19.3	21.2	18.7	20.5	22.4	20.9
DA	2.4	1.5	4.2	3.9	3.6	1.8	2.0	3.5	3.1	3.0	2.1	2.8
DC	1.1	1.7	2.3	1.3	1.0	1.6	1.1	2.2	2.2	2.0	1.0	1.6
DE	1.1	1.3	1.1	2.3	1.3	0.8	2.0	0.9	3.2	3.7	1.3	1.8
DK	3.0	3.7	6.8	7.9	7.5	6.7	5.6	7.7	8.5	7.1	7.1	6.7
EB	0.0	0.0	0.0	0.0	0.0	0.0	0.0	0.0	0.0	0.0	1.8	0.2
ES	1.5	1.3	1.5	1.4	1.9	1.1	1.3	1.2	2.4	3.1	3.2	1.9
EY	3.0	2.4	4.7	4.5	5.2	2.4	2.5	2.0	2.0	2.1	3.1	2.9
GM	7.8	10.6	9.1	6.8	10.1	9.2	10.3	9.7	8.2	8.8	8.0	8.9
HD	5.7	8.8	4.5	4.3	6.7	5.7	4.5	4.1	5.3	3.3	4.3	5.0
HG	1.1	0.8	0.8	0.8	1.0	1.1	2.4	1.9	0.9	0.9	2.3	1.3
HL	12.6	8.6	10.6	16.6	15.1	11.2	10.1	12.0	13.7	12.3	9.0	11.9
LM	0.4	0.5	0.4	0.0	0.0	0.4	0.8	0.4	0.4	0.1	0.2	0.3
MH	3.1	4.0	3.8	5.3	4.0	4.9	4.7	4.5	3.2	4.0	4.4	4.2
NR	0.7	1.2	0.2	1.0	0.2	0.4	0.8	0.6	0.5	0.2	0.2	0.5
NS	4.6	3.9	6.6	6.4	3.6	4.2	6.3	5.6	5.2	4.6	5.3	5.1
RR	6.7	4.6	3.8	2.1	2.7	4.4	5.6	4.3	4.2	4.0	3.5	4.2
Shared	0.0	0.0	0.0	0.3	0.2	0.4	2.2	0.1	0.3	0.2	1.7	0.6
Total	100.0	100.0	100.0	100.0	100.0	100.0	100.0	100.0	100.0	100.0	100.0	100.0

continued

TABLE App-A-15 Continued

Shared awards:	1992	1993	1994	1995	1996	1997	1998	1999	2000	2001	2002
AA, RR, RR											6
AG, DA, DA										1	
AG, HD, HD										1	
AR, CA, CA							2				
AR, HD, HD						1					
AR, NR, NR				1							
AT, OD, OD											3
CA, DK, DK									1		
CA, HD, HD							1				
CA, MD, MD											1
CA, RR, RR						1					
DC, MD, MD											5
DC, MH, MH					1						
DC, NR, NR								1			
GM, GM				1							
GM, LM, LM						1					
HD, HD							9		2		
HD, MH, MH							3				
HD, NR, NR							1				
MH, MH											1
Total				2	1	3	16	1	3	2	16

SOURCE: National Institutes of Health.

TABLE App-A-16 New Phase I Awards Made in Response to an RFA, 1992-2005

Fiscal Year	PHASE	Total Number of SBIR Awards (Parent Solicitation and Special PA/RFAs)	Total Amount of SBIR Dollars Awarded (Parent Solicitation and Special PA/RFAs)	Number of RFA SBIR Awards	Amount of RFA SBIR Dollars Awarded	RFA—Percent of All Awards	RFA—Percent of All Award Dollars
1992	Phase I	540	26,571,593	0	0	0.0	0.0
1993	Phase I	591	29,447,790	0	0	0.0	0.0
1994	Phase I	527	39,143,156	37	2,779,027	7.0	7.1
1995	Phase I	619	58,441,688	6	470,612	1.0	0.8
1996	Phase I	523	50,634,759	3	300,000	0.6	0.6
1997	Phase I	741	72,325,975	0	0	0.0	0.0
1998	Phase I	702	69,368,224	1	98,525	0.1	0.1
1999	Phase I	874	94,944,070	5	1,283,242	0.6	1.4
2000	Phase I	944	113,276,949	12	2,976,638	1.3	2.6
2001	Phase I	884	112,650,165	10	2,417,588	1.1	2.1
2002	Phase I	930	124,423,577	65	11,586,853	7.0	9.3
2003	Phase I	1,042	150,880,637	15	3,145,579	1.4	2.1
2004	Phase I	1,023	158,499,201	15	4,595,326	1.5	2.9
2005	Phase I	806	126,216,067	20	3,044,462	2.5	2.4

SOURCE: National Institutes of Health.

TABLE App-A-17 New Phase I Awards Made in Response to Program Announcements, 1992-2005 (PAs)

Fiscal Year	PHASE	Total Number of SBIR Awards (Parent Solicitation and Special PA/RFAs)	Total Amount of SBIR Dollars Awarded (Parent Solicitation and Special PA/RFAs)	Number of PA SBIR Awards	Amount of PA SBIR Dollars Awarded	Percent of Awards	Percent of Dollars
1992	Phase I	540	26,571,593	0	0	0.0	0.0
1993	Phase I	591	29,447,790	0	0	0.0	0.0
1994	Phase I	527	39,143,156	0	0	0.0	0.0
1995	Phase I	619	58,441,688	0	0	0.0	0.0
1996	Phase I	523	50,634,759	0	0	0.0	0.0
1997	Phase I	741	72,325,975	0	0	0.0	0.0
1998	Phase I	702	69,368,224	1	106,906	0.1	0.2
1999	Phase I	874	94,944,070	33	5,470,073	3.8	5.8
2000	Phase I	944	113,276,949	42	9,826,643	4.4	8.7
2001	Phase I	884	112,650,165	70	13,256,620	7.9	11.8
2002	Phase I	930	124,423,577	120	23,269,261	12.9	18.7
2003	Phase I	1,042	150,880,637	194	41,577,617	18.6	27.6
2004	Phase I	1,023	158,499,201	215	48,889,961	21.0	30.8
2005	Phase I	806	126,216,067	160	34,525,730	19.9	27.4

SOURCE: National Institutes of Health.

TABLE App-A-18 Phase I Awards—By Company, 1992-2003

Top 20 Phase I winners 1992-2003

Organization	Number of Awards
PANORAMA RESEARCH, INC.	69
INOTEK PHARMACEUTICALS CORPORATION	63
RADIATION MONITORING DEVICES, INC.	56
LYNNTECH, INC.	51
INFLEXXION, INC.	44
OREGON CENTER FOR APPLIED SCIENCE	44
NEW ENGLAND RESEARCH INSTITUTES, INC.	42
CREARE, INC.	40
INSIGHTFUL CORPORATION	40
HAWAII BIOTECH, INC.	38
PHYSICAL OPTICS CORPORATION	33
BIOMEC, INC.	30
SURMODICS, INC.	30
BIOTEK, INC.	29
SPIRE CORPORATION	28
ONE CELL SYSTEMS, INC.	27
COMPACT MEMBRANE SYSTEMS, INC.	26
OSI PHARMACEUTICALS, INC.	26
PERSONAL IMPROVEMENT COMPUTER SYSTEMS	25
PHYSICAL SCIENCES, INC.	25
Total	766

SOURCE: National Institutes of Health.

TABLE App-A-19 Phase I—Supplementary Awards, 1992-2003

Fiscal Year	1993	1994	1995	1998	1999	2000	2001	2002	2003
Count	2	3	2	10	23	21	20	12	10
Average ($)	16,238	9,328	17,987	43,013	35,894	36,210	45,311	57,363	78,782
Maximum ($)	25,000	15,000	26,331	98,000	98,000	105,853	173,059	287,606	151,607
Minimum ($)	7,475	5,998	9,643	3,400	400	9,813	15,000	12,600	4,029
Sum ($)	32,475	27,984	35,974	430,129	825,553	760,416	906,214	688,356	787,821

SOURCE: National Institutes of Health.

TABLE App-A-20 Phase I Awards—Year Two of Support, 1992-2003

Fiscal Year	1992	1993	1995	1996	1997	1998	1999	2000	2001	2002	2003
Count	1	1	3	2	1	1	2	17	29	56	81
Average ($)	44,848	49,957	78,618	57,718	99,944	100,009	55,858	203,928	216,888	214,773	195,558
Minimum ($)	44,848	49,957	100,000	85,435	99,944	100,009	101,715	344,294	387,813	307,200	376,150
Maximum ($)	44,848	49,957	49,413	30,000	99,944	100,009	10,000	71,238	65,428	15,600	49,930
Sum ($)	44,848	49,957	235,855	115,435	99,944	100,009	111,715	3,466,776	6,289,766	12,027,313	15,840,190

SOURCE: National Institutes of Health.

TABLE App-A-21 Phase II—Year One Awards, 1992-2005

Fiscal Year	Number of Awards	Total Dollars	Average Award Size ($)
1992	278	64,634,293	232,497
1993	360	82,904,116	230,289
1994	351	82,130,205	233,989
1995	370	106,153,197	286,901
1996	390	124,699,140	319,741
1997	468	163,756,939	349,908
1998	541	182,404,280	337,161
1999	539	199,696,146	370,494
2000	587	223,656,320	381,016
2001	683	274,218,417	401,491
2002	797	330,503,121	414,684
2003	788	343,893,012	436,412
2004	792	362,710,289	457,968
2005	774	396,764,618	512,616

SOURCE: National Institutes of Health.

TABLE App-A-22 Phase II—Requests for Applications (RFAs), 1992-2005

Fiscal Year	PHASE	Total Number of SBIR Awards (Parent Solicitation and Special PA/RFAs)	Total Amount of SBIR Dollars Awarded (Parent Solicitation and Special PA/RFAs)	Number of RFA SBIR Awards	Amount of RFA SBIR Dollars Awarded	RFA—Percent of All Awards	RFA—Percent of All Award Dollars
1992	Phase II	139	34,653,136	0	0	0.0	0.0
1993	Phase II	215	52,537,980	0	0	0.0	0.0
1994	Phase II	134	36,697,824	0	0	0.0	0.0
1995	Phase II	212	73,059,839	0	0	0.0	0.0
1996	Phase II	172	60,806,200	0	0	0.0	0.0
1997	Phase II	279	104,817,711	0	0	0.0	0.0
1998	Phase II	224	85,283,655	0	0	0.0	0.0
1999	Phase II	278	114,811,423	1	507,041	0.4	0.4
2000	Phase II	231	102,407,911	0	0	0.0	0.0
2001	Phase II	343	148,866,724	0	0	0.0	0.0
2002	Phase II	336	154,925,573	0	0	0.0	0.0
2003	Phase II	327	156,101,955	0	0	0.0	0.0
2004	Phase II	298	153,544,979	5	2,594,326	1.7	1.7
2005	Phase II	312	163,695,822	5	2,659,999	1.6	1.6

SOURCE: National Institutes of Health.

TABLE App-A-23 Phase II—Program Announcements (PAs)—New Awards, 1992-2005

Fiscal Year	PHASE	Total Number of SBIR Awards (Parent Solicitation and Special PA/RFAs)	Total Amount of SBIR Dollars Awarded (Parent Solicitation and Special PA/RFAs)	Number of PA SBIR Awards	Amount of PA SBIR Dollars Awarded	PA—Percent of All Awards	PA—Percent of All Award Dollars
1992	Phase II	139	34,653,136	0	0	0.0	0.0
1993	Phase II	215	52,537,980	0	0	0.0	0.0
1994	Phase II	134	36,697,824	0	0	0.0	0.0
1995	Phase II	212	73,059,839	0	0	0.0	0.0
1996	Phase II	172	60,806,200	0	0	0.0	0.0
1997	Phase II	279	104,817,711	0	0	0.0	0.0
1998	Phase II	224	85,283,655	0	0	0.0	0.0
1999	Phase II	278	114,811,423	2	1,256,508	0.7	1.1
2000	Phase II	231	102,407,911	1	370,000	0.4	0.4
2001	Phase II	343	148,866,724	8	4,398,827	2.3	3.0
2002	Phase II	336	154,925,573	8	4,980,270	2.4	3.2
2003	Phase II	327	156,101,955	30	17,637,373	9.2	11.3
2004	Phase II	298	153,544,979	32	23,569,833	10.7	15.4
2005	Phase II	312	163,695,822	43	30,719,886	13.8	18.8

SOURCE: National Institutes of Health.

TABLE App-A-24 Phase II Extra-large Awards, 1992-2005

NIH Phase II SBIR Awards Over 375K

Fiscal Year	PHASE	Total Number Funded	Year Two		Year Three		Year Four		Year Five	
1992	Phase II	139	1	0.7%	0	0.0%	0	0.0%	0	0.0%
1993	Phase II	215	5	2.3%	0	0.0%	0	0.0%	0	0.0%
1994	Phase II	134	7	5.2%	1	0.7%	0	0.0%	0	0.0%
1995	Phase II	212	70	33.0%	0	0.0%	0	0.0%	0	0.0%
1996	Phase II	172	68	39.5%	0	0.0%	0	0.0%	0	0.0%
1997	Phase II	279	131	47.0%	1	0.4%	0	0.0%	0	0.0%
1998	Phase II	224	106	47.3%	0	0.0%	0	0.0%	0	0.0%
1999	Phase II	278	169	60.8%	2	0.7%	0	0.0%	0	0.0%
2000	Phase II	231	155	67.1%	1	0.4%	0	0.0%	0	0.0%
2001	Phase II	343	240	70.0%	1	0.3%	0	0.0%	0	0.0%
2002	Phase II	336	238	70.8%	2	0.6%	0	0.0%	0	0.0%
2003	Phase II	327	224	68.5%	8	2.4%	0	0.0%	1	0.3%
2004	Phase II	298	194	65.1%	15	5.0%	2	0.7%	1	0.3%
2005	Phase II	312	205	65.7%	15	4.8%	14	4.5%	2	0.6%

NIH Phase II SBIR Awards Over 500K

Fiscal Year	PHASE	Total Number Funded	Year Two		Year Three		Year Four		Year Five	
1992	Phase II	139	0	0.0%	0	0.0%	0	0.0%	0	0.0%
1993	Phase II	215	0	0.0%	0	0.0%	0	0.0%	0	0.0%
1994	Phase II	134	1	0.7%	1	0.7%	0	0.0%	0	0.0%
1995	Phase II	212	6	2.8%	0	0.0%	0	0.0%	0	0.0%
1996	Phase II	172	5	2.9%	0	0.0%	0	0.0%	0	0.0%
1997	Phase II	279	13	4.7%	0	0.0%	0	0.0%	0	0.0%
1998	Phase II	224	7	3.1%	0	0.0%	0	0.0%	0	0.0%
1999	Phase II	278	34	12.2%	0	0.0%	0	0.0%	0	0.0%
2000	Phase II	231	52	22.5%	1	0.4%	0	0.0%	0	0.0%
2001	Phase II	343	72	21.0%	1	0.3%	0	0.0%	0	0.0%
2002	Phase II	336	80	23.8%	1	0.3%	0	0.0%	0	0.0%
2003	Phase II	327	83	25.4%	4	1.2%	0	0.0%	1	0.3%
2004	Phase II	298	91	30.5%	11	3.7%	2	0.7%	1	0.3%
2005	Phase II	312	90	28.8%	11	3.5%	13	4.2%	1	0.3%

continued

TABLE App-A-24 Continued

		NIH Phase II SBIR Awards Over 750K									
Fiscal Year	PHASE	Total Number Funded	Year Two		Year Three		Year Four		Year Five	
1992	Phase II	139	0	0.0%	0	0.0%	0	0.0%	0	0.0%
1993	Phase II	215	0	0.0%	0	0.0%	0	0.0%	0	0.0%
1994	Phase II	134	0	0.0%	0	0.0%	0	0.0%	0	0.0%
1995	Phase II	212	0	0.0%	0	0.0%	0	0.0%	0	0.0%
1996	Phase II	172	0	0.0%	0	0.0%	0	0.0%	0	0.0%
1997	Phase II	279	0	0.0%	0	0.0%	0	0.0%	0	0.0%
1998	Phase II	224	0	0.0%	0	0.0%	0	0.0%	0	0.0%
1999	Phase II	278	8	2.9%	0	0.0%	0	0.0%	0	0.0%
2000	Phase II	231	8	3.5%	0	0.0%	0	0.0%	0	0.0%
2001	Phase II	343	9	2.6%	1	0.3%	0	0.0%	0	0.0%
2002	Phase II	336	23	6.8%	0	0.0%	0	0.0%	0	0.0%
2003	Phase II	327	18	5.5%	4	1.2%	0	0.0%	1	0.3%
2004	Phase II	298	24	8.1%	6	2.0%	1	0.3%	1	0.3%
2005	Phase II	312	27	8.7%	8	2.6%	11	3.5%	1	0.3%

SOURCE: National Institutes of Health.

TABLE App-A-25 Phase I, Year One Supplements, 1992-2003

Fiscal Year	1992	1993	1994	1995	1996	1997	1998	1999	2000	2001	2002	2003
Number of Awards	4	6	9	7	8	15	24	16	31	33	33	30
Average ($)	67,138	64,067	47,101	25,314	43,988	100,688	131,038	96,728	118,537	134,049	163,440	95,711
Maximum ($)	142,502	150,000	121,237	82,000	118,886	214,957	530,198	375,000	574,670	600,000	713,490	370,349
Minimum ($)	26,731	8,112	2,780	6,779	6,417	15,470	9,414	634	10,268	10,000	9,809	2,200
Total ($)	268,550	384,402	423,913	177,196	351,906	1,510,314	3,144,900	1,547,643	3,674,641	4,423,627	5,393,530	2,871,315

SOURCE: National Institutes of Health.

TABLE App-A-26 Phase II, Year One Grants, 1992-2003

All Phase II, year one grants

Fiscal Year	1992	1993	1994	1995	1996	1997
Number of Grants	137	215	131	210	173	280
Average ($)	249,361	244,363	269,611	345,070	349,374	375,304

Amended Phase II, year one grants

Fiscal Year	1992	1993	1994	1995	1996	1997
Number of Grants	29	55	22	48	51	89
Percent of all	21.2	25.6	16.8	22.9	29.5	31.8
Average ($)	238,893	247,948	255,816	354,890	336,933	358,916
Percent all av.	95.8	101.5	94.9	102.8	96.4	95.6
Maximum ($)	286,292	500,000	326,203	647,634	495,215	611,874
Minimum ($)	129,427	112,898	2,000	93,848	3,646	7,666
Sum ($)	6,927,902	13,637,125	5,627,959	17,034,712	17,183,579	31,943,518

SOURCE: National Institutes of Health.

1998	1999	2000	2001	2002	2003	
237	289	267	381	358	349	
379,972	410,335	448,377	437,573	461,216	495,834	

1998	1999	2000	2001	2002	2003	
45	60	62	100	88	82	
19.0	20.8	23.2	26.2	24.6	23.5	
377,707	385,556	391,078	402,757	462,258	461,427	
99.4	94.0	87.2	92.0	100.2	93.1	
514,581	864,785	591,768	822,158	1,152,230	1,230,000	
48,387	12,869	9,782	31,418	23,719	204,950	
16,996,826	23,133,378	24,246,844	40,275,684	40,678,701	37,837,035	

TABLE App-A-27 Phase II Contracts, 1992-2000

Fiscal Year	1992	1993	1994	1995	1996	1997	1998	1999	2000
Number of Awards	27	34	21	26	23	24	28	21	28
Average ($)	205,473	222,926	210,814	337,065	304,616	377,640	359,969	414,538	324,965
Maximum ($)	477,019	499,085	500,000	739,693	750,000	1,184,286	750,000	798,773	749,868
Minimum ($)	14,834	9,660	13,915	10,000	7,050	9,793	7,253	4,245	87,992
Total ($)	5,547,783	7,579,467	4,427,086	8,763,693	7,006,158	9,063,353	10,079,133	8,705,294	9,099,027

SOURCE: National Institutes of Health.

TABLE App-A-28 Phase II Awards by IC, 1992-2003

IC	Number of Awards													Percent of Total	Number of Applications	Success Rate (%)
	1992	1993	1994	1995	1996	1997	1998	1999	2000	2001	2002	2003	Awards			
AA	3	3	1	1	2	5	2	4	5	4	4	2	36	1.2	92	39.1
AG	8	10	4	14	9	10	12	11	10	15	19	14	136	4.5	300	45.3
AI	16	23	17	18	20	28	27	30	23	30	26	22	280	9.3	667	42.0
AR	4	4	2	3	1	7	3	6	6	7	10	8	61	2.0	251	24.3
AT									2		3	1	6	0.2	10	60.0
CA	15	40	16	45	31	53	45	46	49	56	54	51	501	16.6	1,378	36.4
DA	5	13	3	7	6	11	9	5	9	14	14	15	111	3.7	232	47.8
DC	2	4	4	4	4	4	6	4	4	9	5	6	56	1.9	151	37.1
DE	3	6	2	3	4	4	3	4	1	4	9	8	51	1.7	175	29.1
DK	9	13	8	12	13	18	14	20	16	23	26	24	196	6.5	450	43.6
EB											6	6	12	0.4	25	48.0
ES	1	4	2	3	3	3	6	4	4	6	8	11	55	1.8	135	40.7
EY	5	6	1	5	4	9	9	9	11	13	6	12	90	3.0	242	37.2
GM	16	13	17	19	16	25	17	26	23	37	20	26	255	8.4	567	45.0
HD	6	11	10	12	9	15	15	14	21	26	25	21	185	6.1	464	39.9
HG	2	4	1	4	1	4	4	6	6	5	3	7	47	1.6	76	61.8
HL	23	35	18	27	21	42	28	45	31	59	47	53	429	14.2	1,018	42.1
LM										2		2	4	0.1	20	20.0
MD															1	0.0
MH	8	9	10	11	11	16	14	20	18	23	13	18	171	5.6	330	51.8
NR		1			2	1	1	2	2	3	3	3	18	0.6	51	35.3
NS	7	12	7	10	9	15	12	17	18	20	31	17	175	5.8	384	45.6
RR	4	4	6	9	7	10	7	11	7	22	19	22	128	4.2	326	39.3
Shared			2	3			3	5	1	3	7		24	0.8		
Total	137	215	131	210	173	280	237	289	267	381	358	349	3,027	100		

SOURCE: National Institutes of Health.

TABLE App-A-29 Conversion Fates of Top 20 Phase II Winners, 1992-2003

Organization	Number of Phase II Awards	Number of Phase I Awards	Conversion Rate (%)
NEW ENGLAND RESEARCH INSTITUTES, INC.	34	42	81.0
INFLEXXION, INC.	32	44	72.7
RADIATION MONITORING DEVICES, INC.	30	56	53.6
OREGON CENTER FOR APPLIED SCIENCE	27	44	61.4
INSIGHTFUL CORPORATION	22	40	55.0
LYNNTECH, INC.	17	51	33.3
PANORAMA RESEARCH, INC.	15	69	21.7
INOTEK PHARMACEUTICALS CORPORATION	14	63	22.2
SOCIOMETRICS CORPORATION	13	16	81.3
ABIOMED, INC.	13	13	100.0
PERSONAL IMPROVEMENT COMPUTER SYSTEMS	13	25	52.0
CREARE, INC.	13	40	32.5
CLEVELAND MEDICAL DEVICES, INC.	13	23	56.5
INDIVIDUAL MONITORING SYS, INC. (IM SYS)	12	14	85.7
BIOTEK, INC.	12	29	41.4
SURMODICS, INC.	11	30	36.7
ADVANCED MEDICAL ELECTRONICS CORPORATION	11	18	61.1
ELECTRICAL GEODESICS, INC.	11	19	57.9
WESTERN RESEARCH COMPANY, INC.	11	20	55.0
PHYSICAL SCIENCES, INC.	11	25	44.0
Total (top 20 award winners)	335	681	49.2
All Awards	3,027		
Top 20 as percent of all Phase II awards	11.1		

SOURCE: National Institutes of Health.

TABLE App-A-30 Phase II Awards by State, 1992-2005

State	Number of Awards by Fiscal Year														Grand Total	Percent of Total
	1992	1993	1994	1995	1996	1997	1998	1999	2000	2001	2002	2003	2004	2005		
ALABAMA	2	2	4	6	4	3	3	3	4	6	7	5	5	3	57	0.7
ARIZONA	3	7	6	2	2	5	7	13	14	8	17	18	10	11	123	1.6
ARKANSAS			1	1								1		2	6	0.1
CALIFORNIA	39	62	53	60	66	94	119	99	94	120	137	137	155	142	1,377	17.8
COLORADO	3	5	5	5	9	10	11	16	17	15	17	17	15	15	160	2.1
CONNECTICUT	8	9	9	12	12	8	7	6	9	9	5	7	12	11	124	1.6
DELAWARE				2	2	2	1	3	4	3	2	2	3	3	25	0.3
DIST OF COL	1	2	4	2	2	6	6	6	14	9	12	8	8	5	85	1.1
FLORIDA	6	13	10	4	3	10	9	4	11	10	9	9	10	11	119	1.5
GEORGIA	1	1	3	3	1	4	6	8	8	6	5	5	7	8	66	0.9
HAWAII	2	2	2	3	2	1		1	2	2	2	1	2	3	25	0.3
IDAHO		1	1					1	1				1	1	6	0.1
ILLINOIS	7	10	9	6	6	11	13	17	17	18	17	19	15	6	171	2.2
INDIANA	4	2		3	5	5	4	4	5	5	7	6	6	9	65	0.8
IOWA	2	1				1	2	2	1	3	4	2	2	1	21	0.3
KANSAS				2	2	2	2		2	3	4	4	4	3	29	0.4
KENTUCKY	1	2	1		1	1		1	1	3	4	1	4	5	25	0.3
LOUISIANA		2	1	1		1	1		1	2	3	3		3	20	0.3
MAINE	1	1	2	2		2	3	2	2	1	3	4	1	1	25	0.3
MARYLAND	23	31	34	30	28	26	37	30	29	42	48	43	52	52	505	6.5
MASSACHUSETTS	57	70	52	60	80	94	86	84	99	118	121	105	100	103	1,229	15.9
MICHIGAN	8	10	7	8	11	15	19	14	10	14	13	19	18	24	190	2.5
MINNESOTA	10	9	9	8	10	16	12	8	8	10	13	16	14	20	163	2.1
MISSISSIPPI												1	1		2	0.0
MISSOURI	2	1	2			1	3	3	6	3	7	7	4	5	44	0.6
MONTANA					1	1	1	2	1	2	2	1	2	2	15	0.2
NEBRASKA	1	1	1	2	2	2	2	1	3	4	2	1	4	3	29	0.4

continued

TABLE App-A-30 Continued

State	Number of Awards by Fiscal Year														Grand Total	Percent of Total
	1992	1993	1994	1995	1996	1997	1998	1999	2000	2001	2002	2003	2004	2005		
NEVADA	1	1	2	2	1	1			1	3	4	2	2		20	0.3
NEW HAMPSHIRE		2	2	6	6	4	6	7	6	7	8	12	12	8	87	1.1
NEW JERSEY	12	7	8	8	7	6	9	19	21	20	27	21	16	19	200	2.6
NEW MEXICO		1	2	2	4	4	2	1	4	5	5	6	5	4	45	0.6
NEW YORK	13	21	23	20	23	26	22	22	27	32	42	40	36	36	383	5.0
NORTH CAROLINA	7	9	13	8	6	14	16	15	13	17	24	23	23	31	219	2.8
NORTH DAKOTA													1	2	3	0.0
OHIO	4	8	8	11	10	14	18	18	17	16	29	29	19	19	220	2.9
OKLAHOMA	1	2	3	1	1		1	1		1	2	1	3	4	21	0.3
OREGON	6	7	8	11	12	14	19	18	21	21	21	20	23	27	228	3.0
PENNSYLVANIA	13	7	7	12	16	13	17	20	18	27	38	34	34	32	288	3.7
PUERTO RICO									1	1					2	0.0
RHODE ISLAND	1	1						2	4	4	6	7	10	8	43	0.6
SOUTH CAROLINA									1	1	6	7	5	6	26	0.3
SOUTH DAKOTA								2	2	2	2	1	1		10	0.1
TENNESSEE	2	2	4	4	1	1	5	5	3	2	2	5	6	4	46	0.6
TEXAS	13	12	12	16	13	15	17	18	23	27	25	32	36	37	296	3.8
UTAH	8	13	9	7	5	2	5	8	9	12	16	9	5	4	112	1.5
VERMONT	2	1	2	4	3	2	1		1	1	1	5	6	4	33	0.4
VIRGINIA	6	10	18	21	15	10	12	14	13	22	24	32	36	23	256	3.3
WASHINGTON	5	11	12	14	14	15	30	34	26	30	42	42	36	35	346	4.5
WISCONSIN	1	1	2	3	4	6	7	6	12	16	10	17	20	17	122	1.6
WYOMING									1	1	1	1	1	1	6	0.1
Grand Total	278	360	351	370	390	468	541	539	587	683	797	788	792	774	7,718	100.0

SOURCE: National Institutes of Health.

TABLE App-A-31 NIH—Target Groups

Size of Affected Populations	In Commercialization		In Use		Total	
	Number	Percent	Number	Percent	Number	Percent
MISSING	1	0.5	0	0.0	1	0.3
Under 10,000 persons	30	16.3	42	26.6	72	21.1
10,000-49,999	23	12.5	21	13.3	44	12.9
50,000-199,999	9	4.9	17	10.8	26	7.6
200,000-499,999	13	7.1	14	8.9	27	7.9
500,000 or more	76	41.3	44	27.8	120	35.1
Not sure	32	17.4	20	12.7	52	15.2
	184	100.0	158	100.0	342	100.0

SOURCE: National Institutes of Health, *National Survey to Evaluate the NIH SBIR Program: Final Report*, July 2003.

TABLE App-A-32 High-impact Projects—By Target Sector

Population	Number of High-impact Respondents	All Respondents (#)	Percent of High-impact	Percent of All
Outpatients	4	28	9.1	11.2
Inpatients	1	22	2.3	8.8
Hospital personnel	4	10	9.1	4.0
Research labs	2	67	4.5	26.9
Diagnostic labs	4	15	9.1	6.0
Medical practitioners	7	24	15.9	9.6
Homecare providers	1	1	2.3	0.4
Other	2	3	4.5	1.2
Other health services	4	12	9.1	4.8
General public	8	26	18.2	10.4
Educators	2	7	4.5	2.8
Worksites	0	1	0.0	0.4
Schools, universities	1	11	2.3	4.4
Other companies, other technologies	2	7	4.5	2.8
Health researchers	1	5	2.3	2.0
MISSING	1	10	2.3	4.0
Total	44	249	100.0	100.0

NOTE: High-impact respondents are those with products in use, who expect to affect more than 500,000 people.

SOURCE: National Institutes of Health, *National Survey to Evaluate the NIH SBIR Program: Final Report*, July 2003.

TABLE App-A-33 NIH—Field of Business

Q2. Field of Business

	Number of Companies	Percent
Biotechnology	175	22.8
Chemical technology	12	1.6
Computer hardware, software	88	11.5
Diagnostics	43	5.6
Engineering, fabrication	13	1.7
Environment, ergonomics	20	2.6
Health care	21	2.7
Information & research	85	11.1
Instrumentation	17	2.2
Medical devices	145	18.9
Medical education, health promotion	65	8.5
Other	25	3.3
Pharmaceuticals	58	7.6
Total companies responding	767	100

Q2. Text:

Which of the following *best* describes this company's *major* field of business?

SOURCE: National Institutes of Health, *National Survey to Evaluate the NIH SBIR Program: Final Report*, July 2003.

TABLE App-A-34 NIH—Other SBIR Awards

How Many Related Phase II Awards?

Number of Awards	NIH Survey		NRC Survey		NIH Survey Breakout—Phase I and Phase II			
					Related Phase I Awards		Related Phase II Awards	
	Number of Companies	Percent	Number of Companies	Percent	Number of Companies	Percent	Number of Companies	Percent
0					2	0.5	143	35.8
1	152	59.4	92	56.8	156	39.1	152	38.1
2	65	25.4	30	18.5	106	26.6	65	16.3
3	19	7.4	11	6.8	68	17.0	19	4.8
4	8	3.1	8	4.9	19	4.8	8	2.0
5	4	1.6	12	7.4	20	5.0	4	1.0
6	2	0.8	0	0.0	13	3.3	2	0.5
7	1	0.4	1	0.6	3	0.8	1	0.3
8	1	0.4	0	0.0	1	0.3	1	0.3
9			2	1.2	1	0.3		
10	3	1.2	0	0.0	2	0.5	3	0.8
11			1	0.6	2	0.5		
12	1	0.4	5	3.1	2	0.5	1	0.3
14					1	0.3		
17					1	0.3		
20					1	0.3		
24					1	0.3		
Total					399	100.0	399	100.0
Number of Responses (excluding companies with 0 awards)	256		162					
Number of Surveyed Companies	768		496					

SOURCE: National Institutes of Health, *National Survey to Evaluate the NIH SBIR Program: Final Report*, July 2003; NRC Phase II Survey.

TABLE App-A-35 NIH—Related Phase II Awards

Q5. Other related SBIR awards

	Number of Responses	Percent
Yes	399	52.0
No	325	42.3
Other	44	5.7
	768	100.0

Q5. Text:

Has the *company* won any other SBIR Phase I or Phase II awards, in addition to the referenced award, for products, processes, or services that are *related to this project*?

Q6. How many related Phase I awards?

Number of Awards	Number of Companies	Percent
0	2	0.5
1	156	39.1
2	106	26.6
3	68	17.0
4	19	4.8
5	20	5.0
6	13	3.3
7	3	0.8
8	1	0.3
9	1	0.3
10	2	0.5
11	2	0.5
12	2	0.5
14	1	0.3
17	1	0.3
20	1	0.3
24	1	0.3
Total	399	100.0

Q6. Text:

How many SBIR *Phase I awards*, that involve products, processes, or services *related to the project* supported by the SBIR award referenced earlier, has the company won?

TABLE App-A-35 Continued

Q7. How Many Related Phase II Awards?

Number of Awards	NIH Survey Number of Companies	Percent	NRC Survey Number of Companies	Percent
0	512	66.7	334	67.3
1	152	19.8	92	18.5
2	65	8.5	30	6.0
3	19	2.5	11	2.2
4	8	1.0	8	1.6
5	4	0.5	12	2.4
6	2	0.3	0	0.0
7	1	0.1	1	0.2
8	1	0.1	0	0.0
10	3	0.4	2	0.4
12	1	0.1	0	0.0
11			1	0.2
18			5	1.0
Total (excluding 0 awards)	256		162	
Total	768		496	

Number of Awards	Related Phase I Awards Number of Companies	Percent	Related Phase II Awards Number of Companies	Percent
0	2	0.5	143	35.8
1	156	39.1	152	38.1
2	106	26.6	65	16.3
3	68	17.0	19	4.8
4	19	4.8	8	2.0
5	20	5.0	4	1.0
6	13	3.3	2	0.5
7	3	0.8	1	0.3
8	1	0.3	1	0.3
9	1	0.3		
10	2	0.5	3	0.8
11	2	0.5		
12	2	0.5	1	0.3
14	1	0.3		
17	1	0.3		
20	1	0.3		
24	1	0.3		
Total	399	100.0	399	100.0

Q7. Text:

How many other SBIR *Phase II awards*, that involve products, processes, or services *related to the project* supported by the SBIR award referenced earlier, has the company won?

SOURCE: National Institutes of Health, *National Survey to Evaluate the NIH SBIR Program: Final Report*, July 2003.

TABLE App-A-36 Impact of SBIR Award on Company Growth

Percent of company growth attributed to SBIR

	Number of Responses	Percent
Less than 25%	132	29.5
25% to 50%	100	22.4
51% to 75%	78	17.4
More than 75%	137	30.6
Total	447	100.0

SOURCE: NRC Firm Survey.

TABLE App-A-37 Company Formation

Company was formed wholly or in part because of the SBIR award	Number of Responses	Percent	Number of founders with academic background	Number of Responses	Percent
No	342	74.8	0	86	
Yes	49	10.7	1	214	
Yes, in part	66	14.4	2	106	
	457	100.0	3	33	
			4	11	
			5	3	
			7	2	
				369	
			None	86	18.9
			At least one	369	81.1
			All	455	100.0

SOURCE: NRC Firm Survey.

TABLE App-A-38 Project Impacts

Q3YN. Project go without award?

	Number of Responses	Percent
Project go without award	114	14.9
Project no go without award	489	63.8
Not Applicable	164	21.4
Total	767	100.0

Q3YN. Text:

If the SBIR program were *not* available, would the project funded by the referenced award still have been pursued?

NIH—Project impacts

Q8. How important was the SBIR funding to the project?

	Number of Responses	Percent
Very important	668	87.4
Important	82	10.7
Somewhat important	13	1.7
Not important	1	0.1
Not very important	0	0.0
Total	764	100.0

Q8. Text:

How important overall has SBIR support been, or how important will it be, in research and development of this product, process, or service?

NIH—Project impacts

Q9. Specific impacts

	Missing	Yes	No	Not Applicable
Pursuing high-risk ideas	9	667	42	50
Hiring additional personnel	13	616	107	32
Raising additional capital	23	341	305	99
Credibility or visibility for finding partners	18	541	128	81

Q9. Text:

Did the granting of one or more SBIR awards for this product, process, or service have an impact on any of the following activities?

SOURCE: National Institutes of Health, *National Survey to Evaluate the NIH SBIR Program: Final Report*, July 2003.

TABLE App-A-39 FDA Approval

Q11. FDA review required?

	Number of Responses	Percent
Yes	324	42.2
No	444	57.8
	768	100.0

Q11. Text:

Was or is FDA approval required for the product, process, or service selected above?

Q12. Submitted to FDA?

	Number of Responses	Percent
Yes	90	27.8
No	234	72.2
	324	100.0

Q12. Text:

Has this product, process, or service been submitted for FDA review?

Q13. Stage of Review

	Number of Responses	Percent
Applied for approval	8	8.9
Review ongoing	13	14.4
Approved	48	53.3
Not approved	2	2.2
IND; Clinical trials	11	12.2
Other	7	7.8
MISSING	1	1.1
Total	90	100.0

Q13. Text:

In what stage of the FDA approval process is this product, process, or service?

Business Type	FDA Approval Required (#)	All Respondents (#)	Percent
Pharmaceuticals	47	58	81.0
Medical devices	102	145	70.3
Biotechnology	87	175	49.7
Diagnostics	21	43	48.8
All Respondents	323	767	42.1
Other	5	12	41.7
Instrumentation	31	88	35.2
Chemical technology	6	20	30.0
Health care	5	21	23.8
Computer hardware, software	14	85	16.5
Engineering, fabrication	2	17	11.8
Environment, ergonomics	1	13	7.7
Information & research	1	25	4.0
Medical education, health promotion	1	65	1.5

SOURCE: National Institutes of Health, *National Survey to Evaluate the NIH SBIR Program: Final Report*, July 2003.

TABLE App-A-40 Project Focus

Q15B. Project focus:

	Number of Responses	Percent
Preventing disease or disability	84	10.9
Detecting disease or disability	82	10.7
Diagnosing disease or disability	66	8.6
Treating disease or disability	236	30.7
Reducing the cost of medical care	34	4.4
Developing information for healthcare professionals	41	5.3
Developing health information for the general public	28	3.6
Fostering new research collaborations	3	0.4
Improving research tools	146	19.0
Other	7	0.9
Training research investigators	1	0.1
Improving quality of technology, products	7	0.9
Improving quality of life for general public	16	2.1
Missing data	17	2.2
Total	768	100.0

Q15B. Text:

Select the single category that is the most important medical, societal, or technological outcome.

Q16B. Population focus

	Number of Responses	Percent
Outpatients	121	15.8
Inpatients	73	9.5
Hospital personnel	32	4.2
Research labs	150	19.5
Diagnostic labs	47	6.1
Medical practitioners	92	12.0
Homecare providers	4	0.5
Emergency medical services	4	0.5
Military medical services		0.0
Other	5	0.7
Other health services	23	3.0
General public	104	13.5
Educators	13	1.7
Worksites	3	0.4
Schools, universities	29	3.8
Police, fire, other municipal workers	4	0.5
Other companies, other technologies	37	4.8
Health researchers	7	0.9
MISSING	20	2.6
Total	768	100.0

Q16B. Text:

Select the single population that is the most important population.

continued

TABLE App-A-40 Continued

Q17. Projected size of benefiting populations			Q17. Text:
	Number of Responses	Percent	Within the next few years, what is the anticipated size of the *total target populations* that would benefit from or use the product, process, or service being developed under this project?
Under 10,000 persons	142	18.5	
10,000-49,999	81	10.5	
50,000-199,999	74	9.6	
200,000-499,999	59	7.7	
500,000 or more	242	31.5	
Not sure	164	21.4	
MISSING	6	0.8	
	768	100.0	

SOURCE: National Institutes of Health, *National Survey to Evaluate the NIH SBIR Program: Final Report*, July 2003.

TABLE App-A-41 Sales by Source of Data

	NIH		DoD		NRC	
	Number of Responses	Percent	Number of Responses	Percent	Number of Responses	Percent
>$0 and <$1M	140	68.3	179	86.1	147	72.8
$1M-<$5M	42	20.5	22	10.6	39	19.3
$5M-<$50M	22	10.7	6	2.9	15	7.4
$50M +	1	0.5	1	0.5	1	0.5
Sales >$0	205	30.3	208	31.6	202	40.7
No sales yet	472	69.7	450	68.4	294	59.3
Total responses	677		658		496	

Detailed Sales Responses, 3 data sources

	NIH	DoD	NRC
<$50K	35	54	44
$50-<$100K	19	32	13
$100K-<$500K	59	68	51
$500K-<$1M	27	25	19
$1M-<$5M	42	22	29
$5M-<$50M	22	6	12
$50M +	1	1	1
Total	205	208	169

SOURCE: National Institutes of Health, *National Survey to Evaluate the NIH SBIR Program: Final Report*, July 2003; DoD Commercialization Database; and NRC Phase II Survey.

TABLE App-A-42 Sales-related Data

NIH—Sales

Q21. Expectations of sales

	Number of Responses	Percent
Yes	576	85.1
No	101	14.9
	677	100

Q21. Text:

Upon *completion* of the project, were (or are) sales expected? (Include *both sales and sales of licenses.*)

Q22. Sales Results

	Number of Responses	Percent
Sales were realized	224	39.2
Sales are anticipated	340	59.4
No sales	8	1.4
	572	100

Q22. Text:

With regard to sales, which of the following resulted?

Q23. Dollar ranges for cumulative sales

	Number of Responses	Percent
$50,000 or less	35	
$50,000-$99,999	19	
$100,000-$499,999	59	
$500,000-$999,999	27	
$1M to <$5M	42	
$5M to <$50M	22	
$50M +	1	
	205	

Q23. Text:

What is the dollar range of *cumulative* sales related to the product, process, or service developed under this project?

Results for those with no sales to date, but expecting sales

	Number of Responses	Percent
None expected	101	13.4
Sales expected	445	59.3
Sales, <$1M	140	18.6
Sales, $1-5M	42	5.6
Sales, $5-50M	22	2.9
Sales >$50M	1	0.1
	751	100

More than $0 sales, 3 data sources

Source	Percent
NRC	40.7
NIH	30.3
DoD	31.6

continued

TABLE App-A-42 Continued

Yes sales, yes licensing agreement

Dollar Range of Sales	Number of Responses	Percent
$50,000 or less	11	15.1
$50,000-$99,999	4	5.5
$100,000-$499,999	25	34.2
$500,000-$999,999	10	13.7
$1,000,000-$4,999,999	13	17.8
$5,000,000-$49,999,999	10	13.7
	73	100.0

SOURCE: National Institutes of Health, *National Survey to Evaluate the NIH SBIR Program: Final Report*, July 2003; DOD Commercialization Database; and NRC Phase II Survey.

TABLE App-A-43 Sales by Number of Employees

Number of Employees	Number of Responses				Grand Total	Percentage
	<$1M	$1M-<$5M	$5M-<$50M	$50M+		
Other	19	6	2		29	12.9
0-5	54	6			70	31.3
6-10	20	6	2		30	13.4
11-15	18	10	2		30	13.4
16-20	6	2	3		13	5.8
21-25	2	2	1		5	2.2
26-30	5				5	2.2
31-40	4				4	1.8
41-50	3	3	3		9	4.0
51-75	1	2			5	2.2
101-200	5	1	4	1	11	4.9
201-300		1	4		6	2.7
301-500	1	2			3	1.3
500 plus	2	1	1		4	1.8
Grand Total	140	42	22	1	224	100.0
Percentage	62.5	18.8	9.8	0.4	100.0	

SOURCE: National Institutes of Health, *National Survey to Evaluate the NIH SBIR Program: Final Report*, July 2003.

TABLE App-A-44 Employment Patterns

Q CQ24. Current number of employees						Q CQ24. Text:

	NIH Groupings			NRC Regroupings		Q CQ24. Text:
	Number of Responses	Percent		Number of Responses	Percent	What is the *current* number of *total* employees (full-time equivalents) in your company?
5 or Fewer	102	13.6	0-5	259	34.4	
6-10	106	14.1	6-10	124	16.5	
11-15	86	11.5	11-15	67	8.9	
16-20	89	11.9	16-20	45	6.0	
21-25	112	14.9	21-25	35	4.7	
26-30	73	9.7	26-30	22	2.9	
31-50	90	12.0	31-40	31	4.1	
51-75	93	12.4	41-50	19	2.5	
76-100		0.0	51-75	31	4.1	
101-250		0.0	76-100	20	2.7	
251-500		0.0	101-200	37	4.9	
500 or More		0.0	201-300	26	3.5	
MISSING		0.0	301-500	14	1.9	
	751	100.0	>500	22	2.9	
				752	100.0	

emp-adjusted	emp-adj. # responses	Percent of responses	emp_adj total employees	Percent of all employees	Percent of all employees at firms <500 employees
	0		−17		
0-5	259	34.5	783	1.0	2.7
6-10	124	16.5	972	1.3	3.4
11-15	67	8.9	842	1.1	2.9
16-20	45	6.0	804	1.0	2.8
21-25	35	4.7	836	1.1	2.9
26-30	22	2.9	628	0.8	2.2
31-40	31	4.1	1,131	1.5	4.0
41-50	18	2.4	852	1.1	3.0
51-75	31	4.1	1,922	2.5	6.7
76-100	20	2.7	1,859	2.4	6.5
101-200	37	4.9	5,534	7.2	19.4
201-300	26	3.5	6,589	8.6	23.1
301-500	14	1.9	5,825	7.6	20.4
More than 500	22	2.9	48,281	62.8	
	751	100.0	76,858	100.0	

SOURCE: National Institutes of Health, *National Survey to Evaluate the NIH SBIR Program: Final Report*, July 2003.

TABLE App-A-45 Sales by Industry Sector

Number of Sales by Industry, By Size

Industry	MISSING	<$50K	>$50K-<$100K	$100K-<$500K	$500K-<$1M	$1M-<$5M	$5M-<$50M	$50M+	Grand Total
Biotechnology	3	8	2	11	5	7	6		42
Chemical technology	1	1	2	3	2	1	1		11
Computer hardware, software	2	7	4	8	2	11			34
Diagnostics		1	1	4		1	1		8
Environment, ergonomics	1	1		4					6
Healthcare	1		2	1	1		3		8
Information and research			2		1		4		7
Instrumentation	1		2	17	3	7	3		33
Medical devices	6	6	2	4	9	11	4	1	43
Medical education, health promotion	4	11	2	6	3	1			27
Other				1	1				2
Pharmaceuticals						3			3
Grand Total	19	35	19	59	27	42	22	1	224

Percentage of Respondents with Sales, By Industry

Industry	MISSING	<$50K	>$50K-<$100K	$100K-<$500K	$500K-<$1M	$1M-<$5M	$5M-<$50M	$50M+	Grand Total
Biotechnology	7.1	19.0	4.8	26.2	11.9	16.7	14.3	0.0	100
Chemical technology	9.1	9.1	18.2	27.3	18.2	9.1	9.1	0.0	100
Computer hardware, software	5.9	20.6	11.8	23.5	5.9	32.4	0.0	0.0	100
Diagnostics	0.0	12.5	12.5	50.0	0.0	12.5	12.5	0.0	100
Environment, ergonomics	16.7	16.7	0.0	66.7	0.0	0.0	0.0	0.0	100
Healthcare	12.5	0.0	25.0	12.5	12.5	0.0	37.5	0.0	100
Information and research	0.0	0.0	28.6	0.0	14.3	0.0	57.1	0.0	100
Instrumentation	3.0	0.0	6.1	51.5	9.1	21.2	9.1	0.0	100
Medical devices	14.0	14.0	4.7	9.3	20.9	25.6	9.3	2.3	100

Industry									
Medical education, health promotion	14.8	40.7	7.4	22.2	11.1	3.7	0.0	0.0	100
Other	0.0	0.0	0.0	50.0	50.0	0.0	0.0	0.0	100
Pharmaceuticals	0.0	0.0	0.0	0.0	0.0	100.0	0.0	0.0	100
Grand Total	8.5	15.6	8.5	26.3	12.1	18.8	9.8	0.4	100

Estimated Revenues by Industry, By Size

Industry	<$50K	>$50K-<$100K	$100K-<$500K	$500K-<$1M	$1M-<$5M	$5M-<$50M	$50M+	Total Revenues by Industry ($)	Average Revenues Respondents with Sales ($)	Average Revenues, All
Biotechnology	200,000	150,000	2,750,000	3,750,000	17,500,000	165,000,000	0	189,350,000	4,508,333	1,082,000
Chemical technology	25,000	150,000	750,000	1,500,000	2,500,000	27,500,000	0	32,425,000	2,947,727	1,621,250
Computer hardware, software	175,000	300,000	2,000,000	1,500,000	27,500,000	0	0	31,475,000	925,735	370,294
Diagnostics	25,000	75,000	1,000,000	0	2,500,000	27,500,000	0	31,100,000	3,887,500	723,256
Environment, ergonomics	25,000	0	1,000,000	0	0	0	0	1,025,000	170,833	78,846
Healthcare	0	150,000	250,000	750,000	0	82,500,000	0	83,650,000	10,456,250	3,983,333
Information & research	0	150,000	0	750,000	0	110,000,000	0	110,900,000	15,842,857	4,436,000
Instrumentation	0	150,000	4,250,000	2,250,000	17,500,000	82,500,000	0	106,650,000	3,231,818	1,211,932
Medical devices	150,000	150,000	1,000,000	6,750,000	27,500,000	110,000,000	100,000,000	245,550,000	5,710,465	1,693,448
Medical education, health promotion	275,000	150,000	1,500,000	2,250,000	2,500,000	0	0	6,675,000	247,222	102,692
Other	0	0	250,000	750,000	0	0	0	1,000,000	500,000	83,333
Pharmaceuticals	0	0	0	0	7,500,000	0	0	7,500,000	2,500,000	625,000
All	875,000	1,425,000	14,750,000	20,250,000	105,000,000	605,000,000	100,000,000	847,300,000	3,782,589	1,129,733

continued

TABLE App-A-45 Continued

Industry	Percentage of Revenues, By Industry, By Size							Total Revenues by Industry
	<$50K	>$50K-<$100K	$100K-<$500K	$500-<$1M	$1M-<$5M	$5M-<$50M	$50M+	
Biotechnology	0.1	0.1	1.5	2.0	9.2	87.1	0.0	100
Chemical technology	0.1	0.5	2.3	4.6	7.7	84.8	0.0	100
Computer hardware, software	0.6	1.0	6.4	4.8	87.4	0.0	0.0	100
Diagnostics	0.1	0.2	3.2	0.0	8.0	88.4	0.0	100
Environment, ergonomics	2.4	0.0	97.6	0.0	0.0	0.0	0.0	100
Healthcare	0.0	0.2	0.3	0.9	0.0	98.6	0.0	100
Information & research	0.0	0.1	0.0	0.7	0.0	99.2	0.0	100
Instrumentation	0.0	0.1	4.0	2.1	16.4	77.4	0.0	100
Medical devices	0.1	0.1	0.4	2.7	11.2	44.8	40.7	100
Medical education, health promotion	4.1	2.2	22.5	33.7	37.5	0.0	0.0	100
Other	0.0	0.0	25.0	75.0	0.0	0.0	0.0	100
Pharmaceuticals	0.0	0.0	0.0	0.0	100.0	0.0	0.0	100
Grand Total	0.1	0.2	1.7	2.4	12.4	71.4	11.8	100

SOURCE: National Institutes of Health, *National Survey to Evaluate the NIH SBIR Program: Final Report*, July 2003.

TABLE App-A-46 Sales by Company Size (employment)

Company Size	Sum of Responses, By Sales Range							Total Number of Companies	Size Distribution (%)
	<$50K	$50K-<$100K	$100K-<$500K	$500K-<$1M	$1M-<$5M	$5M-<50M	$50M+		
0-5	21	3	23	7	6			60	34.4
6-10	3	5	10	2	6	2		28	16.5
11-15	3	2	6	7	10	2		30	8.9
16-20	3	2	6	3	4	2		20	6.0
21-25	1		3	2	2	3		11	4.7
26-30		1	1		2	1		5	2.9
31-40		2	2	1				5	4.1
41-50	2		2					4	2.5
51-75		1	1	1	3	3		9	4.1
76-100			1		2			3	2.7
101-200		2	1	2	1	4	1	11	4.9
201-300					1	4		5	3.5
301-500				1	2			3	1.9
>500			1	1	1	1		4	2.9
Total Companies	33	18	57	27	40	22	1	198	100.0

Excludes companies reporting no sales.
Excludes companies reporting zero sales or missing data.
Excludes companies not reporting companies size.

Number of Employees	Total Number of Responses	Distribution by Size	Companies with Sales	Distribution by Size	Percent with Sales/ Percent all Companies
0-5	259	34.4	60	30.3	88.0
6-10	124	16.5	28	14.1	85.8
11-15	67	8.9	30	15.2	170.1
16-20	45	6.0	20	10.1	168.8
21-25	35	4.7	11	5.6	119.4
26-30	22	2.9	5	2.5	86.3
31-40	31	4.1	5	2.5	61.3
41-50	19	2.5	4	2.0	80.0
51-75	31	4.1	9	4.5	110.3
76-100	20	2.7	3	1.5	57.0
101-200	37	4.9	11	5.6	112.9
201-300	26	3.5	5	2.5	73.0
301-500	14	1.9	3	1.5	81.4
>500	22	2.9	4	2.0	69.1
	752	100	198	100	100.0

SOURCE: National Institutes of Health, *National Survey to Evaluate the NIH SBIR Program: Final Report*, July 2003.

TABLE App-A-47 Other Company Effects

Q25. Received nonSBIR funding for project

	Number of Responses	Percent
Yes	281	36.6
No	487	63.4
	768	100.0

Q25. Text:

Has your company received any additional *non-SBIR* funding or capital for this project?

Q27. Sources of project funding

	Number of Responses	Percent
Non-SBIR federal funds	95	10.7
Your own company	229	25.8
Other private company	131	14.8
U.S. venture capital	66	7.4
Foreign venture capital	22	2.5
Private individual investor	107	12.1
Personal funds	124	14.0
State or local government funds	63	7.1
College or university	36	4.1
Other	6	0.7
Foundations	8	0.9
	887	100.0

Q27. Text:

Thinking now about the *sources* of additional funding or capital for this project and its outcome (product, service, or process), were or are any of the following sources important?

Q26. Validation effect of SBIR

	Number of Responses	Percent
Yes	214	78.4
No	31	11.4
Not sure	28	10.3
	273	100.0

Q26. Text:

Do you believe that this additional funding or capital is a result of the NIH SBIR funding for the product, process, or service developed under this project?

Q28. Most important source of project funding

	Number of Responses	Percent
None	487	x
Non-SBIR federal funds	19	6.8
Your own company	85	30.6
Other private company	61	21.9
U.S. venture capital	22	7.9
Foreign venture capital	3	1.1
Private individual investor	37	13.3
Personal funds	22	7.9
State or local government funds	15	5.4
College or university	5	1.8
Other	5	1.8
Foundations	4	1.4
	278	100.0

Q28. Text:

Which source has been or is the *most important* source of additional funding or capital?

SOURCE: National Institutes of Health, *National Survey to Evaluate the NIH SBIR Program: Final Report*, July 2003.

TABLE App-A-48 VC Funding and SBIR Awards at NIH, 1992-2005.

NIH Top 200 Award Winners: SBIR and venture funding

Company Name	First Phase II Funded	Start of Latest Phase II	Number of NIH Phase II Awards	Total SBIR Funding ($)
	Phase II Awards			All Awards
Aastrom Biosciences, Inc.	2/1/1993	3/1/1999	5	4,905,444
Abiomed Inc.	3/1/1990	9/30/2000	13	8,924,132
Ambion, Inc.	1/1/1993	9/1/2001	8	8,566,387
Biomedical Development Corporation	5/1/1992	9/1/2000	9	6,967,861
Cambridge Neuroscience, Inc.	9/27/1989	9/30/1997	3	2,267,025
Cengent Therapeutics, Inc.	4/1/1999	9/1/1999	3	2,647,188
Centaur Pharmaceuticals, Inc.	2/1/1994	9/30/1996	4	3,989,316
Conductus, Inc.	4/15/1994	6/15/1994	3	2,904,807
Corixa Corporation	9/15/1994	9/1/2000	8	7,971,063
Cortechs Labs, Inc.	5/1/1998	6/15/2002	5	3,793,553
Cubist Pharmaceuticals, Inc.	4/1/1995	4/1/1999	4	4,758,137
Cytel	4/15/1994	7/20/2001	5	4,026,867
Diversa	9/30/1996	11/1/1997	4	4,228,546
EKOS Corporation	6/1/1998	12/1/2000	4	3,350,438
Electro-Optical Sciences, Inc.	6/23/1993	4/19/2001	4	3,262,764
Epoch Biosciences	9/1/1990	1/1/1999	4	3,238,220
Exocell, Inc.	7/1/1992	7/1/2003	6	4,352,150
Foster Miller	5/1/90	4/15/01	10	11,827,620
Genaissance Pharmaceuticals, Inc.	9/1/1993	8/29/1997	4	3,581,611
GenPharm International, Inc.	4/1/91	5/1/92	4	1,748,679
Gliatech, Inc.	2/24/1995	9/15/1998	3	3,342,616
Hawaii Biotech, Inc.	7/1/1991	2/15/1999	7	9,643,061
IDEC Pharmaceuticals Corporation	8/15/1991	9/30/1997	3	2,441,576
Illumina, Inc.	2/1/1999	7/1/2000	5	5,715,123
Immusol, Inc	9/30/1996	9/1/1999	4	3,347,984
Inotek Pharmaceuticals Corporation	3/1/1998	8/1/2001	14	29,421,600
Invitrogen Corporation	9/1/1995	9/30/1998	4	4,254,170
Isis Pharmaceuticals	9/1/1991	3/1/1999	4	3,210,263
Martek Bioscience Corporation	9/30/1991	9/30/1995	3	3,220,694
Medical Physics Colorado	5/1/1990	2/15/1992	4	2,698,149
Medimmune, Inc.	3/1/1992	5/1/1996	4	2,912,945
Meridian Instruments, Inc.	2/1/1991	5/1/1993	3	1,746,413
Micronix Corporation	8/1/1992	3/1/2001	6	4,264,334
Neurocrine Biosciences, Inc.	9/30/1994	1/1/2001	8	10,349,174
Nimbus Medical, Inc.	9/1/1990	2/1/1995	4	3,147,990

continued

TABLE App-A-48 Continued

	Phase II Awards			All Awards
Company Name	First Phase II Funded	Start of Latest Phase II	Number of NIH Phase II Awards	Total SBIR Funding ($)
OSI Pharmaceuticals, Inc.	5/1/1990	2/1/1997	10	8,563,722
Photon Imaging Corporation	4/1/1995	7/1/2001	7	7,889,307
Physical Optics Corporation	2/15/1993	9/30/2001	10	9,782,753
Physical Sciences, Inc.	3/1/1990	3/15/2003	11	8,175,374
Progenics Pharmaceuticals, Inc.	9/1/1991	9/30/1996	5	14,190,507
RiboGene, Inc.	8/1/1994	9/30/1996	4	3,727,127
Scios Nova, Inc.	1/5/1992	11/1/1997	4	2,704,476
Spencer Technologies	9/1/1992	8/1/2000	4	3,334,165
Spire Corporation	9/25/1989	2/15/2002	8	7,514,150
State of The Art, Inc.	8/23/1993	9/14/2000	8	8,231,063
Stratagene Cloning Systems	7/1/1991	4/7/1997	6	4,754,214
Talaria Holdings, LLC	7/12/1996	7/1/2001	8	7,068,657
Third Wave Technologies, Inc.	4/1/1993	9/25/2001	5	3,778,257
Transoma Medical, Inc.	4/1/1996	*8/1/1999*	3	3,384,761
Valentis	3/26/1993	3/1/1997	4	3,350,068
Volumetrics Medical Imaging	8/1/1993	6/1/2000	3	2,432,255
			271	272,332,457

	VC Funding				
Company Name	1st Round	Most Recent Round	Number of Rounds	Total Funding ($)	
Aastrom Biosciences, Inc.	8/18/1989	10/30/2002	8	36,385,000	(FKA: Ann Arbor Stromal, Inc.)
Abiomed, Inc.	12/1/1984	12/1/1984	1	3,000,000	(FKA: Applied Biomedical Corp.)
Ambion, Inc.	5/1/2003	5/1/2003	1	10,500,000	
Biomedical Development Corporation	10/1/1987	10/1/1987	1	150,000	
Cambridge Neuroscience, Inc.	1/1/1986	6/21/1997	9	35,879,000	(FKA: Synax, Inc.)
Cengent Therapeutics, Inc.	1/11/1996	11/30/2000	5	47,350,000	

TABLE App-A-48 Continued

| Company Name | VC Funding | | | |
	1st Round	Most Recent Round	Number of Rounds	Total Funding ($)
Centaur Pharmaceuticals, Inc.	12/1/1992	**11/2/2001**	7	26,561,000
Conductus, Inc.	9/1/1987	**3/27/2002**	6	45,700,000
Corixa Corporation	12/2/1994	10/2/1997	3	59,330,000
Cortechs Labs, Inc.	1/9/1987	11/23/1992	4	51,000,000
Cubist Pharmaceuticals, Inc.	9/1/1992	9/23/1998	5	36,283,000
Cytel	8/1/1987	11/22/1991	4	68,000,000
Diversa	12/1/1994	**2/14/2000**	5	210,200,000 (FKA: Recombinant BioCatalysis, Inc.)
EKOS Corporation	10/1/1996	**8/30/2001**	5	42,900,000
Electro-Optical Sciences, Inc.	1/15/1986	**6/20/2003**	8	32,440,000
Epoch Biosciences	3/1/1986	7/1/1993	13	29,980,000 (FKA: MicroProbe Corporation)
Exocell, Inc.	3/1/1988	3/1/1988	1	900,000
Foster Miller	1/1/80	1/1/80	1	750,000
Genaissance Pharmaceuticals, Inc.	4/1/1998	5/22/2000	7	73,522,000
GenPharm International, Inc.	12/3/88	**4/1/95**	9	40,100,000
Gliatech, Inc.	7/1/1988	6/1/1995	7	32,596,000
Hawaii Biotech, Inc.	6/7/2002	**6/6/2003**	2	7,300,000
IDEC Pharmaceuticals Corporation	5/1/1986	2/1/1990	4	43,870,000
Illumina, Inc.	11/30/1998	11/1/1999	2	36,567,000
Immusol, Inc	6/1/2001	**9/24/2003**	2	23,500,000
Inotek Pharmaceuticals Corporation	3/31/2004	**3/31/2004**	1	20,000,000
Invitrogen Corporation	6/20/1997	6/20/1997	1	15,000,000
Isis Pharmaceuticals	2/1/1989	8/11/1994	6	17,490,000
Martek Bioscience Corporation	1/15/1986	11/23/1993	5	22,750,000
Medical Physics Colorado	12/30/1991	12/30/1991	1	20,000
Medimmune, Inc.	5/1/1988	12/9/1991	5	143,850,000

continued

TABLE App-A-48 Continued

Company Name	VC Funding			
	1st Round	Most Recent Round	Number of Rounds	Total Funding ($)
Meridian Instruments, Inc.	5/1/1983	10/1/1993	4	3,557,000
Micronix Corporation	5/1/1981	7/1/1987	15	76,598,000
Neurocrine Biosciences, Inc.	9/25/1992	5/23/1996	12	43,000,000
Nimbus Medical, Inc.	7/1/1986	3/1/1987	2	5,598,000
OSI Pharmaceuticals, Inc.	3/1/1988	3/1/1988	1	4,000,000
Photon Imaging Corporation	9/1/1983	9/1/1983	1	750,000
Physical Optics Corporation	8/1/1987	8/1/1990	4	3,337,000
Physical Sciences, Inc.	5/9/1995	7/1/1995	2	492,000
Progenics Pharmaceuticals, Inc.	1/1/1995	12/1/1995	1	5,670,000
RiboGene, Inc.	1/1/1990	2/1/1997	14	43,577,000
Scios Nova, Inc.	6/1/1982	6/1/1982	1	5,425,000
Spencer Technologies	7/1/1997	7/1/1997	1	435,000
Spire Corporation	11/1/1979	1/1/1987	3	3,750,000
State of The Art, Inc.	9/1/1983	1/1/1986	2	3,400,000
Stratagene Cloning Systems	4/1/1987	12/31/1992	2	1,873,000
Talaria Holdings, LLC	1/1/2001	4/1/2001	2	28,673,000
Third Wave Technologies, Inc.	6/30/1995	7/26/2000	5	78,064,000
Transoma Medical, Inc.	2/5/2002	2/5/2002	1	12,075,000
Valentis	8/12/1993	10/1/2002	6	47,405,000
Volumetrics Medical Imaging	1/1/1995	6/27/2003	6	10,706,000
			224	1,592,258,000

Legend

34		First VC funding before first SBIR P2
17	**bold**	Last VC funding after start of latest SBIR P2
6	*italics*	First VC funding after last start date for SBIR P2

SOURCE: National Institutes of Health; Thomson Financial, VentureSource, and RDNA databases.

TABLE App-A-49 Knowledge Effects

Q32—Patents

Number of Patents	Number of Responses	Percent
1	158	55.1
2	63	22.0
3-5	44	15.3
6-10	18	6.3
11-20	3	1.0
>21	1	0.3
	287	100.0

Q32 Text:

Number of Patents related to this project

752
287

Q32—Copyrights

Number of Copyrights	Number of Responses	Percent
1	86	57.7
2	17	11.4
3-5	33	22.1
6-10	9	6.0
11-20	3	2.0
>21	1	0.7
	149	100.0

Q32 Text:

Number of copyrights related to this project

SOURCE: National Institutes of Health, *National Survey to Evaluate the NIH SBIR Program: Final Report*, July 2003.

TABLE App-A-50 Further Related Investments by Size of Investment

	Number of Investments by Size of Investment	Total Investment by Size of Investment ($)
	DoD data	DoD data
$50M+	1	77,000,000
$5M-<$50M	3	32,329,122
$1M-<$5M	36	80,492,819
<$1M	180	38,637,715
None	721	0
Total investments	220	228,459,656
Percent of all respondents	24.4	
Average (all)		253,562
Average (with investment)		1,038,453

SOURCE: DoD Commercialization Database.

TABLE App-A-51 Distribution of Sales Responses

Sales	Number of Responses	Percent
<$50K	54	26.0
$50K-<$100K	32	15.4
$100K-<$500K	68	32.7
$500K-<$1M	25	12.0
$1M-<$5M	22	10.6
$5M-<$50M	6	2.9
$50M+	1	0.5
	584	

SOURCE: DoD Commercialization Database.

TABLE App-A-52 Patents by Size of Company (employees)

Company Size	Sum of Responses, By Number of Patents														Total Responses	Total Patents
	1	2	3	4	5	6	7	8	9	10	14	15	18	38		
0-5	60	25	8	6	1			1							101	172
6-10	25	12	8	2			1					1			49	103
11-15	15	3	1			1	1			1					22	47
16-20	6	3	2	2	1	1				2					17	57
21-25	6	4	1												11	17
26-30	6										1				7	20
31-40	3	1		1			1								6	16
41-50	6	3	1												10	15
51-75	5	1		1											7	11
101-200	9	2	3					2					1	1	18	94
201-300	5	2	2		1	1				1					12	38
301-500		1		1											2	7
500 plus	1	1	1	1			1								5	19
Total responses	147	58	27	12	4	3	4	3	1	4	1	1	1	1	267	616

Company Size	All Companies (#)	Number of Patenting Companies	Number of Patents	Patents per Company
0-5	259	101	172	0.7
6-10	124	49	103	0.8
11-15	67	22	47	0.7
16-20	35	17	57	1.6
21-25	22	11	17	0.8
26-30	31	7	20	0.6
31-40	19	6	16	0.8
41-50	31	10	15	0.5
51-75	20	7	11	0.6
101-200	37	18	94	2.5
201-300	26	12	38	1.5
301-500	14	2	7	0.5
500 plus	22	5	19	0.9
All	707	267	616	0.9

NOTE: Company = Award in this case.

SOURCE: National Institutes of Health, *National Survey to Evaluate the NIH SBIR Program: Final Report*, July 2003

TABLE App-A-53 Customer Satisfaction Survey

Number of Responses

	Completely Satisfied	Mostly Satisfied	Mixed	Mostly Dissatisfied	Completely Dissatisfied	Not Applicable	Total Responses
Obtaining information	446	268	37	2		6	759
Instructions	368	305	69	8	2	6	758
Review process	240	320	159	31	5	4	759
Award process	316	319	91	18	7	5	756
Post-award admin	377	269	75	19	5	11	756
Staff helpfulness			missing data				
Business support			missing data				
Other			missing data				

Percentage of Total Responses

	Completely Satisfied	Mostly Satisfied	Mixed	Mostly Dissatisfied	Completely Dissatisfied	Not Applicable	Total Responses
Obtaining information	58.8	35.3	4.9	0.3	0.0	0.8	100.0
Instructions	48.5	40.2	9.1	1.1	0.3	0.8	100.0
Review process	31.6	42.2	20.9	4.1	0.7	0.5	100.0
Award process	41.8	42.2	12.0	2.4	0.9	0.7	100.0
Post-award admin	49.9	35.6	9.9	2.5	0.7	1.5	100.0
Staff helpfulness			missing data				
Business support			missing data				
Other			missing data				

SOURCE: National Institutes of Health, *National Survey to Evaluate the NIH SBIR Program: Final Report*, July 2003.

TABLE App-A-54 FDA Approval

Approval Stage	Percent of Responding Projects which Require FDA Approval
Applied for approval	5.0
Review ongoing	3.0
Approved	38.5
Not Approved	6.5
IND: Clinical trials	16.0
Other	32.0

SOURCE: NRC Phase II Survey.

TABLE App-A-55 Sales by Dependence on SBIR

Firm Revenues	Percent of Firm Revenues that Come from SBIR (Number of responses in each percent range)						
	0	1-10%	11-25%	26-50%	51-75%	76-100%	Total Responses
0	31	41	2	2	2	14	92
<$100K	12	15	7	16	14	29	93
$100K-<$500K	10	16	7	9	18	25	85
$500K-<$1M	28	33	17	28	21	32	159
$1M-<$5M	13	26	3	10	20	2	74
$5M-<$20M	2	21	5	2	2	0	32
$20M-<$100M	0	3	0	1	0	0	4
$100M+	2	2	0	0	0	0	4
Total	98	157	41	68	77	102	543

NOTE: These data are for all agencies, not NIH-specific.

SOURCE: NRC Firm Survey.

TABLE App-A-56 Company-level Activities

Activities	U.S. Companies/ Investors			Foreign Companies/ Investors		
	Finalized (%)	Ongoing (%)	Total (%)	Finalized (%)	Ongoing (%)	Total (%)
Licensing Agreement(s)	19	16	35	9	6	15
Sale of Company	1	4	5	0	1	1
Partial sale of Company	2	4	6	0	1	1
Sale of technology rights	6	7	13	1	1	2
Company merger	0	3	3	0	1	1
Joint Venture agreement	3	9	12	1	3	4
Marketing/distribution agreement(s)	21	10	31	12	6	18
Manufacturing agreement(s)	7	4	11	2	2	4
R&D agreement(s)	15	11	26	4	3	7
Customer alliance(s)	8	10	18	3	1	4
Other Specify_____	2	2	4	0	1	1

SOURCE: NRC Phase II Survey.

TABLE App-A-57 Sales by Size of Reported Revenues

Revenues	NRC Responses	Percent of Those with Sales	Percent of All Responses
<$1M	147	72.8	29.6
>$1M to <$5M	39	19.3	7.9
>$5M to <$10M	8	4.0	1.6
>$10M to <$50M	7	3.5	1.4
>$50M	1	0.5	0.2
Reporting Sales Total	202		
All Responses	496		
Percent Reporting Sales	40.7		

SOURCE: NRC Phase II Survey.

TABLE App-A-58 Change in Employment Caused by SBIR

Change in Employment	Number of Responses	Percent
<0	63	39.7
0	40	11.6
1-10	123	6.1
11-25	36	1.9
26-50	19	1.0
50-75	6	4.8
76-99	3	1.3
101-250	15	0.3
251-500	4	100.0
+>500	1	0.0
	310	

SOURCE: NRC Phase II Survey.

TABLE App-A-59 Sales Expectations

Sales Expectations	Number of Responses	Percent of Responses
Sales expected	95	19.2
Sales not expected	32	6.5
Sales	226	45.6
No response	143	28.8
Total responses		353
All responses to survey		496

SOURCE: NRC Phase II Survey.

TABLE App-A-60 Sales by Licensees

Revenues Reported for Licensee	Number of Responses
<$1M	39
$1M-<$5M	5
$5M-<$50M	5
$50M+	3
Total	52
All responses	496

NOTE: These data are as reported by the recipient, not the licensee.

SOURCE: NRC Phase II Survey.

TABLE App-A-61 Additional Investment Dollars (For projects receiving additional investment).

Source of Investment	Total Investment ($)	Percent	Number of Investments	Percent	Average Investment ($)
Private investment from U.S. venture capital	154,617,045	33.9	15	3.9	10,307,803
Private investment from other private equity	141,992,212	31.1	40	10.4	3,549,805
Private investment from foreign investment	39,616,075	8.7	12	3.1	3,301,340
Private investment from other domestic private company	21,624,866	4.7	31	8.1	697,576
Your own company	82,118,851	18.0	188	49.1	436,802
State or local government	6,290,000	1.4	23	6.0	273,478
Personal funds	9,850,408	2.2	67	17.5	147,021
College or universities	236,500	0.1	7	1.8	33,786
Total	456,345,957	100.0	383	100.0	1,191,504

SOURCE: NRC Phase II Survey.

TABLE App-A-62 The "Go" Decision

In the absence of this SBIR award, would your company have undertaken this project?

Definitely yes	5%
Probably yes	8%
Uncertain	14%
Probably not	28%
Definitely not	46%

SOURCE: NRC Phase II Survey.

TABLE App-A-63 Patents and Publications

Patent applications and awards

	Applications		Awarded	
	Number	Percent	Number	Percent
No	249	58.7	280	66.0
Yes	175	41.3	144	34.0
Total Responses	424	100.0	424	100.0

Publications

Number of Publications	Number of Responses	Total Number of Publications
1	72	72
2	52	104
3	32	96
4	19	76
5	15	75
6-10	15	133
11-30	9	146
30+	7	420
Totals	236	1,122

SOURCE: NRC Phase II Survey.

TABLE App-A-64 Time to Market

Number of Years to Market	Number of Projects	Percent of Projects
−11	1	0.4
−7	1	0.4
−4	1	0.4
−3	4	1.8
−2	6	2.7
−1	9	4.0
0	18	8.0
1	29	12.9
2	50	22.3
3	48	21.4
4	26	11.6
5	17	7.6
6	8	3.6
7	3	1.3
8	1	0.4
9	2	0.9
	224	

Award Year	All Respondents	Sales	No Sales Yet	Years Since Award
1992	21	10	11	13
1993	34	14	20	12
1994	27	10	17	11
1995	32	17	15	10
1996	35	15	20	9
1997	59	25	34	8
1998	57	29	28	7
1999	81	44	37	6
2000	63	26	37	5
2001	87	34	53	4

NOTE: Negative answers are possible if the research represents enhancement of an existing product.

SOURCE: NRC Phase II Survey.

Appendix B

NRC Phase II and Firm Surveys

The first section of this appendix describes the methodology used to survey Phase II SBIR awards (or contracts.) The second part presents the results—first of the awards (NRC Phase II Survey) and then of the NRC Firm Survey. (Appendix C presents the NRC Phase I Survey.)

ABOUT THE SURVEYS

Starting Date and Coverage

The survey of SBIR Phase II awards was administered in 2005, and included awards made through 2001. This allowed most of the Phase II awarded projects (nominally two years) to be completed, and provided some time for commercialization. The selection of the end date of 2001 was consistent with a GAO study, which in 1991, surveyed awards made through 1987.

A start date of 1992 was selected. The year 1992 for the earliest Phase II project was considered a realistic starting date for the coverage, allowing inclusion of the same (1992) projects as the DoD 1996 survey, and of the 1992, and 1993 projects surveyed in 1998 for SBA. This adds to the longitudinal capacities of the study. The 10 years of Phase II coverage spanned the period of increased funding set-asides and the impact of the 1992 reauthorization. This time frame allowed for extended periods of commercialization and for a robust spectrum of economic conditions. Establishing 1992 as the cut-off date for starting the survey helped to avoid the problem that older awards suffer from several problems, including meager early data collection as well as potentially irredeemable data loss; the fact that some firms and PIs are no longer in place; and fading memories.

Award Numbers

While adding the annual awards numbers of the five agencies would seem to define the larger sample, the process was more complicated. Agency reports usually involve some estimating and anticipation of successful negotiation of selected proposals. Agencies rarely correct reports after the fact. Setting limitations on the number of projects to be surveyed from each firm required knowing how many awards each firm had received from all five agencies. Thus, the first step was to obtain all of the award databases from each agency and combine them into a single database. Defining the database was further complicated by variations in firm identification, location, phone numbers, and points of contact within individual agency databases. Ultimately, we determined that 4,085 firms had been awarded 11,214 Phase II awards (an average of 2.7 Phase II awards per firm) by the five agencies during the 1992-2001 timeframe. Using the most recent awards, the firm information was updated to the most current contact information for each firm.

Sampling Approaches and Issues

The Phase II survey used an array of sampling techniques, to ensure adequate coverage of projects to address a wide range of both outcomes and potential explanatory variables, and also to address the problem of skew. That is, a relatively small percentage of funded projects typically account for a large percentage of commercial impact in the field of advanced, high-risk technologies.

- **Random samples.** After integrating the 11,214 awards into a single database, a random sample of approximately 20 percent was sampled. Then a random sample of 20 percent was ensured for each year; e.g., 20 percent of the 1992 awards, of the 1993 awards, etc. Verifying the total sample one year at a time allowed improved ability to adapt to changes in the program over time, as otherwise the increased number of awards made in recent years might dominate the sample.

- **Random sample by agency.** Surveyed awards were grouped by agency; additional respondents were randomly selected as required to ensure that at least 20 percent of each agency's awards were included in the sample.

- **Firm surveys.** After the random selection, 100 percent of the Phase IIs that went to firms with only one or two awards were polled. These are the hardest firms to find for older awards. Address information is highly perishable, particularly for earlier award years. For firms that had more than two awards, 20 percent were selected, but no less than two.

- **Top performers.** The problem of skew was dealt with by ensuring that all Phase IIs known to meet a specific commercialization threshold (total of $10 million in the sum of sales plus additional investment) were surveyed (derived from the DoD commercialization database). Since 56 percent of all awards were in the random and firm samples described above, only 95 Phase IIs were added in this fashion.

- **Coding.** The project database tracks the survey sample, which corresponds with each response. For example, it is possible for a randomly sampled project from a firm that had only two awards to be a top performer. Thus, the response could be analyzed as a random sample for the program, a random sample for the awarding agency, a top performer, and as part of the sample of single or double winners. In addition, the database allows examination of the responses for the array of potential explanatory or demographic variables.

- **Total number of surveys.** The approach described above generated a sample of 6,410 projects, and 4,085 firm surveys—an average of 1.6 award surveys per firm. Each firm receiving at least one project survey also received a firm survey. Although this approach sampled more than 57 percent of the awards, multiple award winners, on average, were asked to respond to surveys covering about 20 percent of their projects.

Administration of the Survey

The questionnaire drew extensively from the one used in the 1999 National Research Council assessment of *SBIR at the Department of Defense, SBIR: An Assessment of the Department of Defense Fast Track Initiative.*[1] That questionnaire in turn built upon the questionnaire for the 1991 GAO SBIR study. Twenty-four of the 29 questions on the earlier NRC study were incorporated. The researchers added 24 new questions to attempt to understand both commercial and noncommercial aspects, including knowledge base impacts, of SBIR, and to gain insight into impacts of program management. Potential questions were discussed with each agency, and their input was considered. In determining questions that should be in the survey, the research team also considered which issues and questions were best examined in the case studies and other research methodologies. Many of the resultant 33 Phase II Award survey questions and 15 Firm Survey questions had multiple parts.

The surveys were administered online, using a Web server. The formatting,

---

[1]National Research Council, *The Small Business Innovation Research Program: An Assessment of the Department of Defense Fast Track Initiative*, Charles W. Wessner, ed., Washington, DC: National Academy Press, 2000.

encoding and administration of the survey was subcontracted to BRTRC, Inc. of Fairfax, VA.

There are many advantages to online surveys (including cost, speed, and possibly response rates). Response rates become clear fairly quickly, and can rapidly indicate needed follow up for nonrespondents. Hyperlinks provide amplifying information, and built-in quality checks control the internal consistency of the responses. Finally, online surveys allow dynamic branching of question sets, with some respondents answering selected subsets of questions but not others, depending on prior responses.

Prior to the survey, we recognized two significant advantages of a paper survey over an online one. For every firm (and thus every award), the agencies had provided a mailing address. Thus, surveys could be addressed to the firm president or CEO at that address. That senior official could then forward the survey to the correct official within the firm for completion. For an online survey we needed to know the email address of the correct official. Also, each firm needed a password to protect its answers. We had an SBIR Point of Contact (POC) and email address and password for every firm, which had submitted for a DoD SBIR 1999 survey. However, we had only limited email addresses and no passwords for the remainder of the firms. For many, the email addresses that we did have were those of Principal Investigators rather than an official of the firm. The decision to use an online survey meant that the first step of survey distribution was an outreach effort to establish contact with the firms.

Outreach by Mail

This outreach phase began with the establishing a NAS registration Web site which allowed each firm to establish a POC, email address and password. Next, the Study Director, Dr. Charles Wessner, sent a letter to those firms for which email contacts were not available. Ultimately only 150 of the 2,080[2] firms provided POC/email after receipt of this letter. Six hundred fifty of those letters were returned by the post office as invalid addresses. Each returned letter required thorough research by calling the agency provided phone number for the firm, then using the Central Contractor Registration database, <Business.com> (powered by Google) and Switchboard.com to try to find correct address information. When an apparent match was found, the firm was called to verify that it was in fact the firm, which had completed the SBIR. Two hundred thirty-seven of the 650 missing firms were so located. Another ten firms were located which had gone out of business and had no POC.

Two months after the first mailing, a second letter from the Study Director went to firms whose first letter had not been returned, but which had not yet

---

[2]The letter was also erroneously sent to an additional 43 firms that had received only STTR awards.

registered a POC. This letter also went to 176 firms, which had a POC email, but no password, and to the 237 newly corrected addresses. The large number of letters (277) from this second mailing that were returned by the postal service, indicated that there were more bad addresses in the first mailing than indicated by its returned mail. (If the initial letter was inadvertently delivered, it may have been thrown away.) Of the 277 returned second letters, 58 firms were located using the search methodology described above. These firms were asked on the phone to go to the registration Web site to enter POC/email/password. A total of 93 firms provided POC/email/password on the registration site subsequent to the second mailing. Three additional firms were identified as out of business.

The final mailing, a week before survey, was sent to those firms that had not received either of the first two letters. It announced the study/survey and requested support of the 1,888 CEOs for which we had assumed good POC/email information from the DoD SBIR submission site. That letter asked the recipients to provide new contact information at the DoD submission site if the firm information had changed since their last submission. One hundred seventy-three of these letters were returned. We were able to find new addresses for 53 of these, and ask those firms to update their information. One hundred fifteen firms could not be found and five more were identified as out of business.

The three mailings had demonstrated that at least 1,100 (27 percent) of the mailing addresses were in error, 734 of which firms could not be found, and 18 were reported to be out of business.

Outreach by Email

We began Internet contact by emailing the 1,888 DoD Points of Contact (POCs) to verify their email and give them opportunity to identify a new POC. Four hundred ninety-four of those emails bounced. The next email went to 788 email addresses that we had received from agencies as PI emails. We asked that the PI have the correct company POC identify themselves at the NAS Update registration site. One hundred eighty-eight of these emails bounced. After more detailed search of the list used by NIH to send out their survey, we identified 83 additional PIs and sent them the PI email discussed above. Email to the POCs not on the DoD submission site resulted in 110 more POC/email/password being registered on the NAS registration site.

We began the survey at the end of February with an email to 100 POCs as a beta test and followed that with another email to 2,041 POCs (total of 2,141) a week later.

Survey Responses

By August 5, 2005 five months after release of the survey, 1,239 firms had begun and 1,149 firms had completed at least 14 of 15 questions on the firm survey. Project surveys were begun on 1,916 Phase II awards. Of the 4,085 firms that received Phase II SBIR awards from DoD, NIH, NASA, NSF, or DOE from 1992 to 2001, an additional seven firms were identified as out of business (total of 25) and no email addresses could be found for 893. For an additional 500 firms, the best email addresses that were found were also undeliverable. These 1,418 firms could not be contacted, thus had no opportunity to complete the surveys. Of these firms, 585 had mailing addresses known to be bad. The 1,418 firms that could not be contacted were responsible for 1,885 of the individual awards in the sample.

Using the same methodology as the GAO had used in the 1992 report of their 1991 survey of SBIR, undeliverables and out-of-business firms were eliminated prior to determining the response rate. Although 4,085 firms were surveyed, 1,418 firms were eliminated as described. This left 2,667 firms, of which 1,239 responded, representing a 46 percent response rate by firms,[3] which could respond. Similarly when the awards, which were won by firms in the undeliverable category, were eliminated (6,408 minus 1,885), this left 4,523 projects, of which 1,916 responded, representing a 42 percent response rate. Table App-B-1 displays by agency the number of Phase II awards in the sample, the number of those awards, which by having good email addresses had the opportunity to respond, and the number that responded.[4] Percentages displayed are the percentage of awards with good addresses, the percentage of the sample that responded and the responses as a percentage of awards with the opportunity to respond.

The NRC Methodology report had assumed a response rate of about 20 percent. Considering the length of the survey and its voluntary nature, the rate achieved was relatively high and reflects both the interest of the participants in the SBIR program and the extensive follow-up efforts. At the same time, the possibility of response biases that could significantly affect the survey results must be recognized. For example, it may be possible that some of the firms that could not be found have been unsuccessful and folded. It may also be possible that unsuccessful firms were less likely to respond to the survey.

---

[3]Firm information and response percentages are not displayed in Table App-B-1, which displays by agency, since many firms received awards from multiple agencies.

[4]The average firm size for awards, which responded, was 37 employees. Nonresponding awards came firms that averaged 38 employees. Since responding Phase IIs were more generally more recent than nonresponding, and awards have gradually grown in size, the difference in average award size ($655,525 for responding and $649,715 for nonresponding) seems minor.

TABLE App-B-1 NRC Phase II Survey Responses by Agency, August 4, 2005

Agency	Phase II Sample Size	Awards with Good Email Addresses	Percent of Sample Awards with Good Email Addresses	Answered Survey as of August 4, 2005	Surveys as a Percent of Sample	Surveys as a Percent of Awards Contacted
DoD	3,055	2,191	72	920	30	42
NIH	1,680	1,127	67	496	30	44
NASA	779	534	69	181	23	34
NSF	457	336	74	162	35	48
DoE	439	335	76	157	36	47
Total	6,408	4,523	70	1,916	30	42

NRC Phase II Survey Results For NIH

NOTE: SURVEY RESPONSES APPEAR IN BOLD, AND EXPLANATORY NOTES ARE IN TYPEWRITER FONT.

Project Information 496 respondents answered the first question. Since respondents are directed to skip certain questions based on prior answers, the number that responded varies by question. Also some respondents did not complete their surveys. 444 completed all applicable questions. For computation of averages, such as average sales, the denominator used was 496, the number of respondents who answered the first question. Where appropriate, the basis for calculations is provided in typewriter font after the question.

PROPOSAL TITLE:
AGENCY: NIH
TOPIC NUMBER:
PHASE II CONTRACT/GRANT NUMBER:

Part I. Current status of the Project

1. What is the current status of the project funded by the referenced SBIR award? *Select the one best answer.* Percentages are based on the 496 respondents who answered this question.
    a.   **7%**   Project has not yet completed Phase II. *Go to question 21.*
    b. **19%**   Efforts at this company have been discontinued. No sales or additional funding resulted from this project. *Go to question 2.*
    c.   **8%**   Efforts at this company have been discontinued. The project did result in sales, licensing of technology, or additional funding. *Go to question 2.*
    d. **22%**   Project is continuing post Phase II technology development. *Go to question 3.*
    e. **13%**   Commercialization is underway. *Go to question 3.*
    f. **31%**   Products/Processes/Services are in use by target population/customer/consumers. *Go to question 3.*

2.   Did the reasons for discontinuing this project include any of the following? ***PLEASE SELECT YES OR NO FOR EACH REASON AND NOTE THE ONE PRIMARY REASON.***
137 projects were discontinued. The % below are the percent of the discontinued projects that responded with the indicated response.

	Yes	No	Primary Reason
a. Technical failure or difficulties	34%	66%	18%
b. Market demand too small	49%	51%	20%
c. Level of technical risk too high	22%	79%	3%
d. Not enough funding	41%	59%	7%
e. Company shifted priorities	54%	46%	18%
f. Principal investigator left	13%	87%	3%
g. Project goal was achieved (e.g., prototype delivered for federal agency use)	29%	71%	1%
h. Licensed to another company	14%	86%	7%
i. Product, process, or service not competitive	28%	22%	4%
j. Inadequate sales capability	26%	74%	3%
k. Other (please specify): _____	20%	80%	15%

The next question to be answered depends on the answer to question 1. If c, go to question 3. If b, skip to question 16.

Part II. Commercialization activities and planning.

Questions 3-7 concern actual sales to date resulting from the technology developed during this project. Sales includes all sales of a product, process, or service, to federal or private sector customers resulting from the technology developed during this Phase II project. A sale also includes licensing, the sale of technology or rights etc.

3. Has your company and/or licensee had any actual sales of products, processes, services or other sales incorporating the technology developed during this project? *Select all that apply.* This question was not answered for those projects still in Phase II (6%) or for projects, which were discontinued without sales or additional funding (19%). The denominator for the percentages below is all projects that answered the survey. Only 73% of all projects, which answered the survey, could respond to this question.

   a. **19%** No sales to date, but sales are expected. *Skip to question 8.*
   b. **6%** No sales to date nor are sales expected. *Skip to question 11.*
   c. **41%** Sales of product(s)
   d. **4%** Sales of process(es)
   e. **13%** Sales of services(s)
   f. **9%** Other sales (e.g., rights to technology, licensing, etc.)

   From the combination of responses 1b, 3a and 3b, we can conclude that 24% had no sales and expect none, and that 19% had no sales but expect sales.

4. For your company and/or your licensee(s), when did the first sale occur, and what is the approximate amount of total sales resulting from the technology developed during this project? If multiple SBIR awards contributed to the ultimate commercial outcome, report only the share of total sales appropriate to this SBIR project. *Enter the requested information for your company in the first column and, if applicable and if known, for your licensee(s) in the second column. Enter approximate dollars. If none, enter 0 (zero).*

Your Company   Licensee(s)

a. Year when first sale occurred.

45% reported a year of first sale. 63% of these first sales occurred in 2000 or later. 21% reported a licensee year of first sale. 59% of these first sales occurred in 2001 or later.

b. Total Sales Dollars of Product (s) Process(es) **$684,359 $678,785** or Service(s) to date. (Average of 496 survey respondents)

Although 224 reported a year of first sale, only 194 reported sales >0. Their average sales were $1,749,703. Over half of the total sales dollars were due to 4 projects, each of which had $15,000,000 or more in sales. The highest reporting project had $100,000,000 in sales. Similarly of the 103 projects that reported a year of first licensee sale, only 52 reported actual licensee sales >0. Their average sales were $6,474,565. 50% of the total sales dollars were due to 2 projects, each of which had $70,000,000 or more licensee sales. The highest reporting project had $100,000,000 in licensee sales.

c. Other Total Sales Dollars (e.g., Rights to **$65,855 $74,012** technology, Sale of spinoff company, etc.) to date. (Average of 496 survey respondents)

Combining the responses for b and c, the average for each of the 496 projects that responded to the survey is thus sales of over three-quarter million dollars by the SBIR company and over one and one-half million dollars in sales by licensees.

Display this box for Q 4 & 5 if project commercialization is known.
Your company reported sales information to DoD as a part of an SBIR proposal or to NAS as a result of an earlier NAS request. This information may be useful in answering the prior question or the next question. You reported as of *(date)*: DoD sales *($ amount)*, Other Federal Sales *($ amount)*, Export Sales *($ amount)*, Private Sector sales *($ amount)*, and other sales *($ amount)*.

5. To date, approximately what percent of total sales from the technology developed during this project have gone to the following customers? *If none enter 0 (zero). Round percentages. Answers should add to about 100%.*[5]
    496 firms responded to this question as to what percent of their sales went to each agency or sector.

Domestic private sector	**56%**
Department of Defense (DoD)	**1%**
Prime contractors for *DoD or NASA*	**0%**
NASA	**0%**
Agency that awarded the Phase II	**2%**
Other federal agencies *(Pull down)*	**0%** Sales to NIH 4%, DoE 2%, NSF 1%, other federal SBIR agencies 3%. These agencies were customers of 10% of the projects, but such sales represented only 4% of total sales.
State or local governments	**16%**
Export Markets	**19%**
Other (Specify)_____	**6%**

The following questions identify the product, process, or service resulting from the project supported by the referenced SBIR award, including its use in a fielded federal system or a federal acquisition program.

6. Is a Federal System or Acquisition Program using the technology from this Phase II?
    If yes, please provide the name of the federal system or acquisition program that is using the technology. **1% reported use in a federal system or acquisition program.**

7. Did a commercial product result from this Phase II project? **41% reported a commercial product.**

8. If you have had no sales to date resulting from the technology developed during this project, what year do you expect the first sales for your company or its licensee? Only firms that had no sales but answered that they expected sales got this question.

---

[5]Please note: If a NASA SBIR award, the Prime contractors line will state "Prime contractors for NASA." The "Agency that awarded the Phase II" will only appear if it is not DoD or NASA. The Name of the actual awarding agency will appear.

13% expected sales. The year of expected first sale is ☐☐☐☐☐
87% of those expecting sales expected sales to occur before 2009.

9. For your company and/or your licensee, what is the approximate amount of total sales expected between now and the end of 2006 resulting from the technology developed during this project? *If none, enter 0 (zero).* This question was seen by those who already had sales and those w/o sales who reported expecting sales.

    a. Total sales dollars of product(s), process(es) or   **$559,622**
       services(s) expected between now
       and the end of 2006. (Average of 496 projects)

    b. Other Total Sales Dollars (e.g., rights to technology,   **$88,857**
       sale of spinoff company, etc.) expected between now
       and the end of 2006. (Average of 496 projects)

    c. Basis of expected sales estimate. *Select all that apply.*
        a. **21%**   Market research
        b. **18%**   Ongoing negotiations
        c. **43%**   Projection from current sales
        d. **5%**   Consultant estimate
        e. **33%**   Past experience
        f. **41%**   Educated guess

10. How did you (or do you expect to) commercialize your SBIR award?
    a. **2%**   No commercial product, process, or service was/is planned.
    b. **37%**   As software
    c. **35%**   As hardware (final product, component, or intermediate hardware product)
    d. **14%**   As process technology
    e. **14%**   As new or improved service capability
    f. **3%**   As a drug
    g. **6%**   As a biologic
    h. **35%**   As a research tool
    i. **19%**   As educational materials
    j. **10%**   Other, please explain _____

11. Which of the following, if any, describes the type and status of marketing activities by your company and/or your licensee for this project? *Select one for each marketing activity.* This question answered by 340 firms,

which completed Phase II and have not discontinued the project, w/o sales or additional funding.

	Marketing activity	Planned	Need Assistance	Underway	Completed	Not Needed
a.	Preparation of marketing plan	8%	9%	18%	38%	27%
b.	Hiring of marketing staff	10%	10%	7%	25%	48%
c.	Publicity/advertising	16%	9%	25%	26%	24%
d.	Test marketing	11%	8%	12%	28%	41%
e.	Market Research	7%	14%	16%	33%	31%
f.	Other *(Specify)*	2%	1%	2%	1%	26%

Part III. Other outcomes

12. As a result of the technology developed during this project, which of the following describes your company's activities with other companies and investors? *Select all that apply.* Percentage of the 339 who answered this question.

	Activities	U.S. Companies/Investors		Foreign Companies/Investors	
		Finalized Agreements	Ongoing Negotiations	Finalized Agreements	Ongoing Negotiations
a.	Licensing Agreement(s)	19%	16%	9%	6%
b.	Sale of company	1%	4%	0%	1%
c.	Partial sale of company	2%	4%	0%	1%
d.	Sale of technology rights	6%	7%	1%	1%
e.	Company merger	0%	3%	0%	1%
f.	Joint Venture agreement	3%	9%	1%	3%
g.	Marketing/distribution agreement(s)	21%	10%	12%	6%
h.	Manufacturing agreement(s)	7%	4%	2%	2%
i.	R&D agreement(s)	15%	11%	4%	3%
j.	Customer alliance(s)	8%	10%	3%	1%
k.	Other *Specify*_____	2%	2%	0%	1%

13. In your opinion, in the absence of this SBIR award, would your company have undertaken this project?
    (Select one.) Percentage of the 339 who answered this question.
    a. **5%** Definitely yes
    b. **8%** Probably yes *If selected a or b , go to question 14.*
    c. **14%** Uncertain
    d. **28%** Probably not
    e. **45%** Definitely not *If c, d or e, skip to question 16.*

14. If you had undertaken this project in the absence of SBIR, this project would have been Questions 14 and 15 were answered only by the 13% who responded that they definitely or probably would have undertaken this project in the absence of SBIR.
    a.   **5%**   Broader in scope
    b. **44%**   Similar in scope
    c. **51%**   Narrower in scope

15. In the absence of SBIR funding, *(Please provide your best estimate of the impact)*
    a.   The start of this project would have been delayed about **an average of 8** months.
         44% of the 43 firms expected the project would have been delayed. 35% (15 firms) expected the delay would be at least 12 months. 31% anticipated a delay of at least 24 months
    b.   The expected duration/time to completion would have been
                   1) **63%**   longer
                   2) **23%**   the same
                   3) **2%**   shorter
                     12%   No response
    c.   In achieving similar goals and milestones, the project would be
                   1)   **5%**   ahead
                   2) **26%**   the same place
                   3) **56%**   behind
                     14%   No response

16. Employee information. (Enter number of employees. You may enter fractions of full-time effort (e.g., 1.2 employees). Please include both part-time and full-time employees, and consultants, in your calculation.)

Number of employees (if known) when Phase II proposal was submitted	**Ave = 22** **3% report 0** **44% report 1-5** **35% report 6-20** **8% report 21-50** **6% report >100**
Current number of employees	**Ave = 58** **4% report 0** **28% report 1-5** **36% report 6-20** **17% report 2-50** **12% report >100**
Number of current employees who were hired as a result of the technology developed during this Phase II project.	**Ave = 2.7** **42% report 0** **50% report 1-5** **6% report 6-20** **0% report report >20**
Number of current employees who were re- tained as a result of the technology developed during this Phase II project	**Ave = 2.2** **43% report 0** **51% report 1-5** **5% report 6-20** **1% report 1% report >20**

17. The Principal Investigator for this Phase II Award was a (check all that apply)
    a. **22%** Woman
    b. **8%** Minority
    c. **73%** Neither a woman or minority

18. Please give the number of patents, copyrights, trademarks and/or scientific publications for the technology developed as a result of this project. *Enter numbers. If none, enter 0 (zero).* Results are for 426 respondents to this question.

Number Applied For/Submitted		Number Received/Published
430	Patents	**305**
262	Copyrights	**258**
195	Trademarks	**170**
1,172	Scientific Publications	**1122**

Part IV. Other SBIR funding

19. How many SBIR awards did your company receive prior to the Phase I that led to this Phase II?
    a. Number of previous Phase I awards. Average of 4. **37% had no prior Phase I and another 47% had 5 or less prior Phase I.**
    b. Number of previous Phase II awards. Average of 2. **55% had no prior Phase II and another 38% had 5 or less prior Phase II.**

20. How many SBIR awards has your company received <u>that are related to the project/technology</u> supported by this Phase II award ?
    a. Number of related Phase I awards   **Average of two awards 45% had no prior related Phase I and another 45% had 5 or less prior related Phase I.**
    b. Number of related Phase II awards   **Average of one award. 62% had no prior related Phase II and another 34% had 5 or less prior related Phase II.**

Part V. Funding and other assistance

21. Prior to this SBIR Phase II award, did your company receive funds for research or development of the technology in this project from any of the following sources? Of 457 respondents.
    a. **16%** Prior SBIR *Excluding the Phase I, which proceeded this Phase II.*
    b. **5%** Prior non-SBIR federal R&D
    c. **6%** Venture Capital
    d. **7%** Other private company
    e. **10%** Private investor
    f. **34%** Internal company investment (including borrowed money)
    g. **5%** State or local government
    h. **4%** College or University
    i. **7%** Other *Specify* _____

Commercialization of the results of an SBIR project normally requires additional developmental funding. Questions 22 and 23 address additional funding. Additional Developmental Funds include non-SBIR funds from federal or private sector sources, or from your own company, used for further development and/or commercialization of the technology developed during this Phase II project.

---

22. Have you received or invested any additional developmental funding in this project?

    a. **58%** Yes *Continue.*
    b. **42%** No *Skip to question 24.*

23. To date, what has been the total additional developmental funding for the technology developed during this project? Any entries in the **Reported** column are based on information previously reported by your firm to DoD or NAS. They are provided to assist you in completing the **Developmental Funding** column. Previously reported information did not include investment by your company or personal investment. *Please update this information to include breaking out Private investment and Other investment by subcategory. Enter dollars provided by each of the listed sources. If none, enter 0 (zero).)* The dollars shown are determined by dividing the total funding in that category by the 496 respondents who started the survey to determine an average funding. Only 262 of these respondents reported any additional funding.

Source	Reported	Developmental Funding
a. Non-SBIR federal funds	$_ _, _ _ _, _ _ _	$ 91,984
b. Private investment	$_ _, _ _ _, _ _ _	
(1) U.S. venture capital		$311,727
(2) Foreign investment		$ 79,871
(3) Other private equity		$286,274
(4) Other domestic private company		$ 43,598
c. Other sources	$_ _, _ _ _, _ _ _	
(1) State or local governments		$ 12,681
(2) College or Universities		$ 476
d. Not previously reported		
(1) Your own company (including money you have borrowed)		$165,567
(2) Personal funds		$ 19,859
Total average additional developmental funding, all sources, per award		**$1,012,037**

24. Did this award identify matching funds or other types of cost sharing in the Phase II Proposal?[6]
    a. **94%** No matching funds/co-investment/cost sharing were identified in the proposal. *If a, skip to question 26.*
    b. **6%** Although not a DoD Fast Track, matching funds/co-investment/ cost sharing were identified in the proposal.
    c. **0%** Yes. This was a DoD Fast Track proposal.

25. Regarding sources of matching or co-investment funding that were proposed for Phase II, check all that apply. The percentages below are computed for those 28 projects, which reported matching funds.
    a. **79%** Our own company provided funding (includes borrowed funds)
    b. **4%** A federal agency provided non-SBIR funds
    c. **18%** Another company provided funding
    d. **4%** An angel or other private investment source provided funding
    e. **11%** Venture Capital provided funding

26. Did you experience a gap between the end of Phase I and the start of Phase II?
    a. **80%** Yes *Continue.*
    b. **20%** No *Skip to question 29.*
    The average gap reported by 362 respondents was 13 months. 11% of the respondents reported a gap of two or more years.

27. Project history. Please fill in for all dates that have occurred. This information is meaningless in aggregate. It has to be examined project by project in conjunction with the date of the Phase I end and the date of the Phase II award to calculate the gaps.

    Date Phase I ended *Month/year*

    Date Phase II proposal submitted *Month/year*

28. If you experienced funding gap between Phase I and Phase II for this award, *select all answers that apply*
    a. **38%** Stopped work on this project during funding gap.
    b. **53%** Continued work at reduced pace during funding gap.
    c. **8%** Continued work at pace equal to or greater than Phase I pace during funding gap.
    d. **8%** Received bridge funding between Phase I and II.
    e. **5%** Company ceased all operations during funding gap.

---

[6]The words underlined appear only for DoD awards.

29. Did you receive assistance in Phase I or Phase II proposal preparation for this award? Of 380 respondents.
    a. **3%** State agency provided assistance
    b. **2%** Mentor company provided assistance
    c. **0%** Regional association provided assistance
    d. **7%** University provided assistance
    e. **87%** We received no assistance in proposal preparation

    Was this assistance useful?
    a. **75%** Very Useful
    b. **25%** Somewhat Useful
    c. **0%** Not Useful

30. In executing this award, was there any involvement by universities faculty, graduate students, and/or university developed technologies? Of 444 respondents.
    54% Yes
    46% No

31. This question addresses any relationships between your firm's efforts on this Phase II project and any University (ies) or College (s). The percentages are computed against the 444 who answered question 30, not just those who answered yes to question 30.
    (Select all that apply.)
    a. **4%** The Principal Investigator (PI) for this Phase II project was at the time of the project a faculty member.
    b. **7%** The Principal Investigator (PI) for this Phase II project was at the time of the project an adjunct faculty member.
    c. **34%** Faculty member(s) or adjunct faculty member (s) work on this Phase II project in a role other than PI, e.g., consultant.
    d. **15%** Graduate students worked on this Phase II project.
    e. **16%** University/College facilities and/or equipment were used on this Phase II project.
    f. **5%** The technology for this project was licensed from a University or College.
    g. **6%** The technology for this project was originally developed at a University or College by one of the participants in this Phase II project.
    h. **24%** A University or College was a subcontractor on this Phase II project.

In remarks enter the name of the University or College that is referred to in any blocks that are checked above. If more than one institution is referred to, briefly indicate the name and role of each.

32. Did commercialization of the results of your SBIR award require FDA approval? Yes **20%**

    In what stage of the approval process are you for commercializing this SBIR award?
    a. **1.0%**   Applied for approval
    b. **0.6%**   Review ongoing
    c. **7.7%**   Approved
    d. **1.3%**   Not approved
    e. **3.2%**   IND: Clinical trials
    f. **6.4%**   Other

NRC Firm Survey Results

NOTE: ALL RESULTS APPEAR IN BOLD. RESULTS ARE REPORTED FOR ALL 5 AGENCIES (DOD, NIH, NSF, DOE, AND NASA).

1,239 firms began the survey. 1,149 completed through question 14. 1,108 completed all questions.

If your firm is registered in the DoD SBIR/STTR Submission Web site, the information filled in below is based on your latest update as of September 2004 on that site. Since you may have entered this information many months ago, you may edit this information to make it correct. In conjunction with that information, the following additional information will help us understand how the SBIR program is contributing to the formation of new small businesses active in federal R&D and how they impact the economy. Questions A-G are autofilled from Firm database, when available.

A. Company Name: _____
B. Street Address: _____
C. City: _____ State: ____ Zip: _____
D. Company Point of Contact: _____
E. Company Point of Contact Email: _____
F. Company Point of Contact Phone: (___) ___ - ____ Ext: _____
G. The year your company was founded: _____

1. Was your company founded because of the SBIR Program?
   a. **79%** No
   b. **8%** Yes
   c. **13%** Yes, In part

2. Information on company founders. *Please enter zeros or the correct number in each pair of blocks.*
   a. Number of founders.
   5% unknown
   40% 1
   30% 2
   13% 3
   8% 4
   2% 5
   2% >5
   Average = 2 founders/firm

b. Number of other companies started by one or more
   of the founders.

   5% unknown
   46% started no other firms
   23% started 1 other firm
   13% started 2 other firms
   7% started 3 other firms
   3% started 4 other firms
   3% started 5 or more other firms

   Average number of other firms founded is one.

c. Number of founders who have a business background.

   5% Unknown
   50% No founder known to have business background
   30% One founder with business background
   14% More than one founder with business background

d. Number of founders who have an academic background

   5% Unknown
   29% No founder known to have academic background
   38% One founder with academic background
   28% More than one founder with academic background

3. What was the most recent employment of the company founders prior to
   founding this company? *Select all that apply.* **Total >100% since many
   companies had more than one founder.**

   a. **65%** Other private company
   b. **36%** College or University
   c. **9%** Government
   d. **10%** Other

4. How many SBIR and/or STTR awards has your firm received from the Federal Government?

   a. Phase I: _____ **Average number of Phase I reported was 14.**

      13% **1 Phase I**
      34% **2 to 5 Phase I**
      24% **6 to 10 Phase I**
      14% **11 to 20 Phase I**
      11% **21 to 50 Phase I**
      3% **51 to 100 Phase I**
      2% **>100 Phase I Five firms reported >300 Phase I**

What year did you receive your first Phase I Award? _____
- **3%** **reported 1983 or sooner.**
- **33%** **reported 1984 to 1992.**
- **40%** **reported 1993 to 1997.**
- **24%** **reported 1998 or later.**

b. Phase II: _____ **Average number of Phase II reported was 7**
- **27%** **1 Phase II**
- **44%** **2 to 5 Phase II**
- **15%** **6 to 10 Phase II**
- **8%** **11 to 20 Phase II**
- **5%** **21 to 50 Phase II**
- **1%** **>50 Phase II Four firms reported >100 Phase II**

What year did you receive your first Phase II Award? _____
- **3%** **reported 1983 or sooner.**
- **22%** **reported 1984 to 1992.**
- **35%** **reported 1993 to 1997.**
- **41%** **reported 1998 or later.**

5. What percentage of your company's growth would you attribute to the SBIR program after receiving its first SBIR award?
   a. **31%** Less than 25%
   b. **25%** 25% to 50%
   c. **20%** 51% to 75%
   d. **24%** More than 75%

6. Number of company employees (including all affiliates):
   a. At the time of your company's first Phase II Award: _____
   - **56%** **5 or less**
   - **28%** **6 to 20**
   - **9%** **21 to 50**
   - **8%** **> 50 Fourteen firms 1.3% had greater than 200 employees at time of first Phase.**

   b. Currently: _____
   - **29%** **5 or less**
   - **37%** **6 to 20**
   - **17%** **21 to 50**
   - **13%** **51 to 200**
   - **5%** **> 200 Eleven firms report over 500 current employees.**

7. What Percentage of your Total R&D Effort (Man-hours of Scientists and Engineers) was devoted to SBIR activities during the most recent fiscal year?___%
   22% 0% of R&D was SBIR during most recent fiscal year.
   16% 1% to 10% of R&D was SBIR during most recent fiscal year.
   11% 11% to 25% of R&D was SBIR during most recent fiscal year.
   18% 26% to 50% of R&D was SBIR during most recent fiscal year.
   14% 51% to 75% of R&D was SBIR during most recent fiscal year.
   19% >75% of R&D was SBIR during most recent fiscal year.

8. What was your company's total revenue for the last fiscal year?
   a. **10%** <$100,000
   b. **18%** $100,000-$499,999
   c. **16%** $500,000-$999,999
   d. **33%** $1,000,000-$4,999,999
   e. **14%** $5,000,000-$19,999,999
   f. **6%** $20,000,000-$99,999,999
   g. **1%** $100,000,000+
   h. **0.4%** Proprietary information

9. What percentage of your company's revenues during its last fiscal year is federal SBIR and/or STTR funding (Phase I and/or Phase II)? _____
   30% 0% of revenue was SBIR (Phase I or II) during most recent fiscal year.
   17% 1% to 10% of revenue was SBIR (Phase I or II) during most recent fiscal year.
   11% 11% to 25% of revenue was SBIR (Phase I or II) during most recent fiscal year.
   13% 26% to 50% of revenue was SBIR (Phase I or II) during most recent fiscal year.
   13% 51% to 75% of revenue was SBIR (Phase I or II) during most recent fiscal year.
   13% 76% to 99% of revenue was SBIR (Phase I or II) during most recent fiscal year.
   4% 100% of revenue was SBIR (Phase I or II) during most recent fiscal year.

10. **This question eliminated from the survey as redundant.**

11. Which, if any, of the following has your company experienced as a result of the SBIR Program? *Select all that apply.*

   a. **Fifteen** firms made an initial public stock offering in calendar year

   Seven reported prior to 2000; two in 2000; four in 2004; and one in both 2006 and 2007

   b. **Six** planned an initial public stock offering for 2005/2006.

   c. **14%** Established one or more spin-off companies.

   How many spin-off companies?
   242 Spin-off companies were formed.

   d. **84%** reported None of the above.

12. How many patents have resulted, at least in part, from your company's SBIR and/or STTR awards?
   43%   **reported no patents resulting from SBIR/STTR.**
   16%   **reported one patent resulting from SBIR/STTR.**
   27%   **reported 2 to 5 patents resulting from SBIR/STTR.**
   13%   **reported 6 to 25 patents resulting from SBIR/STTR.**
     1%   **reported >25 patents resulting from SBIR/STTR.**

   A total of over 3,350 patents were reported; an average of almost 3 per firm

---

The remaining questions address how market analysis and sales of the commercial results of SBIR are accomplished at your company.

---

13. This company normally first determines the potential commercial market for an SBIR product, process or service
   a. **66%** Prior to submitting the Phase I proposal
   b. **21%** Prior to submitting the Phase II proposal
   c. **9%** During Phase II
   d. **3%** After Phase II

14. Market research/analysis at this company is accomplished by: *(Select all that apply.)*
    a. **28%**   The Director of Marketing or similar corporate position
    b.   **7%**   One or more employees as their primary job
    c. **41%**   One or more employees as an additional duty
    d. **23%**   Consultants
    e. **53%**   The Principal Investigator
    f. **67%**   The company President or CEO
    g.   **1%**   None of the above

15. Sales of the product(s), process(es) or service(s) that result from commercialising an SBIR award at this company are accomplished by: *Select all that apply.*
    a. **35%**   An in house sales force
    b. **52%**   Corporate officers
    c. **30%**   Other employees
    d. **30%**   Independent distributors or other company(ies) with which we have marketing alliances
    e. **26%**   Other company(ies), which incorporate our product into their own.
    f.   **9%**   Spinoff company(ies)
    g. **26%**   Licensing to another company
    h. **11%**   None of the above

Appendix C

NRC Phase I Survey

SURVEY DESCRIPTION

This section describes a survey of Phase I SBIR awards over the period 1992-2001. The intent of the survey was to obtain information on those which did not proceed to Phase II, although most that did receive a Phase II were also surveyed.

Over that period the five agencies (DoD, DoE, NIH, NASA, and NSF) made 27,978 Phase I awards. Of the total number for the five agencies, 7,940 Phase I awards could be linked to one of the 11,214 Phase II awards made from 1992-2001. To avoid putting an unreasonable burden on the firms which had many awards, we identified all firms which had over ten Phase I awards that apparently had not received a Phase II. For those firms we did not survey any Phase I awards that also received a Phase II. This amounted to 1,679 Phase Is that were not surveyed.

We chose to survey the Principal Investigator (PI) rather than the firm both to reduce the number of surveys that any person would have to complete, and because if the Phase I had not gone on to a Phase II, the PI was more likely to have any memory of it than would the firm officials. There were no PI email addresses for 5,030 Phase I awards, a fact that reduced the number of surveys sent since the survey was conduced by email.

Thus there were 21,269 surveys (27,978 minus 1,679 minus 5,030 = 21,269) emailed to 9,184 Principal Investigators. Many PIs had received multiple Phase I awards. Of these surveys, 6,770 were bounced (undeliverable) email. This left possible responses of 14,499. Of these, there were 2,746 responses received. The responses received represented 9.8 percent of all Phase I awards for the five-agencies, or 12.9 percent of all surveys emailed, and 18.9 percent of all possible responses.

The agency breakdown, including NRC Phase I Survey results, is given in Table App-C-1.

TABLE App-C-1 Agency Breakdown for NRC Phase I Survey

Phase I Project Surveys By Agency	Number of Phase I Awards, 1992-2001	Answered Survey (Number)	Answered Survey (%)
DoD	13,103	1,198	9
DoE	2,005	281	14
NASA	3,363	303	9
NIH	7,049	716	10
NSF	2,458	248	10
TOTAL	27,978	2,746	10

SURVEY PREFACE

This survey is an important part of a major study commissioned by the U.S. Congress to review the SBIR program as it is operated at various federal agencies. The assessment, by the National Research Council (NRC), seeks to determine both the extent to which the SBIR programs meet their mandated objectives, and to investigate ways in which the programs could be improved. Over 1,200 firms have participated earlier this year in extensive survey efforts related to firm dynamics and Phase II awards. This survey attempts to determine the impact of Phase I awards that do not go on to Phase II. We need your help in this assessment. We believe that you were the PI on the listed Phase I.

We anticipate that the survey will take about 5-10 minutes of your time. If this Phase I resulted in a Phase II, this survey has only three questions; if there was not a Phase II; there are 14 questions. Where $ figures are requested (sales or funding), please give your best estimate. Responses will be aggregated for statistical analysis and not attributed to the responding firm/PI, without the subsequent explicit permission of the firm.

Since you have been the PI on more than one Phase I from 1992 to 2001, you will receive additional surveys. These are not duplicates. Please complete as many surveys for those Phase I that did not result in a Phase II as you deem to be reasonable.

Further information on the study can be found at <*http://www7.national academies.org/sbir*>. BRTRC, Inc., is administering this survey for the NRC. If you need assistance in completing the survey, call 877-270-5392. If you have questions about the assessment more broadly, please contact Dr. Charles Wessner, Study Director, NRC.

Project Information
Proposal Title:
Agency:
Firm Name:
Phase I Contract / Grant Number:

NRC PHASE I SURVEY RESULTS

NOTE: RESULTS APPEAR IN BOLD. RESULTS ARE REPORTED FOR ALL 5 AGENCIES (DoD, NIH, NSF, DoE, AND NASA). EXPLANATORY NOTES ARE IN TYPEWRITER FONT.

2,746 responded to the survey. Of these 1,380 received the follow on Phase II. 1,366 received only a Phase I.

1. Did you receive assistance in preparation for this Phase I proposal?

Phase I only			**Received Phase II**	
95%	No	Skip to Question 3	**93%**	No
5%	Yes	Go to Question 2	**7%**	Yes

2. If you received assistance in preparation for this Phase I proposal, put an X in the first column for any sources that assisted and in the second column for the most useful source of assistance. Check all that apply. Answered by 74 Phase I only and 91 Phase II who received assistance.

	Phase I only Assisted/Most Useful	**Received Phase II** Assisted/Most Useful
State agency provided assistance	**10/3**	**11/10**
Mentor company provided assistance	**15/9**	**21/15**
University provided assistance	**31/17**	**34/22**
Federal agency SBIR program managers or technical representatives provided assistance	**16/8**	**25/19**

3. Did you receive a Phase II award as a sequential direct follow on to this Phase I award? (If yes, please check yes. Your survey would have been automatically submitted with the HTML format. Using this Word format, you are done after answering this question. Please email this as an attachment to *jcahill@brtrc.com*, or fax to Joe Cahill 703-204-9447. Thank you for you participation.) 2,746 responses

     50% No. We did not receive a follow on Phase II after this Phase I.
     50% Yes. We did receive the follow on Phase II after this Phase I.

4. Which statement correctly describes why you did not receive the Phase II award after completion of your Phase I effort. (Select best answer) All questions which follow were answered by those 1,366 who did not receive the follow on Phase II. % based on 1,366 responses.

    33% The company did not apply for a Phase II. Go to question 5.
    63% The company applied, but was not selected for a Phase II. Skip to question 6.
     1% The company was selected for a Phase II, but negotiations with the government failed to result in a grant or contract. Skip to question 6.
     3% Did not respond to question 4.

5. The company did not apply for a Phase II because: Select all that apply. % based on 446 who answered "The company did not apply for a Phase II" in question 4.

    38% Phase I did not demonstrate sufficient technical promise.
    11% Phase II was not expected to have sufficient commercial promise.
     6% The research goals were met by Phase I. No Phase II was required.
    34% The agency did not invite a Phase II proposal.
     3% Preparation of a Phase II proposal was considered too difficult to be cost effective.
     1% The company did not want to undergo the audit process.
     8% The company shifted priorities.
     5% The PI was no longer available.
     6% The government indicated it was not interested in a Phase II.
    13% Other—explain:

6. Did this Phase I produce a non-commercial benefit? Check all responses that apply. % based on 1,366.

    59% The awarding agency obtained useful information.
    83% The firm improved its knowledge of this technology.
    27% The firm hired or retained one or more valuable employees.
    17% The public directly benefited or will benefit from the results of this Phase I. (Briefly explain benefit.)
    13% This Phase I was essential to founding the firm or to keeping the firm in business.
     8% No

7. Although no Phase II was awarded, did your company continue to pursue the technology examined in this Phase I? Select all that apply. % based on 1,366.

46% The company did not pursue this effort further.

22% The company received at least one subsequent Phase I SBIR award in this technology.

14% Although the company did not receive the direct follow on Phase II to the this Phase I, the company did receive at least one other subsequent Phase II SBIR award in this technology.

12% The company received subsequent federal non-SBIR contracts or grants in this technology.

9% The company commercialized the technology from this Phase I.

2% The company licensed or sold their rights in the technology developed in this Phase I.

16% The company pursued the technology after Phase I, but it did not result in subsequent grants, contracts, licensing or sales.

Part II. Commercialization

8. How did you, or do you, expect to commercialize your SBIR award? (Select all that apply) % based on 1,366.

33% No commercial product, process, or service was/is planned.

16% As software

32% As hardware (final product component or intermediate hardware product)

20% As process technology

11% As new or improved service capability

15% As a research tool

4% As a drug or biologic

3% As educational materials

9. Has your company had any actual sales of products, processes, services or other sales incorporating the technology developed during this Phase I? (Select all that apply.) % based on 1,366.

5% Although there are no sales to date, the outcome of this Phase I is in use by the intended target population.

65% No sales to date, nor are sales expected. Go to question 11.

15% No sales to date, but sales are expected. Go to question 11.

9% Sales of product(s)

1% Sales of process(es)

6% Sales of services(s)
2% Other sales (e.g., rights to technology, sale of spin of company, etc.)
2% Licensing fees

10. For you company and/or your licensee(s), when did the first sale occur, and what is the approximate amount of total sales resulting from the technology developed during this Phase I? If other SBIR awards contributed to the ultimate commercial outcome, estimate only the share of total sales appropriate to this Phase I project. (Enter the requested information for your company in the first column and, if applicable and if known, for your licensee(s) in the second column. Enter dollars. If none, enter 0 (zero), leave blank if unknown.)

	Your Company	Licensee(s)
a. Year when first sale occurred	**89 of 147 after 1999**	**11 of 13 after 1999**
b. Total Sales Dollars of Product(s) Process(es) or Service(s) to date		
(Sale Averages)	**$84,735**	**$3,947**
Top 5 Sales	1. **$20,000,000**	
Accounts for 43% of all sales	2. **$15,000,000**	
	3. **$5,600,000**	
	4. **$5,000,000**	
	5. **$4,200,000**	
c. Other Total Sales Dollars (e.g., Rights to technology, Sale of spin off company, etc.) to date		
(Sale Averages)	**$1,878**	**$0**

Sale averages determined by dividing totals by 1,366 responders.

11. If applicable, please give the number of patents, copyrights, trademarks and/ or scientific publications for the technology developed as a result of Phase I. (Enter numbers. If none, enter 0 (zero); leave blank if unknown.)

Applied For or Submitted / # Received/Published

319 / 251	Patent(s)
50 / 42	Copyright(s)
52 / 47	Trademark(s)
521 / 472	Scientific Publication(s)

12. In your opinion, in the absence of this Phase I award, would your company have undertaken this Phase I research? (Select only one lettered response. If you select c, and the research, absent the SBIR award, would have been different in scope or duration, check all appopriate boxes.) Unless otherwise stated, % are based on 1,366.

5%   Definitely yes
7%   Probably yes, similiar scope and duration
16%   Probably yes, but the research would have been different in the following way

    % based on 218 who responded probably yes, but research would have . . .

    75%   Reduced scope
    4%   Increased scope
    21%   No Response to scope
    5%   Faster completion
    51%   Slower completion
    44%   No Response to completion rate
14%   Uncertain
40%   Probably not
16%   Definitely not
4%   No Response to question 12

Part III. Funding and other assistance

Commercialization of the results of an SBIR project normally requires additional developmental funding. Questions 13 and 14 address additional funding. Additional developmental funds include non-SBIR funds from federal or private sector sources, or from your own company, used for further development and/or commercialization of the technology developed during this Phase I project.

13. Have you received or invested any additional developmental funding in this Phase I? % based on 1,366.

    25% Yes. Go to question 14.
    72% No. Skip question 14 and submit the survey.
     3% No response to question 13.

14. To date, what has been the approximate total additional developmental funding for the technology developed during this Phase I? (Enter numbers. If none, enter 0 (zero; leave blank if unknown).

Source	# Reporting that source	Developmental Funding (Average Funding)
a. Non-SBIR federal funds	**79**	**$72,697**
b. Private Investment		
(1) U.S. Venture Capital	**13**	**$4,114**
(2) Foreign investment	**8**	**$4,288**
(3) Other Private equity	**20**	**$7,605**
(4) Other domestic private company	**39**	**$8,522**
c. Other sources		
(1) State or local governments	**20**	**$1,672**
(2) College or Universitie	**6**	**$293**
d. Your own company (Including money you have borrowed)	**149**	**$21,548**
e. Personal funds of company owners	**54**	**$4,955**

Average Funding determined by dividing totals by 1,366 responders.

Appendix D

Case Studies

Advanced Brain Research, Inc.[1]

Robin Gaster
North Atlantic Research

EXECUTIVE SUMMARY

ABM is a small company whose research has been funded almost entirely by a series of successful SBIR awards. Currently, ABM is poised to enter Phase III, and is seeking the funding needed to do so successfully.

The company was founded on SBIR awards in 1997, and expanded based on Phase II awards in 1999. It received additional SBIR awards in 2002, and some additional funding from DARPA, during the development of two complementary products: home sleep diagnosis products, and an initial sleep disorder screening product for use in office or other settings.

ABM has received six Phase II NIH awards, and seven Phase I NIH awards, and has been supported almost entirely by $6.3 million in SBIR awards and $700,000 from DARPA.

Primary Outcomes:

- One product with FDA clearance and a second that has been submitted for clearance, both entering Phase III.
- Six patents.
- Publications.
- Additional employment.
- Partnerships: Possible pilot program with Waste Management, Inc.

Key SBIR issues:

- Failure of Fast Track.
- Better program manager accountability.
- Commercialization/Phase III support.
- Commercialization review.
- Review quality and oversight.

Key recommendations:

- Optional training program for reviewers.

---

[1] Interview: In Carlsbad, CA, at Advanced Brain Monitoring, Inc., with Daniel Lebedowski, Chief Scientific Officer, and Chris Berkas, CEO. Both are co-founders.

- Accelerate shift to electronic submissions. Consider using DoD submission system.
- Improved program manager assessment using report cards during the Final Report and/or Edison submission processes.
- Review. Improve commercialization reviews, possibly by instituting two-phase screening system.
- Phase III. Improve electronic matchmaking by improving online tools at NIH Web site.

BACKGROUND

Advanced Brain Monitoring, Inc., was founded in January 1997 to create low cost, easy-to-use, portable systems to monitor and interpret physiological signals indicating brain activity, and has developed patented data acquisition technology with automated analysis software to measure the brain's electrical activity (EEG), oxygen levels in the blood and cardiac activity.

ABS used a Phase I award as a founding grant. It opened in 1997 with two full-time and two part-time employees. Phase I awards took the company to January 1999, when it received three Phase II awards. This allowed all three founders to go full time, funded the company's move to Carlsbad, and paid for three EEG technicians who were hired in June 1999.

The founders have invested about $400,000 on the company, funding primarily used for FDA 510k filings and patent filings, which cannot be delayed while more funding is found. Overall, the company has received more than $6 million

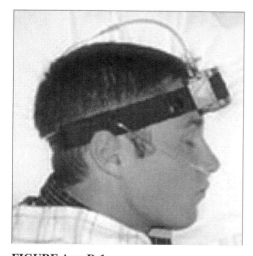

FIGURE App-D-1

SOURCE: Advanced Brain Research.

from NIH in SBIR awards and an additional $700,000 from DARPA under the Augmented Cognition program. ABM has worked with Honeywell and Lockheed in the context of its DARPA-sponsored research.

All current awards will end in March 2005. Company is currently seeking ongoing capital for product rollout.

PRODUCTS

ABM is currently focused entirely on bringing products to market. It has two products that are ready for pilot sales:

(1) The Apnea Risk Evaluation System (ARES™) integrates physiological data acquired in-home with clinical history and anthropomorphic data to quantify level of risk for Obstructive Sleep Apnea (OSA). ARES has three components:

- **ARES Unicorder:** a battery powered, self-applied, single site (forehead) physiological recorder that acquires and stores nocturnal data for use in the diagnosis of OSA.
- **ARES Questionnaire (ARES Q):** designed to assess pre-existing risk factors for OSA, including age, gender, body mass index (BMI), neck circumference, daytime drowsiness, frequency and intensity of snoring, observed apneas, and history of hypertension, diabetes and cardiovascular disease.
- **ARES Insight Software:** automated software to recognize and quantify abnormal respiratory events.

The ARES received FDA clearance in October 2004, and its CE mark in February 2005. It must be ordered by a prescription.

ABM sells the AREA system through two channels:

- **Directly to primary care physicians and industrial customers (employers) (as prescribed by a physician).**
- Licensed to larger users. This service includes the technology and training for user staff, and is designed for larger facilities such as hospitals or other bulk purchasers.

(2) Alertness and Memory Profiling System (AMP™). The AMP simultaneously acquires data on brain function and cognitive performance during vigilance, attention and memory tests. Its components can be used together or separately:

- The patented **Sensor Headset** addresses many of the technical concerns with EEG recordings, including ease of use, comfort, cosmetic acceptability for the workplace, and high quality data acquisition in challenging environments.

FIGURE App-D-2

SOURCE: Advanced Brain Research.

- **B-Alert® Software.** The patented B-Alert software identifies and decontaminates artifacts, monitors changes in the EEG on a second-by-second basis, and classifies each second of brain activity on a continuum from highly vigilant to sleep onset.
- **Neurocognitive test battery.** A battery of vigilance, attention, and memory tests that assess and quantify alertness and memory.

The Sensor Headset has been submitted for FDA clearance in March 2005, and it received its CE mark in February 2005. The medical application must be ordered by a prescription. There are numerous nonmedical applications for the EEG system.

MARKETS

ABM is addressing two markets:

- The traditional market for sleep diagnostics, where its lower cost and easier to use system has competitive advantages.
- New industrial markets for undiagnosed OSA, where companies need better knowledge about employees operating critical equipment.

According to NHLBI, approximately 20 million (6.6 percent) Americans who suffer from OSA, approximately 90 percent are currently undiagnosed.[2] The general market is therefore substantial. More specifically, companies whose employees operate critical machinery—e.g., trucks, air traffic controls, trains, etc.—are a very likely market.

---

[2]National Sleep Foundation.

ABM faces some significant challenges in marketing its products, even though they address important problems. The ARES system is essentially designed to replace current sleep diagnosis procedures, substituting inexpensive and relatively convenient home diagnosis for expensive and inconvenient sleep studies currently performed in hospitals.

Existing sleep diagnosis labs—potentially a major source of customers—are firmly opposed to in-home studies because it will reduce their own income. Insurance reimbursement for in-home unattended studies is inconsistent. Managed care groups reimburse. The PPOs follow CMS' lead and either don't reimburse or at a very low rate. CMS had a review of in-home unattended studies, and—according to ABM—after substantial lobbying of the sleep labs, chose not to categorically reimburse for these studies.

ABM is in discussions with two sleep labs to establish pilot projects that augment rather than cannibalize the sleep labs revenue. ABM has a meeting scheduled with CMS at the end of April to present the results of its study that was funded by NIH (the largest study of its kind for in-home unattended studies).

The AMP system also faces substantial marketing challenges. ABM has established a relationship with Waste Management, Inc., one the country's largest employers of commercial truck drivers. The pilot—which was to be implemented using a Fast Track since rejected by NIH—involved using the ARES and AMP on Waste Management drivers to 1) determine the level of undiagnosed OSA, and 2) develop a model for incorporating sleep apnea screening into the biannual fitness for duty physicals. The rejected application defunded the pilot, and ABM is now seeking other mechanisms to implement this program. More generally, addressing the problem of undiagnosed sleep apnea potentially opens companies such as Waste Management to significant liability issues. This problem has not yet been resolved.

Despite these difficulties, it is clear that ABM has successfully completed the initial research phase for two complementary products, and is now entering Phase III with both. Its current emphasis is acquiring the funding necessary to implement its marketing strategy.

PATENTS

The company has been awarded 6 patents, funded primarily from founder's investment and the 7 percent fixed fee received from SBIR awards. All the patents are based on work developed under the NIH SBIR program.

REGULATORY APPROVAL

Both of the company's products have received the FDA CE mark after completing FDA clinical trials.

PROGRAM MANAGEMENT
DIFFERENT ICS

ABM has had dramatically different experiences at different ICs, which it believes are entirely due to the capabilities and approaches of the different program managers. ABM has had a very positive experience with one program manager, but had problems with another who they believe has been, at best, unsupportive, and does not provide the support that reflects NIH guidelines on collaboration between program managers and companies. Short of changing its products and research goals, ABM has not found a way around this program manager, and no way to generate improvement.

ABM's experience highlights the problem of using program managers as gatekeepers without any tools in place to monitor their effectiveness, or in some cases apparently to train them in relation to new programs.

FAST TRACK

ABM was encouraged by presentations made by Jo Anne Goodnight and started submitting Fast Track applications almost from the start of the program, but has had very mixed experiences at best:

(1) Fast Track Application 1. The application received a very high quality review, which recommended splitting the application into Phase I and Phase II. ABM agreed and did so, receiving first a Phase I and then $1.2 million for Phase II, where ABM noted the extensive help from the relevant program manager in preparing a justification for the extra-sized funding.

(2) Fast Track 2. This award ran into major administrative problems. The Fast Track was approved in March 2003. The Phase I work was completed in August and a "streamlined noncompeting award process" (SNAP) report was submitted (a short version report designed for projects that are not subject to further competition). This is standard procedure for a Fast Track award and was provided by the program manager in his/her instructions to ABM. However, several problems developed:

- The total amount of the award was reduced by 5 percent by the review committee because of their opinion that a key consultant was not needed. After discussion with the program manager, the company submitted justification for the payment but the program manager said the review committee's suggestion was final. If the company needed to pay the consultant, they would have to rebudget form other areas.
- Even though the program is designed to avoid a gap in funding between Phase I-Phase II, review of the Phase I report was delayed until after October because the Institute needed the new fiscal year to begin in order to have funds for Phase II.

- According to ABM, the program manager and the Institute conducted an internal review of the Phase I and turned down the Phase II award due to insufficient detail on what was accomplished in Phase I. (The investigators could easily have written a full Phase I final report but instead provided the amount of information required by the SNAP submission as instructed.) This notification occurred in November, approximately 2.5 months after ABM had notified their program manager that they began the Phase II work that that the pre-award authorization would be used to re-capture the funds. The program manager felt that was appropriate because at the time the only delay was due to the new fiscal year. The company wanted to push forward toward commercialization and since the award was noncompetitive and because the company had met its Phase I goals, there was no reason to expect this financial commitment might jeopardize the company's future.
- After much negotiations with the NIH program coordinator (which included reviewing with the program coordinator that he/she provided instructions to the company to submit the SNAP, preparation of a full Phase I report, and subsequent re-review), this error was eventually reversed. Because the company had to stop work in November, approximately 12 of the subjects being studied had to be dropped and there was a gap in funding from August when the Phase I ended until the following February.
- Although the funding was delayed and it interrupted some of the studies, there was no compromise on the part of the program officer about the number of subjects and other research issues. The net result was money was allocated in a manner that reduced the benefits of the large study and reduced the power of the data needed for commercialization.

(3) Fast Track 3. An application to take the technology developed during earlier SBIR awards and apply it in to the needs of the trucking industry. An agreement for a pilot implementation program was made with Waste Management, Inc., one of the largest operators of commercial trucks.

- An initial score of 320 meant substantial revisions were needed.
- ABM resubmitted and was awarded a priority score of 274. Key criticisms included some scientific objections, privacy concerns, issues to do with drivers (social issues), and the lack of women in the study. To address the concern of inadequate female representation, the company had to rewrite the proposal to impose enormous potential costs on ABM including test sites right across the country to increase the number of women in the study. The percentage of female drivers at Waste Management is less that 2 percent of 35,000 drivers. This stringent guideline applied to this unique situation was, in the company's view, mindless adherence to new

guidelines designed to ensure that projects are not based on male-only research (guidelines which ABM supports in general).

- ABM resubmitted the application a third time, but in a new year and with an entirely new panel. This time ABM's review was so poor, it did not receive a priority score at all. One of the lead reviewers simply said that he did not believe that sleep apnea was a widespread medical problem. Because this was the third submission of the application, ABM was forced to give up on this SBIR application.

The lessons from this experience seem to be that the Fast Track application is not very well implemented, or at minimum people were not trained prior to implementation. ABM endorses the concept of the Fast Track program. Given the likelihood of obtaining a Fast Track award vs. Phase I and II, the fact that the Phase II dollars are not set aside at the beginning, and misunderstandings about the Fast Track, the company has decided to avoid this program in the future.

REVIEW PROCESS

ABM identified some substantial problems in the review process. The company has noted apparent changes at NIH in how priority scores are calculated, and in the nature of reviewers—notably a pronounced shift toward quasi-commercial concerns. Specifically—

- Beginning in 2003, the company noticed that reviewer comments ("pink sheets") no longer tracked closely with the scores.
- ABM believes that in recent panels, business people may have been over-influencing panel reviews, even when they are not the primary reviewer. The impact of business-based reviews may help to explain the apparent disconnect betweens cores (generated form the panel as whole) and pink sheets (generated primarily from lead reviewers).
- Study sections often suffer from substantial confusion between the functions and objectives of RO1s and R44s (SBIR awards). Section members who are used to reviewing RO1s are often not prepared for the application-heavy focus of ABM's applications.
- Reviewers are sometimes not properly briefed. In one case, for example, a Phase I proposal was sharply criticized for not having a commercialization plan—even though no such plan is required for Phase I.
- Lead reviewers are sometimes not properly monitored. There appears to be no process for assessing major biases (e.g., the second resubmission on the pilot study).
- Panel memberships. Letters seeking to affect participants in study sections do not work. ABM knows that in one case it explicitly asked for specific

reviewers to be excluded for conflict of interest—and two of those reviewers was the lead reviewer for their application.
- In a recent review, of both RO1s and R44s, the Committee gave ABM the third highest priority score of 270. The best score was less than 200 and the second highest score was between 200 and 270, both R01s. ABM had the highest R44 score. Over 65 percent of the grants received no priority score.

COMMERCIALIZATION TRAINING

ABM has been a long-time participant in the San Diego Regional Technology Alliance (SDRTA), and is now participating in the NIH commercialization program operated by LARTA. Initial events were not especially helpful, but ABM will be participating in a major technology showcase organized by LARTA in May 2005, for which it has substantial expectations.

LARTA is currently funding a few hours a month from three business consultants, all of whom are viewed fairly positively by ABM, and they have provided some useful market research as well as a contact with Innovex, which provides turn-key national sales forces to sell to physicians, although none has yet provided a real potential partner—which is their primary assigned role.

ABM has also presented posters at the NIH annual conference twice, but in neither case did any business connections result.

PHASE III

SBIR does not permit use of funds for marketing or market research, which makes the transition to Phase III very difficult. ABM did receive CAL-TIP (state) funding of $175,000, which the company said was crucial for the market research necessary to get toward product launch.

AWARD FUNDING LEVELS

ABM's experience is that applications for more than $1 million get reduced during review.

PROGRAM MANAGEMENT

ABM believes that funding can be delayed when submitting in the April funding cycle: This inevitably means getting caught up in delays in the review process due to summer vacations and the end-of-fiscal year problems at NIH. From a standpoint of counting on an SBIR grant to meet payroll, delay of funding until October can be a significant disruption to a small company that is reliant on the SBIR program as a primary funding source.

However, this contradicts points made in interviews at other companies, who noted that while funding is delayed to October, it does become available as soon as the appropriation is passed, in contrast to funding allocated toward the end of the fiscal year where there may be a liquidity crunch.

SBIR AND VENTURE CAPITAL

ABM has experienced mixed reviews of its SBIR awards from venture capitalists. Some write it off, others view the peer review process as a prohibitive indication of research quality. Receiving more than $6 million in funding from NIH gives ABM immediate legitimacy in discussions with funders, although VCs always discount this funding in the course of valuation.

RECOMMENDATIONS

- Training program for reviewers (e.g., one-day, on a regional basis). This would not only encourage a more standardized approach, perhaps based on a standard curriculum. It could also encourage some potential participants who might otherwise feel unqualified to become reviewers (e.g., Mr. Levendowski, an MBA with scientific training).
- Accelerate shift to electronic submissions. ABM is very favorably impressed by the DoD electronic submission process, in comparison to NIH.
- Improved program manager assessment. ABM felt strongly that final reports and/or Edison submissions should include a report card for the program manager concerned, and that NIH should have review processes in place to improve or eliminate underperforming managers.
- Review. Commercialization reviews are a problem.
  - ○ ABM suggested that an online questionnaire might help companies answer key commercialization questions, and would also highlight obvious problem areas.
  - ○ ABM supported two phase reviews, with an initial screening by study sections focused entirely on science, and a second level screening of commercialization plans for Phase II. Problems at the second level could then be fixed within a single funding cycle, or applicants could be asked to resubmit for commercialization review only, substantially shortening the entire application process for many awards while improving quality and eliminating many of the current problems with commercialization review.
- Commercialization. NIH could do much more electronic matchmaking. Recommended in particular that NIH implement technology that would permit companies to update their own listings and identify information that is available for review (e.g., business plans, results from Phase I or II, patent applications, etc). Current listings are usually out of date and hence not used much by potential partners.

ADVANCED BRAIN RESEARCH—ANNEX

TABLE App-D-1 Advanced Brain Research NIH SBIR Awards-I

Fiscal Year	Phase Type	Award Size ($)	Project Title	Funding Institute-Center
1996	Phase I	99,980	Ambulatory, battery powered, physiological recording	NS
1999	Phase II	543,000	Ambulatory brain monitoring device	NS
2000	Phase II	204,167	Ambulatory brain monitoring device	NS
1997	Phase I	99,940	Alertness quantification system using normative indices	NS
1999	Phase II	798,773	Alertness quantification system using normative indices	NS
2000	Phase II	276,228	Alertness quantification system using normative indices	NS
1997	Phase I	99,400	Portable self-applying drowsiness detection device	NS
1999	Phase II	365,994	Portable drowsiness monitoring device	NS
2000	Phase II	41,556	Portable drowsiness monitoring device	NS
2000	Phase II	347,762	Portable drowsiness monitoring device	NS
2001	Phase I	125,306	In-home sleep apnea risk evaluation system	HL
2002	Phase II	838,890	Validation of In-Home Sleep Apnea Risk Evaluation System	HL
2003	Phase II	318,195	Validation of In-Home Sleep Apnea Risk Evaluation System	HL
2002	Phase I	139,428	Biobehavioral Measurements of Alertness in Sleep Apnea	HL
2003	Phase II	691,925	Automated Detection of Sleep Disordered Breathing	HL
2003	Phase II	683,352	Biobehavioral Measurements of Alertness in Sleep Apnea	HL
2001	Phase I	99,991	Drowsiness Detection: Effects of Feedback Based on EEG	MH
2001	Phase I	99,994	Novel systems to evaluate sleep apnea and vigilance	HL

NOTE: For a list of codes for National Institutes of Health institutes and centers, see Box App-A-1.

SOURCE: Advanced Brain Research.

TABLE App-D-2 Advanced Brain Research NIH SBIR Awards-II

ABM Grant Name	Description	Award Size ($)	Start Date	Funding Institute-Center or Agency
ABMD—I	Solid state digital recorder	99,831	11/1/1996	NS
Drowsy—I	Quantify sleep onset	99,648	5/1/1997	NS
Alertness—I	Quantify states of alertness	99,940	3/1/1988	NS
Drowsy—II	Quantify sleep onset—large clinical study	755,312	1/1/1999	NS
ABMD—II	Wireless EEG system	747,167	5/1/1999	NS
Alertness—II	Clinical and validation studies	1,075,001	6/1/1999	NS
ARES-A—II	Prototype ARES and baseline AMP for OSA	99,994	1/1/2001	HLB
CAPTIP	Commercialization and EEG sensor production	175,000	1/1/2001	HLB
ARES-B—I	ARES Questionnaire	125,306	8/1/2001	HLB
DMD—I	Assess real-time recognition of sleep onset	99,991	8/1/2001	MH
ARES-A—II	Development of ARES & clinical studies	1,157,083	1/1/2002	HLB
AMP—I	Development of AMP	139,428	3/1/2002	HLB
AMP—II	AMP Clinical Studies	968,669	4/1/2003	HLB
ARES-B—II	Enclosure, Nasal Pressure, clinical studies	978,327	7/1/2003	HLB
	Total NIH	6,620,697		
DARPA—I	Assess workload	50,715	5/1/2002	DARPA
DARPA—II.a	Assess workload	100,000	1/1/2003	DARPA
DARPA II.b	Assess workload	250,000	1/1/2004	DARPA
	Total DoD	400,715		
	Total grants awarded	7,021,412		

NOTE: For a list of codes for National Institutes of Health institutes and centers, see Box App-A-1.

SOURCE: Advanced Brain Research.

Advanced Targeting System, Inc.[3]

Robin Gaster
North Atlantic Research

EXECUTIVE SUMMARY

Background

ATS is a small biotech company located in San Diego. Unusually, it has had a strong product line since inception in 1994, and currently offers more than 40 products for sale on its Web site.

The company is based on the application of targeted toxins to neuroscience, where the selective approach offered by what the company calls Molecular Neurosurgery offers obvious advantages if successful.

Initial products have been sold to other research companies, but the company is now reaching the clinical trials stage for products aimed at addressing chronic pain. The American Chronic Pain Association (ACPA) estimates that one in three Americans (approximately 50 million people) suffers from some type of chronic pain.

Future research will focus on ways of enhancing the company's current approach, so as to permit cell modification via cytotoxins, beyond the current tools which allow selective elimination of cells only.

ATS is currently seeking partnerships/funding for clinical trials of chronic pain technologies, partly through participation in the commercialization assistance program operated for NIH by LARTA.

SBIR History and Status

ATS has used SBIR since its inception. It has received two long running Phase II awards, one new Phase II in 2003, and also one of the first CCAs in 2003. ATS has received three additional Phase I awards.

Key Utilization of SBIR

ATS has funded its research primarily through SBIR awards. New funding is now being used for the toxicology/safety testing phase of FDA approval process.

---

[3]Interview: At ATS in San Diego, February 24, 2005, with Dr. Douglas Lappi, President and Chief Scientific Officer. Dr. Lappi is a co-founder of ATS.

Outcomes:

- Numerous products (more than 40).
- Current research supported by SBIR: focused on chronic pain, a major quality of life issue for one in three Americans.
- Many scientific papers.
- Two patents.
- Partnerships with major medical research centers and academics.
- "Profound addition" to knowledge in the field of chronic pain.

Key issue/concern: resolving the Phase III funding problem.

Recommendations:

- Phase III: the neuroscience-funding institutes at NIH should collectively fund a research hospital for clinical trials, similar to that funded by NCI.
- Size/duration. No additional funding for Phase I.
- Funding cycles. Eliminate the 2-year window for Phase I winners to apply for Phase II.
- Direct to Phase II. Companies should be allowed to compete directly for Phase II without previous Phase I.

BACKGROUND

ATS was founded in April 1994 by Douglas Lappi, Ph.D. and Ronald Wiley, M.D., Ph.D. (Scientific Advisor), initially for commercial development of ideas and products developed in their academic labs.

ATS is located in an R&D hub (San Diego), is not woman- or minority-owned, is small (9 employees), has been funded internally and by SBIR, has won several SBIR awards but is not a top-20 award winner, and has reached market with its products. ATS has received SBIR awards only from NIH.

FOUNDER/COMPANY HISTORY

Douglas Lappi began work in the field in the 1970s. He worked on targeted toxins focused on cancer, but could not interest previous employers in his ideas for targeting toxins on the brain. He has extensive experience in laboratory work. His two partners are Ron Wylie, Chief of Neurology at the Veteran's Administration Medical Center, and Professor of Neurology and Pharmacology at Vanderbilt University, Nashville, TN, and Denise Higgins (VP Business Development), previously at the Salk Institute with Lappi.

Wylie's key insight, according to Lappi, was that "cancer people could learn nothing from us, but we could learn a lot from them." Essentially, there were many possibilities for applying the science of targeted toxins from cancer to neurological research. The field of targeted toxins and the brain was largely ignored by mainstream research, and the company was started in 1994 with a specific focus on selling targeted agents, especially for neurological research.

Company products were immediately well received by biotech companies. In 1994, at the Society for Neuroscience annual meeting, the ATS poster and booth showed the first use of the 192-Saporin by an independent researcher Dr.Waits at Rutgers University. Lappi says that this was a "thunderbolt" in the field, as researchers had been waiting for this capability for years.

As a result, the first month of product release in 1994 generated "huge" sales, which the company surpassed on a monthly basis only recently.

ATS does not disclose revenues.

TECHNICAL FOCUS

ATS is focused on implementing known techniques in neuroscience by means of new mechanisms. Lesioning of a region by surgical means and observing the effects is a well known and widely used technique in neuroscience research and medicine. ATS aims to provide similar outcomes by application of specific cytotoxins (essentially, using chemistry instead of surgery, with—if successful—much greater specificity and control).

ATS calls the new technique Molecular Neurosurgery (MN). The first ATS MN product (192-Saporin) is now in use in laboratories world-wide.

PRODUCTS

The ATS product line includes targeted toxins, antibodies, and custom services for assisting neuroscientists in studying nervous system function, and brain-related diseases and disorders.

ATS has had products on the market from its first month of operation, and was first to market with cytotoxin research reagents, which are sold to other biotech researchers primarily in the neurosciences. This is a niche market, and as presumably of limited interest to larger companies, although ATS understands a larger company could enter the market at any time. Its original partner, Chemi-Con did compete for a while but has left the market. ATS has protected its position through two patents.

ATS currently offers a large number of products through its online catalog, which lists 20 targeted toxins, four control conjugates, six secondary conjugates, eight proteins and peptides, and four fluorescent conjugates, and more than 25 other neuroscience products, as of March 2005.

EMPLOYMENT

The company began with one full-time employee. The first Phase II award allowed ATS to hire one additional person. Currently, ATS has nine staff members.

COMPANY STRATEGY

ATS is emphatically not a pharmaceutical company. It is too small, and the company does not intend to grow rapidly into a large enough company to pursue drug development on its own.

Accordingly, ATS is seeking development partners, a process in which it has been supported by CAP and the NIMH program officer.

CURRENT AND FUTURE PROJECTS

The general research strategy is to identify a target cell type, place a bioactive molecule inside the cell, determine whether it functions, and then track the results. This is the basis for all products to date.

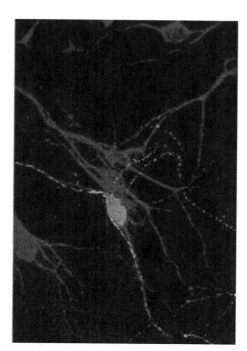

Substance P-Saporin targets delivery of a toxic compound to eliminate those nerve cells that transmit pain messages up the spinal cord to the brain.

This precise method allows chronic pain to be permanently stopped without affecting other neurons.

FIGURE App-D-3 Substance P and targeted cell death.

SOURCE: Advanced Targeting System.

Current Research

The company is working on SP-SAP—a patented chemical conjugate composed of the neuropeptide Substance P, and the ribosome-inactivating protein saporin. The project is aimed addressing the problem of chronic pain, a very difficult area to understand, with high levels of complexity and multiple areas of research, including the spinal cord and neuron receptors within the brain.

ATS research has addressed a central question in chronic pain research, namely whether chronic pain can be defined in terms of a unique pathway within the nervous system, or whether it results from some characteristic of the system as whole. ATS determined that chronic pain is related to specific and unique pathways, which could be both identified and disrupted. Lappi claims that this was an enormous breakthrough in the field of chronic pain, and that it is at a minimum a "profound addition" to the field.

Initial research in this area was followed by more work on other screens starting in 1999, developing more chronic pain models in rats. All of these were successful.

This work has potentially profound implications for millions of people who currently endure chronic pain. On the ATS Web site, Lappi notes that—

> Many people suffering from intractable chronic pain have exhausted all of their options. Their quality of life is diminished. We envision, in the not too distant future, offering a one-time injection that will end the pain. Chronic pain sufferers won't need to take a pill every day. Advanced Targeting Systems has excellent preclinical data that leads us to believe that SP-SAP will be safe and effective and compels us to develop SP-SAP for clinical use.

Researchers at UCSD have now completed preliminary toxicology studies with SP-SAP in one of the FDA-required large animal models (funded under 2001 SBIR award from NIMH), and will carry out the full toxicology studies required to address safety issues.[4]

ATS has submitted its work on chronic pain to the FDA, and its preclinical data have been accepted. Clinical trials are the next step. Toxicology/safety testing will be funded by the SBIR CCA award, and ATS is now actively seeking partnerships or venture funding for this expensive stage of development.

The FDA has advised ATS that SP-SAP may best be developed as an orphan drug for treatment of pain in patients with terminal cancer.

---

[4]Including Dr. Tony Yaksh, Professor of Anesthesiology and Pharmacology at the University of California, San Diego School of Medicine, a leading expert in spinal cord delivery of experimental agents.

Future Directions

So far, the company's lead compound, 192-Saporin, has been used to kill selective cells. ATS is now interested in seeing whether a similar delivery mechanism can be used to modify the behavior of cells, instead of just killing them.

Working with an academic—Bob Solveter—the company is working on epilepsy-related problems. Epilepsy is triggered by reduced activity/loss of inhibitor neurons in the hippocampus. This is a fairly well established hypothesis. Animal experiments have shown that the destruction of inhibitor neurons does result in status epilectucus.

The problem then is to increase the activity of inhibitor neurons. ATS is working on using its established delivery systems for inducing enhanced activity among inhibitor cells.

FUNDING

ATS was bootstrapped on the basis of its products and SBIR funding, with no outside funding and minimal investment from founders. It formed an initial partnering agreement with Chemi-Con, which both marketed the first product—192-Saporin—and provided ATS with office and laboratory space.

Subsequently, ATS rented space on favorable terms from Invitrogen, which was seeking further intellectual cross-fertilization at the time.

ATS sought venture capital funding in 2003, but the timing was too difficult. The company is now seeking funding or a corporate partner under NIH auspices via the LARTA program. It is however too soon to tell whether this will lead to anything substantive.

ATS does not see a significant halo with respect to private funding. It believes that while an SBIR award with peer review may help a company get through the door to see a venture capitalist, the latter are focused on economics, not science, and an SBIR award says little about that.

SBIR HISTORY

ATS applied for its first SBIR immediately after being founded in 1994, received a Phase I in 1995, and another in 1996. Both became long-running Phase II awards, with the second running all the way to the first round of CCAs in 2003—the fifth year of support for this project. Three of the six ATS Phase I awards have resulted in Phase IIs.

CCA

ATS received one of the first CCA awards in October 2003—providing $2.4 million over three years—designed to allow ATS to complete toxicology studies and to prepare clinical-grade material for use in human trials.

This CCA award was part of a Program Announcement at NIMH, "Competing Continuation Awards of SBIR Phase II Grants for Pharmacologic Agents and Drugs for Mental Disorders." ATS notes that "For small businesses like Advanced Targeting Systems, this latest expansion of the SBIR program provides important support at a time when alternative funding is expensive and difficult to find."

IMPACT OF SBIR

Without SBIR, ATS says that it would have followed much the same development and research trajectory, but at a much slower pace. The existence of markets for its first products would have generated the funding necessary for this lower level of effort. However, the path itself would probably have been different, especially in relation to its relationship with outside academics, who would have been much harder to fund, and whose work with ATS has clearly been critical to the company's strategy and to its success so far.

SBIR suits ATS for several reasons, according to Lappi:

- Competition is fair because only other small companies are involved.
- Many of the other applications are not all that good.
- SBIR is focused on the same goal as ATS—making a product.
- Overall, a "tremendously appealing program. SBIR has been a great help, and we appreciate it tremendously."

PUBLICATIONS

ATS staff have a long history of publications. Lappi himself has more than 70 scientific publications to his credit, and Lappi and Wiley have a book on targeted toxins due out in March 2005.[5] However, ATS prefers to focus on the publications generated by other researchers—especially academics—who are using their tools.

The ATS Web site cross references more than 65 papers for 2004 alone that used technologies sold by ATS. Lappi sees scientific publications as "the highest form of advertising."

Publications have also had an important impact on the company. In the course of its research on chronic pain, ATS submitted a Phase II application that was originally not funded with a score of 220. After publication of the first article on 192-Saporin in *Science*, the resubmitted application scored 121 (the highest score so far encountered in NRC research at NIH). Reviewers also increased the budget above that originally requested. This research has subsequently received

---

[5]Ronald G. Wiley and Douglas A. Lappi, *Molecular Neurosurgery with Targeted Toxins*, Humana Press, 2005.

several additional rounds of funding from NIH under this SBIR award, leading to the CCA award described above.

PATENTS

ATS has received patents on its first two molecules, and has several other patent applications pending. However, applications are expensive (ATS estimates $25,000 per patent), and the company is careful in selecting targets for patenting. Both current patents are based on work funded by SBIR.

UNIVERSITY PARTNERSHIPS

Partnerships appear to have a played a very important role. Academics have acted as first adopters for ATS technology. Subsequent academic/scientific publications then provide the validation necessary for market acceptance, and for further funding from NIH.

This approach is demonstrated in the context of the research on chronic pain currently under way at ATS. Initial pain model studies showing efficacy in rats were performed in the laboratories of University of Minnesota pain expert Dr. Patrick Mantyh. Results from these studies were published in *Science* in 1997,[6] providing enormous validation to the ATS approach. Mantyh's laboratory published a second *Science* article in 1999,[7] demonstrating the long-term elimination of chronic pain with SP-SAP.

Dr. Tony Yaksh, Professor of Anesthesiology and Pharmacology at the University of California, San Diego, is a leading expert on the administration and pharmacology of drugs in the spinal cord and spinal fluid. His associate, Dr. Jeff Allen, completed preliminary toxicology studies with SP-SAP in one of the FDA-required large animal models. UCSD will carry out the full toxicology studies with funding from the grant awarded to Advanced Targeting Systems.

Management Issues

Phase I-Phase II Gap. Acknowledged by ATS, Lappi notes that, "you can't run a company on SBIR awards." He does not see this as a criticism. In fact, he suggests that the gap acts as a kind of guarantee that there is a real business here: that the government is not supporting a business, but is helping an established business do product development.

---

[6]M. L. Nichols, B. J. Allen, S. D. Rogers, J. R. Ghilardi, P. Honore, J. Li, D. A. Lappi, D.A. Simone, and P. W. Mantyh, "Transmission of chronic nociception by spinal neurons expressing the substance P receptor," *Science*, 286:1558-1561.

[7]P. W. Mantyh, S. Rogers, P. Honore, B. Allen, J. R. Ghilardi, J. Li, R. S. Daughters, S. R. Vigna, D. A. Lappi, R. G. Wiley, D. A. Simone, "Inhibition of hyperalgesia by ablation of lamina I spinal neurons expressing the substance P receptor," *Science*, 278:275-279.

Differences between ICs.   ATS sees significant differences, but based more on scientific prejudices than program management. ATS simply cannot get funded at NCI, despite its strong track record at NIMH. Lappi sees this as stemming from scientific bias—targeted toxicology has not historically worked well in cancer, and as a result reviewers are biased against this approach.

Commercialization assistance program (CAP).   ATS is currently involved in the new CAP being operated by LARTA. The company was present at a February 2005 meeting with potential funders in Newport Beach.

RECOMMENDATIONS

- Size/duration (Phase I): No additional funding for Phase I, even though current levels mean that no business can afford the risk of hiring someone just on the basis of a Phase I Award.
- Funding cycles. Eliminate the 2-year window for Phase I winners to apply for Phase II. Serves no useful purpose, and sometimes a Phase II application must wait for more data (Lappi has examples).
- Build a hospital for clinical trials. ATS suggests resolving the Phase III problem in neurological sciences by building a hospital for clinical trials, with the costs shared by multiple ICs. Claims that this approach has been adopted at NCI.
- Direct to Phase II. Companies should be allowed to compete directly for Phase II without previous Phase I.

ADVANCED TARGETING SYSTEMS—ANNEX

Table App-D-3 Advanced Targeting Systems NIH SBIR awards

Fiscal Year	Phase Type	Award	Institute-Center	Project Title
1995	Phase I	$96,844.00	NS	Specific tool for modeling neuronal degeneration
2001	Phase II	$282,235.00	NS	A specific tool for targeting neurodegeneration
2002	Phase II	$237,928.00	NS	A specific tool for targeting neurodegeneration
2003	Phase II	$245,067.00	NS	A specific tool for targeting neurodegeneration
		$862,074.00		
1999	Phase I	$113,227.00	MH	Mabs to target specific neuronal populations
2003	Phase II	$341,281.00	MH	Monoclonal antibodies to target neuronal populations
		$454,508.00		
1996	Phase I	$99,997.00	MH	New tool for basic neurobiological research
1998	Phase II	$404,567.00	MH	New tools for basic neurobiological research
1999	Phase II	$344,750.00	MH	New tools for basic neurobiological research
2001	Phase II	$397,984.00	MH	Toxicology/safety studies of a chronic pain therapeutic
2002	Phase II	$276,089.00	MH	New tools for basic neurobiological research
2003	Phase II	$799,709.00	MH	Drug development of a chronic pain therapeutic
		$2,323,096.00		
1999	Phase I	$100,000.00	NS	Tools for the dissection of pain transmission pathways
2000	Phase I	$100,000.00	DA	A tool to study the diverse behavior effects of galanin
2001	Phase I	$133,547.00	DE	Targeting neurons involved in chronic pain transmission
Total		$3,973,225.00		

NOTE: For a list of codes for National Institutes of Health institutes and centers, see Box App-A-1.

SOURCE: Advanced Targeting Systems.

Bioelastics Research, Ltd.

Paula Stephan
Georgia State University

DESCRIPTION OF THE FIRM

Bioelastics Research, Ltd. (Bioelastics) was founded by Dan W. Urry, Ph.D., at the University of Alabama (UAB) Birmingham in 1989. The company suspended operations October 31, 2004. At the time the firm was founded, Dr. Urry was a professor of molecular biophysics at UAB. Dr. Urry currently is on the faculty of University of Minnesota (department of chemical engineering and material science) teaching courses from January to April each year, having retired from UAB. After retiring from UAB, and prior to the suspension of operations, Dr. Urry continued to spend time in Birmingham each year. Before it suspended operations in the fall of 2004, the firm had four employees. Its average annual revenue was under $700,000. At its largest, the firm employed eight people. The firm resided in incubator space at UAB. During the first few years, the firm paid rent that was under market but in subsequent years the firm paid the market rate.

The company's technology evolved around polymers made from elastic sequences in the body. This technology provided a basis for producing materials that would prevent adhesion, deliver drugs, and provide acoustic prevention (sound deadening) as well as a number of other applications including tissue reconstruction that is applicable to urinary incontinence and spinal injuries. Initial work on this technology was done at University of Alabama Birmingham in the lab of the founder, Dr. Dan W. Urry.

The firm received an initial investment of $333,000 from three local investors. UAB provided funding for initial patent applications before Bioelastics was founded. The firm was required to pay back the amount UAB spent and Bioelastics has born all the patent cost forward since then. In addition, Bioelastics was required to pay a $50,000 per year due-diligence payment to UAB. Patent costs are still being generated and are to be paid by the current holding company or whoever eventually acquires the technology.

The initial impetus to found the firm was the availability of cost-sharing funds for small businesses from the Office of Naval Research (ONR) which Dr. Urry found out about while working on research funded by ONR in the late 1980s at UAB. The firm was started using the $333,000 investment money noted above to set up the firm and then applied to ONR for a project directed towards wound repair technology.

INPUTS

The firm had 18 SBIRs from NIH: 12 Phase I and 6 Phase II. Almost all of the SBIR awards were to facilitate the development of products based on bioelastic material. Examples include: ingestible implants to correct urinary incontinence; materials for strabismus and retinal surgery, development of an artificial pericardia.

SBIRs did not play a role in the founding of the company, but the company began applying for SBIR awards at an early stage, and from 1991 on "SBIRs were very important to the company's financial position." The company received no funding from the state of Alabama. In addition to the SBIR awards, and the ONR award referred to above, the company received federal funding from DoD and DARPA as well as some other, non-SBIR funds, from ONR.

The only angel/VC money that the company received was the initial $333,000 noted above. The inability to attract other funding arguably relates to the Birmingham location of the firm. There was very little VC money in Birmingham at the time and what was available had been invested in several high profile companies that "soaked up the local money and provided no return." This hindered Bioleastics ability to raise local venture funding. Moreover there was little interest at the time Bioleastics started up from VC companies operating out of state. "You don't get the interest from the Carolinas and Atlanta to come over here. They [VCs] didn't really like having it in Birmingham." Moving, which could have opened up the opportunity for VC, was problematic for the company. Some of the primaries did not want to leave the Birmingham area and the initial investors wanted the company to stay. There was, to quote Mr. Parker, a hometown spirit of "make this benefit Alabama."[8]

The SBIR awards helped the company to raise many of the contracts that the company had with firms, allowing the company to expand the research that it was doing and grow the research to a particular application. By way of example, one of the SBIR awards dealt with adhesions. As a result of this research, Bioelastics established a research contract with a firm to develop a product in which the firm was potentially interested. While they were able to do that successfully, the stumbling block was the lack of a production facility. This "chicken-and-egg" problem plagued the company throughout its entire tenure and eventually played a key role in causing the company to suspend operation. In essence, the contracting companies did not want to take on the initial expense, estimated at approximately $10 million, of creating a facility to produce the polymers but would have readily bought the polymers if they were available and reasonably priced for the application.

The company had four employees when it received its first SBIR award in the early 1990s. At its largest, the company employed eight individuals. SBIR

---

[8]Venture capital funds were not sought after the first several years. Existence depended on contracts and grants.

awards impacted company hiring in the sense that in several instances individuals were hired on after the grant was awarded. For example, Dr. Asima Pattanaik was hired as a result of an SBIR award and remained with the company for eight years, becoming a PI on several SBIR awards. To quote Mr. Parker, "SBIR funding allowed the company to maintain a "core group."

KEY OUTCOMES FROM SBIR

Commercial Outcomes

The company never had a product that generated revenue. It did, as noted above, have research contracts with commercial companies to explore the development of bioelastic material and viable products were produced. The key constraint in producing a viable commercial product was the lack of a production facility capable of producing the material. According to Mr. Parker, the initial investors did not believe that they needed to have a production company. They "expected someone to walk up and pick it [the company] up for a big chunk of change and move on with it but no one wants to put that type of capital into a company that . . . needed a production company and a huge initial investment." Stated somewhat differently, as long as the firm focused on research it was successful. But when the initial investors pushed for commercialization and got control of the company, problems emerged. "The firm was good at doing research, but when they [the initial investors] turned it over to someone else to move it from a research company to a production company the transition was not successful."

The company never licensed a product but in several instances option payments were obtained from firms interested in taking a look at the company's technology.

Noncommercial Outcomes

The company has been awarded ten patents. The last patent was awarded in March of 2004; it had been applied for in April 2001. A few of these patents were based on research that was funded by SBIR grants. Most of the SBIR grants furthered the advance of technology for which a patent had already been applied.

Scientists working at the firm published a considerable number of papers. A list of publications is provided at the end of this paper, representing the scholarly work accomplished during the existence of Bioelastics. A number of these papers can be attributed to SBIR awards. Some of the publications represent basic research papers; some proposed the basis for what became an SBIR proposal from work done with other agencies and some resulted from work done with other companies.

Dr. Urry has established a considerable scientific reputation based on his work in the field of bioelastics and is a frequent speaker at international conferences and symposia. The SBIRS helped open up a whole new frontier: "We now believe we can understand how the body works—the reactions of how proteins change in the body. How they will release, conform and reattach and the nanoscale processes they undergo. Also, Dr. Urry can now explain how the internal motors work and describe the forces that drive them. No one else in the world developed or conceived of this. It's really a Nobel Prize type application."

IMPACT OF SBIR ON THE FIRM

The SBIR program was very important to the company's financial position from 1991 on. It provided a good deal of the funding for the research that the company did. The SBIR program also influenced the research direction of the company to the extent that the company would respond to special SBIR initiatives. The SBIR program allowed the company to hire additional researchers and maintain a core group. But the company could only maintain a core group and that was a problem. "You cannot go out and hire new people if there is only six months of funding; for two years you can afford to go out and hire people but there is not a six-month job market. If you are going to grow on a Phase II you better have a clear exit strategy."

Participation in the SBIR program affected the firm's commercialization strategy by allowing the firm to take a longer view of commercialization. "A problem the company had from inception is that it didn't have a rock solid business plan. It was a spin out—first company to spring out from UAB. In so doing, the university put a lot of restrictions on the inventor. . . . It required an annual due diligence payment in order to keep the technology of $50,000 at minimum; it held a 7 percent ownership stake and a 50 percent stake in any patent that developed in the company." This made it more difficult to commercialize. "SBIRs were not a constraint; the business plan, the structure of the company were the big constraints."

SBIR PROGRAM ADMINISTRATION

The firm first became aware of SBIRs as a result of a program solicitation that was distributed in the early 1990s. At the time, Mr. Parker was working in Dr. Urry's lab at UAB.

The firm managed the delay between Phase I and Phase II awards by having multiple concurrent ongoing projects.

Mr. Parker found the size of the SBIR awards to be "decent." But the time frame to be "short." Unless you are working on a project full force, for example, it is hard to accomplish Phase I research in a six month period. They learned to live with the lag time between submission and receipt of funding. But, if it "took a year for a project to be funded it could be hard to keep all of that [personnel and equipment] together. It really takes an established company that has a product to support bringing in money during that period of time."

Mr. Parker answered affirmatively in terms of whether it would be beneficial to increase the duration of the award, adding that if "not increasing duration, being careful when you fund people with a broad scope, which has trouble fitting in a six month period." He added that broad scope projects also have difficulty fitting into the time allowed for a Phase II. "Six month SBIRs turned into a year; Phase IIs turned into three years."

The company did not find the paperwork to be severe. "It took some time but it was not overly demanding." Mr. Parker went on to say that "I think it would be useful if they provided a few more guidelines as to what they want in reports. It might be helpful to them to have more uniform reporting."

The primary recommendation for change that Mr. Parker offered was to demand a better plan as to how the research will be commercialized. "If a company goes in with the sole purpose of doing research and getting the information, that's fine . . . but also need to have a plan of what to do with it once they get it. Will they be able to market it? Will they be able to get additional funding? Is additional funding available?" He believes that one would see a higher success rate coming out of SBIR funding if the agency paid more attention to the business plan.

Mr. Parker also recommended that NIH consider requiring grantees to submit an annual report that is directed at sharing information with other SBIR awardees. He compared the current NIH reporting requirements to reporting required by DoD for certain non-SBIR awards. "DoD requires a PowerPoint presentation, summing up your program and stating where you are going. . . . This is presented and you have to at least develop it to the point where you can get somebody's attention with it." These DoD reports, presented annually, brought people up to speed concerning what was going on and drove collaborations. The experience made participants aware that they were "part of a group; you all are striving to overcome something. It brings the community together."

CROSS-CUTTING RESEARCH QUESTIONS

Firms are inhibited from getting an SBIR award if they are not well established, having neither facilities, equipment, nor personnel in place. He saw a number of companies in the incubator space at UAB that were not as established as Bioelastics. "One person in one room with one piece of equipment. They got a Phase I and they scaled up. But there was no way to survive beyond except to enter into the Phase II."

One benefit that a firm gains from an SBIR award that is not available through many other programs is the ability to "take advantage of connections with other people" by going to meetings organized for SBIR recipients at NIH. The company found attending such meetings and interacting with other researchers to be helpful in building collaborations that helped to move projects forward.

"Especially in small companies it is the collaborations that strengthen you and builds you . . . it's these meetings that really bring you together."

Note: the interview was conducted February 18, 2005 by Paula Stephan, with Tim Parker, in Pell City, Alabama. Mr. Parker was Manager of Research for Bioelastics and had worked with the company for thirteen years. Prior to joining the company, Mr. Parker worked for five years in Dr. Urry's lab at the University of Alabama Birmingham.

REFERENCES

Alkalay, Ron N., David H. Kim, Dan W. Urry, Jie Xu, Timothy M. Parker, and Paul A. Glazer. 2003. "Prevention of Postlaminectomy Epidural Fibrosis Using Bioelastic Materials." *Spine*. 28:1659-1665.

Daniell, H., C. Guda, D. T. McPherson, X. Zhang, and D. W. Urry. 1996. "Hyper Expression of a Synthetic Protein Based Polymer Gene." Pp. 359-371 in *Methods in Molecular Biology Vol. 63: Recombinant Proteins: Protocol Detection and Isolation*. R. Tuan, ed. Totowa, NJ: Humana Press, Inc.

Gowda, D. Channe, Timothy M. Parker, R. Dean Harris, and Dan W. Urry. 1994. "Synthesis, Characterizations and Medical Applications of Bioelastic Materials." Pp. 81-111 in *Peptides: Design, Synthesis, and Biological Activity*. Channa Basava and G. M. Anantharamaiah, eds. Boston: Birkhäuser.

Gowda, D. C., T. M. Parker, C. M. Harris, R. D. Harris and D. W. Urry. 1994. "Design and Synthesis of Poly-tricosapeptides to Enhance Hydrophobic-induced pKa Shifts." Pp. 940-943 in *Peptides: Chemistry, Structure and Biology*. R. S. Hodges and J. A. Smith, eds., Proceedings of the Thirteenth American Peptide Symposium, Edmonton, Alberta, Canada.

Gowda, D. Channe, Chi-Hao Luan, Raymond L. Furner, ShaoQing Peng, Naijie Jing, Cynthia M. Harris, Timothy M. Parker, and Dan W. Urry. 1995. "Synthesis and Characterization of Human Elastin W_4 Sequence." *International Journal of Peptide and Protein Research*. 46:453-463.

Guda, C., X. Zhang, D. T. McPherson, J. Xu, J. H. Cherry, D. W. Urry, and H. Daniell. 1995. "Hyper Expression of an Environmentally Friendly Synthetic Polymer Gene." *Biotechnology Letters*. 17, 745-750.

Herzog, R. W., N.K. Singh, D. W. Urry, H. Daniell. 1997. "Expression of a Synthetic Protein-based Polymer (Elastomer) Gene in *Aspergillus Nidulans*." *Applied Microbiology & Biotechnology*. 47:368-372.

Hoban, Lynne D., Marissa Pierce, Jerry Quance, Isaac Hayward, Adam McKee, D. Channe Gowda, Dan W. Urry, and Taffy Williams. 1994. "The Use of Polypenta-peptides of Elastin in the Prevention of Postoperative Adhesions." *Journal of Surgical Research*. 56:179-183.

Jing, Naijie. Kari U. Prasad. and Dan W. Urry, "The Determination of Binding Constants of Micellar-packaged Gramicidin A by 13C and 23Na NMR,"Biochem. Biophys. Acta, 1238, 1-11, 1995.

Jing, Naijie and Dan W. Urry. 1995. "Ion Pair Binding of Ca^{++} and Cl^- Ions in Micellar-packaged Gramicidin A." *Biochem. Biophys. Acta*. 1238(1):12-21.

Kemppainen, B. W., D. W. Urry, C.-X. Luan, J. Xu, S. F. Swaim and S. Goel. 2004. "In vitro skin penetration of dazmegrel delivered with a bioelastic matrix." *International Journal of Pharmaceutics*. 271:301-303.

Kemppainen, B., N.-Z. Wang, S. Swaim, D. W. Urry, C.-X. Luan, J. Xu, E. Sartin, R. Gillette, S. Hinkle, and S. Coolman. 2004. "Bioelastic Membranes for Topical Application of Thromboxane Synthetase Inhibitor for Protection of Skin from Pressure Injury: A Preliminary Study." *Wound Repair and Regeneration*. 12.

Luan, Chi-Hao and Dan W. Urry. 1999. "Elastic, Plastic, and Hydrogel Protein-based Polymers." Pp. 78-89 in *Polymer Data Handbook*. James. E. Mark, ed. New York: Oxford University Press.

Manno, M., A. Emanuele, V. Martorana, P. L. San Biagio, D. Bulone, M. B. Palma-Vitorelli, D. T. McPherson, J. Xu, T. M. Parker, and D. W. Urry. 2001. "Interaction of processes on different time scales in a bioelastomer capable of performing energy conversion." *Biopolymers.* 59:51-64.

McPherson, David T., Jie Xu, and Dan W. Urry. 1996. "Product Purification by Reversible Phase Transition Following *E. coli* Expression of Genes Encoding up to 251 Repeats of the Elastomeric Pentapeptide GVGVP." *Protein Expression and Purification* 7:51-57.

Nicol, Alastair, D. Channe Gowda, Timothy M. Parker, and Dan W. Urry. 1994. "Cell Adhesive Properties of Bioelastic Materials Containing Cell Attachment Sequences." Pp. 95-113 in *Biotechnol. Bioactive Polym.* Charles G. Gebelein and Charles E. Carraher, Jr., eds. New York: Plenum Press.

Nicol, Alastair, D. Channe Gowda, Timothy M. Parker, and Dan W. Urry. 1993. "Elastomeric Polytetrapeptide Matrices: Hydrophobicity Dependence of Cell Attachment from Adhesive, $(GGIP)_n$, to Non-adhesive, $(GGAP)_n$, Even in Serum." *J. Biomed. Mater. Res.* 27:801-810.

Patkar, Anant, Natarajan Vijayasankaran, Dan W. Urry, and Friedrich Srienc. 2002. "Flow Cytometry as a Useful Tool for Process Development: Rapid Evaluation of Expression Systems." *Journal of Biotechnology.* 93:217-229.

Strzegowski, Luke A., Manuel Bueno Martinez, D. Channe Gowda, Dan W. Urry, and David A. Tirrell. 1994. "Photomodulation of the Inverse Temperature Transition of a Modified Elastin Poly (Pentapeptide)," *Journal of the American Chemical Society.* 116:813-814.

Urry, D. W., A. Nicol, D. C. Gowda, L. D. Hoban, A. McKee, T. Williams, D. B. Olsen, and B. A. Cox. 1993. "Medical Applications of Bioelastic Materials." Pp. 82-103 in *Biotechnological Polymers: Medical, Pharmaceutical and Industrial Applications*. Charles G. Gebelein, ed. Atlanta: Technomic Publishing Company, Inc.

Urry, D. W., D. C. Gowda, S. Q. Peng, T. M. Parker, and R. D. Harris. 1992. "Design at Nanometric Dimensions to Enhance Hydrophobicity-induced pKa Shifts." *Journal of the American Chemical Society.* 114:8716-8717.

Urry, Dan W., Larry C. Hayes, D. Channe Gowda, Cynthia M. Harris, and R. Dean Harris. 1992. "Reduction-driven Polypeptide Folding by the ΔT_t Mechanism." *Biochem. Biophys. Res. Comm.* 188:611-617.

Urry, Dan W. 1993. "Bioelastic Materials as Matrices for Tissue Reconstruction." Pp. 199-206 in *Tissue Engineering: Current Perspectives*. Eugene Bell, ed. New York: Birkhäuser Boston, Div. Springer-Verlag.

Urry, Dan W., Larry C. Hayes, Timothy M. Parker and R. Dean Harris. 1993. "Baromechanical Transduction in a Model Protein by the ΔT_t Mechanism." *Chem. Phys. Letters.* 201:336-340.

Urry, Dan W., ShaoQing Peng and Timothy M. Parker. 1993. "Delineation of Electrostatic-and Hydrophobic-Induced pKa Shifts in Polypentapeptides: The Glutamic Acid Residue." *Journal of the American Chemical Society.* 115:7509-7510.

Urry, Dan W., D. Channe Gowda, Betty A. Cox, Lynne D. Hoban, Adam McKee and Taffy Williams. 1993. "Properties and Prevention of Adhesions Applications of Bioelastic Materials." *Mat. Res. Soc. Symp. Proc.* 292:253-264.

Urry, Dan W., ShaoQing Peng, Timothy M. Parker, D. Channe Gowda, and Roland D. Harris. 1993. "Relative Significance of Electrostatic- and Hydrophobic-Induced pK_a Shifts in a Model Protein: The Aspartic Acid Residue." *Angew. Chem.* (German). 105:1523-1525; *Angew. Chem. Int. Ed. Engl.* 32:1440-1442.

Urry, Dan W., D. Channe Gowda, Cynthia M. Harris and R. Dean Harris. 1994. "Bioelastic Materials and the ΔT_t-Mechanism in Drug Delivery." Pp. 15-28 in *Polymeric Drugs and Drug Administration*. Raphael M. Ottenbrite, ed., *American Chemical Society Symposium Series* 545:15-28.

Urry, Dan W., D. Channe Gowda, ShaoQing Peng, Timothy M. Parker, Naijie Jing, and R. Dean Harris. 1994. "Nanometric Design of Extraordinary Hydrophobicity-induced pKa Shifts for Aspartic Acid: Relevance to Protein Mechanisms." *Biopolymers.* 34:889-896.

Urry, Dan W., Shaoqing Peng, D. Channe Gowda, Timothy M. Parker, and R. Dean Harris. 1994. "Comparison of Electrostatic- and Hydrophobic-induced pKa Shifts in Polypentapeptides: The Lysine Residue." *Chemical Physics Letters.* 225:97-103.

Urry, Dan W. 1994. "Conversion of Available Energy Forms into Desired Forms by a Biologically Accessible Mechanism." Pp. 629-636 in *Nondestructive Characterization of Materials VI.* Robert E. Green, Jr., ed. Proceedings of the Sixth International Symposium on Nondestructive Characterization of Materials, Oahu, Hawaii. New York: Plenum Press.

Urry, Dan W. 1994. "Biophysics of Energy Converting Model Proteins." *Mat. Res. Soc. Symp. Proc.* 321-332.

Urry, Dan W., Alastair Nicol, David T. McPherson, Jie Xu, Peter R. Shewry, Cynthia M. Harris, Timothy M. Parker, and D. Channe Gowda. 1995. "Properties, Preparations and Applications of Bioelastic Materials." Pp. 1619-1673 in *Encyclopedic Handbook of Biomaterials and Bioengineering—Part A—Materials, Vol. 2.* New York: Marcel Dekker, Inc.

Urry, Dan W., David T. McPherson, Jie Xu, D. Channe Gowda, and Timothy M. Parker. 1995. "Elastic and Plastic Protein-based Polymers: Potential for Industrial Uses," Pp. 259-281 in *Industrial Biotechnological Polymers.* Chas. Gebelein and Chas. E. Carraher, Jr., eds. Lancaster, PA: Technomic Publshing Co.

Urry, D. W. 1994. "Postulates for Protein (Hydrophobic) Folding and Function." *International Journal of Quantum Chemistry.* 21:3-15.

Urry, Dan W. and Chi-Hao Luan. 1995. "A New Hydrophobicity Scale and Its Relevance to Protein Folding and Interactions at Interfaces." Pp. 92-110 in *Proteins at Interfaces 1994.* Thomas A. Horbett and John L. Brash, eds. American Chemical Society Symposium Series: Washington, DC.

Urry, Dan W. 1995. "Elastic Biomolecular Machines: Synthetic chains of amino acids, patterned after those in connective tissue, can transform heat and chemical energy into motion." *Scientific American.* January, 64-69.

Urry, Dan W., Larry C. Hayes, and D. Channe Gowda. 1994. "Electromechanical Transduction: Reduction-driven Hydrophobic Folding Demonstrated in a Model Protein to Perform Mechanical Work." *Biochemical and Biophysical Research Communications.* 204:230-237.

Urry, Dan W., Chi-Hao Luan, and ShaoQing Peng. 1995. "Molecular Biophysics of Elastin Structure, Function and Pathology." Pp. 4-30 in *Proceedings of The Ciba Foundation Symposium No. 192. The Molecular Biology and Pathology of Elastic Tissues.* Sussex, UK: John Wiley & Sons, Ltd.

Urry, Dan W., D. T. McPherson, J. Xu, H. Daniell, C. Guda, D. C. Gowda, Naijie Jing, and T. M. Parker. 1996. "Protein-Based Polymeric Materials: Syntheses and Properties" Pp. 7263-7279 in *The Polymeric Materials Encyclopedia: Synthesis, Properties and Applications.* Boca Raton, FL: CRC Press.

Urry, D. W., C.-H. Luan, C. M. Harris, and T. Parker. 1997. "Protein-based Materials with a Profound Range of Properties and Applications: The Elastin T_t Hydrophobic Paradigm." Pp. 133-177 in *Proteins and Modified Proteins as Polymeric Materials.* Kevin McGrath and David Kaplan, eds. Birkhauser Press.

Urry, D. W., Cynthia M. Harris, Chi Xiang Luan, Chi-Hao Luan, D. Channe Gowda, Timothy M. Parker, ShaoQing Peng, and Jie Xu. 1997. "Transductional Protein-based Polymers as New Controlled Release Vehicles," Part VI: New Biomaterials for Drug Delivery. Pp. 405-437 in *Controlled Drug Delivery: The Next Generation.* (Kinam Park, ed. Am. Chem. Soc. Professional Reference Book.

Urry, Dan W., D. Channe Gowda, ShaoQing Peng, and Timothy M. Parker. 1995. "Non-linear Hydrophobic-induced pKa Shifts: Implications for Efficiency of Conversion to Chemical Energy." *Chemical Physics Letters.* 239:67-74.

Urry, Dan W., Larry C. Hayes, D. Channe Gowda, ShaoQing Peng, and Naijie Jing. 1995. "Electro-chemical Transduction in Elastic Protein-based Polymers." *Biochem. Biophys. Res. Commun.* 210:1031-1039.

Urry, Dan W. and ShaoQing Peng. 1995. "Non-linear Mechanical Force-induced pKa Shifts: Implications for Efficiency of Conversion to Chemical Energy." *Journal of the American Chemical Society* 8478-8479.

Urry, Dan W. and Chi-Hao Luan. 1995. "Proteins: Structure, Folding and Function." Pp. 105-182 in *Bioelectrochemistry: Principles and Practice.* Giorgio Lenaz, ed. Basel, Switzerland: Birkhäuser Verlag AG.

Urry, Dan W., Asima Pattanaik, Mary Ann Accavitti, Chi-Xiang Luan, David T. McPherson, Jie Xu, D. Channe Gowda, Timothy M. Parker, Cynthia M. Harris, and Naijie Jing. 1997. "Transductional Elastic and Plastic Protein-based Polymers as Potential Medical Devices." Pp. 367-386 in *Handbook of Biodegradable Polymers.* Domb, Kost, and Wiseman, eds. Chur, Switzerland: Harwood Academic Publishers.

Urry, Dan W., Larry C. Hayes, and Shao Qing Peng. 1996. "Designing for Advanced Materials by the T_t-Mechanism." SPIE—The International Society for Optical Engineering Smart Structures and Materials. San Diego, CA: Smart Materials Technologies and Biomimetics 2716, 343-346.

Urry, Dan W. 1996. "Engineers of Creation." Pp. 39-42 in *Chemistry in Britain.*

Urry, Dan W., ShaoQing Peng, Jie Xu, and David T. McPherson. 1997. "Characterization of Waters of Hydrophobic Hydration by Microwave Dielectric Relaxation." *Journal of the American Chemical Socitey.* 119:1161-1162.

Urry, Dan W. 1997. "On the Molecular Structure, Function and Pathology Of Elastin: The Gotte Stepping Stone." Pp. 11-22 in *The Structure, Function and Pathology of Elastic Tissue.* Potenza, Italy: Tip. Mario Armento & Co. Publisher.

Urry, Dan W. and Asima Pattanaik. 1997. "Elastic Protein-based Materials in Tissue Reconstruction." *Annals of the New York Academy Sciences.* 831:32-46.

Urry, D. W., S. Q. Peng, L. C. Hayes, D. T. McPherson, Jie Xu, T. C. Woods, D. C. Gowda, and A. Pattanaik. 1998. "Engineering Protein-based Machines to Emulate Key Steps of Metabolism (Biological Energy Conversion)." *Biotechnology and Bioengineering.* 58:175-190.

Urry, Dan W. 1997. "Physical Chemistry of Biological Free Energy Transduction as Demonstrated by Elastic Protein-based Polymers." *J. Phys. Chem.* 101:11007-11028.

Urry, Dan W. 1999. "Five Axioms for Protein Engineering: Keys for Understanding Protein Structure/Function?" Pp. 75-78 in *Proceedings of the Fifteenth American Peptide Symposium, Peptides: Frontiers of Peptide Science.* James P. Tam and Pravin T. P. Kaumaya, eds. Boston: Kluwer Academic Publishers.

Urry, Dan W., ShaoQuing Peng, Chi-Hao Luan, Chi-Xiang, Luan, Asima Pattanaik, Jie Xu, David T. McPherson. 1998. "Hydrophobic Hydration in Protein Models for Muscle Contraction: Calcium Ion, Thermal, Stretch and pH Activation." *Scanning Microscopy International.*

Urry, Dan W. 1998. "Five Axioms for the Functional Design of Peptide-Based Polymers as Molecular Machines and Materials: Principle for Macromolecular Assemblies." *Biopolymers (Peptide Science)* 47:167-178.

Urry, Dan W., Asima Pattanaik, Jie Xu, T. Cooper Woods, David T. McPherson, and Timothy M. Parker. 1998. "Elastic Protein-based Polymers in Soft Tissue Augmentation and Generation." *J. Biomater. Sci. Polymer Edn.* 9:1015-1048.

Urry, Dan W. 1999. "Elastic Molecular Machines in Metabolism and Soft Tissue Restoration." TIBTECH. 17:249-257.

Urry, Dan W., Larry Hayes, Chixiang Luan, D. Channe Gowda, David McPherson, Jie Xu, and Timothy Parker. 2001. "ΔT_t-Mechanism in the Design of Self-Assembling Structures." Pp. 323-340 in *Self Assembling Peptide Systems in Biology, Medicine and Engineering.* Amalia Aggeli and Neville Boden, eds. Netherlands: Kluwer Academic Publishers.

Urry, D. W., T. Hugel, M. Seitz, H. Gaub, L. Sheiba, J. Dea, J. Xu and T. Parker. 2002. "Elastin: A Representative Ideal Protein Elastomer." *Phil. Trans. R. Soc. Lond. B.* 357:169-184.

Urry, D. W., T. Hugel, M. Seitz, H. Gaub, L. Sheiba, J. Dea, J. Xu, L. Hayes, and T. Parker. 2002. "Ideal Protein Elasticity: The Elastin Model." In P. Shewry and A. Bailey, eds. Cambridge: Cambridge University Press.

Urry D. W. and T. M. Parker. 2002. "Mechanics of Elastin: Molecular Mechanism of Biological Elasticity and its Relationship to Contraction, Special Issue: Mechanics of Elastic Biomolecules." Journal of Muscle Research and Cell Motility, 23, 543-559, 2002.

Urry, D. W., T. C. Woods, L. C. Hayes, J. Xu, D. T. McPherson, M. Iwama, M. Furuta, T. Hayashi, M. Murata, and T. M. Parker. 2004. "Elastic Protein-Based Biomaterials: Elements of Basic Science, Controlled Release and Biocompatiblity." Pp. 31-54 in *Tissue Engineering and Novel Delivery Systems*. New York: Marcel Dekker, Inc.

Urry, D. W., J. Xu, W. Wang, L. Hayes, F. Prochazka, and T. M. Parker. 2003. "Development of Elastic Protein-based Polymers as Materials for Acoustic Application." *Mat. Res. Soc. Symp. Proc.* 774:81-92.

Wang, N. Z., D. W. Urry, S. F. Swaim, R. L. Gillette, C. E. Hoffman, S. H. Hinkle, S. L. Coolman, C.-X. Luan, J. Xu, and B. W. Kemppainen. 2004. "Skin concentrations of thromboxane synthetase inhibitor after topical application with bioelastic membrane." *J. Vet. Pharmacol. Therap.* 27:37-43.

Zhang, X., C. Guda, R. Datta, R. Dute, D. W. Urry, and H. Daniell. 1995. "Nuclear Expression of an Environmentally Friendly Synthetic Protein-based Polymer Gene in Tobacco Cells." *Biotech Letters.* 17(12):1279-1284.

Zhang, X., D. W. Urry, and H. Daniell. 1996. "Expression of an Environmentally Friendly Synthetic Protein-based Polymer Gene in Transgenic Tobacco Plants." *Plant Cell Reports.* 16:174-179.

Cambridge NeuroScience[9]

Paula Stephan
Georgia State University

DESCRIPTION OF THE FIRM

Cambridge NeruoScience (CNS) was incorporated in Delaware in December 1985 and began operations in January 1986 in Cambridge, MA where it remained until 2000, when the firm relocated to Norwood, MA.

The company's focus involved the discovery and development of pharmaceutical products to treat a variety of severe neurological and psychiatric disorders. Product development was focused in three areas: neuroprotective compounds for the treatment of acute neurological disorders such as stroke and traumatic brain injury, novel antipsychotic compounds for the treatment of mental illnesses such as schizophrenia, and growth factors for the treatment of neurodegenerative diseases such as diabetic peripheral neuropathies, and ALS (Prospectus, Initial Public Offering, June 6, 1991, page 10). As the company grew and advanced its technology, research and development programs in multiple sclerosis and neuropathetic pain were added to the portfolio.

The company had a number of collaborations with academic institutions at the time that it went public in 1991. These included, for example, the Ludwig Institute for Cancer Research in London, the University of Oregon, Cornell University, the Medical College of Virginia, the University of Toledo, Columbia University and the Massachusetts Institute of Technology.

The company raised venture capital from Warburg, Pincus Capital Partners and Aeneas Venture Corporation and, as noted above, made an initial public offering in 1991 which generated $22,320,000 in proceeds for the company. Warburg, Pincus provided the initial funds (approximately $5 million) to start the company.

The company states in its prospectus that it had 44 full-time employees at the time of its initial public offering in 1991; 36 of these employees were engaged in research and development.

The company faced considerable competition both from start-ups and from fully integrated pharmaceutical companies in the neuroscience pharmaceutical market. At the same time, the neurological market has significant growth potential. For example, the number of cases of Alzheimer's disease is rapidly growing

---

[9]Paula Stephan spoke with Mark A. Marchionni, Ph.D., on February 24, 2005, at 4:00 p.m. by phone. Dr. Marchionni worked at Cambridge NeuroScience from 1987 to 2001. His initial position at CNS was that of Group Leader/Staff Scientist II; at the time that the company was closed by CeNeS Pharmaceuticals, he was Vice President, Research. During his time with the company, he was the PI on five funded SBIR grants and the company obtained support from at least 13 SBIR grants.

as life expectancy increases; and the population of individuals suffering from strokes is large and growing. Moreover, there are virtually no products that successfully treat strokes. Strokes are, according to Dr. Marchionni, "a graveyard for compounds in development."

The Cambridge location was chosen because of the location of the three co-founders: Joe Martin, M.D., Ph.D., Chairman of Neurology, Massachusetts General Hospital and Professor of Neurology at Harvard Medical School at the time the company was founded; Howard M. Goodman, Ph.D., Chief of the Department of Molecular Biology at Massachusetts General Hospital and Professor of Genetics at Harvard Medical School; and Rod Moorhead, of Warburg, Pincus Capital Partners. Dr. Goodman was a board member of Warburg, Pincus Capital partners at the time the company was started.

Cambridge NeuroScience was acquired by CeNeS Pharmaceuticals, Inc., a company listed on the London Stock Exchange, in December of 2000; CeNeS closed operations of Cambridge NeuroScience in January of 2002.

INPUTS

The company received at least 13 SBIRs during its 15 years in existence. With but one exception, the company followed up each Phase I with a successful Phase II. In one instance, the company chose not to follow up with a Phase II application due to the limited success in demonstrating feasibility of the approach in Phase I. The first SBIR award was applied for soon after the company was founded. The company was in the process of being awarded an SBIR Phase II at the time that it was acquired by CeNeS but did not receive the actual funding because, by the time it received the Notice of Grant Award, it had been acquired by a foreign-owned company.

Although SBIRs played no role in the founding of the company, "right off the bat" the company applied for SBIR funding. This was facilitated by the fact that from the beginning, with seed money from venture capitalists, the company was able to hire a team of fairly experienced staff scientists. While none had SBIR experience, all had worked in postdoc positions and understood something about the culture of grant writing.

The company received no funding from the state of Massachusetts. In addition to receiving SBIR support from the federal government, one of its scientists, Dr. Stanley Goldin, was able to transfer an R01 from NIH to Cambridge NeuroScience when he came to Cambridge NeuroScience from Harvard Medical School. Since his area of research was of interest to the company, this arrangement was of mutual benefit and he was able to continue his research program with the aid of NIH support.

SBIRs played a role in external partners' decisions to provide funding in two ways. First, the award of an SBIR was treated by the company as a marketing tool, with the company issuing a press release at the time of the award. Second,

and more meaningfully, SBIR grant proposals were often given to potential partners through confidentiality agreements to provide a detailed description of the
science that was being done at Cambridge NeuroScience, allowing the potential
partner to do due diligence on scientific elements. A funded grant, according to
Dr. Marchionni, "says here is a group of disinterested reviewers who have decided to say we are going to invest tax payers money to try to move this forward;
we think it has a reasonable chance of success. The independent endorsement
carries some weight. It's probably worth more than what you get paid to do the
research." It should also be noted that SBIR funding was discussed in the prospectus for the IPO as a source of revenue.

The company had an extremely aggressive hiring plan when it started, and
had 16 to 18 staff scientists by the time it received its first SBIR award.

SBIRs definitely helped the company grow in terms of hiring. Dr. Marchionni, for example, was able to add people to his group because of the grant.
"It helped fortify a group that would have been half the size if it did not have
the grant." Moreover, the grants allowed for "growing" a work force, not only in
size, but also in skill, by permitting the company to bring individuals on board
and then invest in training them. "Because they were on board I could train them
in advanced molecular skills." This meant that when they came to another project
in which the company had a greater corporate investment and which went beyond
the grant, "these people were ready to go." As a result, "we were able to compete
in a high-stakes game; we were able to compete against Genentech in essence.
If I had had to wait until the time that we got into the growth factor work to hire
people, the whole project would have been taken over by Genentech before we
were ever able to get into it."

The hiring and training that result from SBIRs allow the company to take a
longer-run view than it would have taken without such funding. While company
management focuses on immediate returns to shareholders, SBIRs permit the
company to focus on things that don't necessarily have an immediate return to
shareholders. The grant allows the company to "hire on new employees who
will have an impact that goes long beyond the period of the grant." It provides
the company enough flexibility to hire individuals and create a critical mass of
a sort.

KEY OUTCOMES FROM SBIR

Commercial Outcomes

The company never had a product that generated revenue through sales.
However, the ion channel patents were bought by Scion and, as part of the deal,
an SBIR Phase II award relating to ion channels moved, with the PI, from Cambridge NeuroScience to Scion. Scion also bought the chemical library of CNS.

The company had one drug that made it to Phase III Clinical Trials for the treatment of stroke and TBI. The drug (CERESTAT, or aptiganel hydrachloride) did not show efficacy during Phase III and the trials were terminated, both for the treatment of stroke and for TBI. "It was really close . . . it is possible that if the trial were rerun and using more narrow inclusion criteria, it might work." SBIR funding was involved at some early stage of the work that developed CERESTAT.

Many biotech companies choose to focus on niche markets. The stroke market, by way of contrast, is not a niche market and if the trials had proved successful the company could have anticipated considerable commercial success. "We appeared on local television news; we were on the verge of making a major breakthrough in treatment for stroke. It really looked like it was going to work." The company's stock fell by something like 80 percent when the trail was discontinued. "Once that happened, the company died a slow death."

A consistent theme during the interview was that SBIR awards are not appropriate for the "front runner of the company." Those projects should be funded primarily by the company. But SBIR awards permit the company to enhance the projects it is doing, providing the opportunity for greater depth, thus improving the chances of success. SBIRs also allow a company to diversify. Stated somewhat differently, companies are under pressure to convert their dollars to a profit; the pace of research on the "front runner" is too fast to permit for the grant approach. The SBIR program is relatively slow: eight months at least before the money comes in. "You can't wait that long; things happen too fast; you risk losing out to your competition if it is the lead program in the company."

Dr. Marchioni elaborated on this theme, saying that SBIRs provide a relatively successful company two specific benefits. First, SBIR awards allow a company to round out research on the lead program that it cannot fully support. "You write a grant for something you will be needing six months from now; that will require adding things; you can round out your effort through the grant by making it a fuller program with a greater chance of success. But the grant funding is not the main source of support. The grant is not driving the company. You have to have other sources of funding to make it work. You use the grant to embellish and enhance the effort, but it's not the sole source of support." Second, the SBIR program allows relatively successful companies to explore alternatives that the company might otherwise not be able to explore, providing for diversification. For example, the company acquires an option to technology from a university; the university researchers continue to work on it in the laboratory. The company then can begin to participate directly in this technology by applying for SBIR funding. If successful in obtaining this grant support, then the company is in a strong position to convert the option into a license and start working on it internally in a new area. This approach will grow the scientific base of the company's technology and accrue benefit to the university as well—technology originally discovered there can be developed further and eventually be commercialized. Diversification provides the company some hedge against risk and creates jobs.

Venture capitalists, to paraphrase Dr. Marchionni, cannot necessarily fund the complete diversification of a company that needs to diversify. Neither do they want to. "Venture Capitalists want you to be extremely focused and often give you one chance to succeed."

Noncommercial Outcomes

Cambridge NeruoScience had many patents; it is difficult to determine which ones specifically related to SBIR grants. An example of a patent that was clearly based on SBIR research is patent 6,232,061 issued May 15, 2001 for homology cloning. Scientists working at CNS published a considerable number of scientific papers; some were based on research funded through the SBIR program.

IMPACT OF SBIR ON THE FIRM

The SBIR program clearly contributed to the research productivity of Cambridge NeuroScience. It allowed the company to hire and train additional researchers, engage in more in-depth research for front runner programs and diversify its research efforts.

SBIR PROGRAM ADMINISTRATION

Although none of the scientific staff had prior SBIR experience, from the time the company started the research scientists were aware of the SBIR program.

The company did not experience problems related to the delay between Phase I and Phase II awards. Dr. Marchionni indicated that they were aware that the SBIR grant did not pay for everything and that they realized they had to have other sources of income to support research. He also noted that once the project took on "front burner" status in the company, the company was no longer interested in using grants to support the research because the time required to write the grant application detracted from the research effort.

He expressed the view that the perfect SBIR company is a Series A company or a company with angel backing. Such companies have sufficient funds to take advantage of the grant and make full use of it, but not be overly concerned about what is going to happen when they lose it for a period of time. "If SBIR income is the only income you have, SBIR money is being wasted."

Dr. Marchionni viewed much of the review and selection process to be fair. The company chose a strategy of including detailed data in the grant applications. They knew they were competitive and they didn't want to provide an excuse for not getting funding. They also thought that the chances that disclosure would cause a problem were "pretty minimal." "In the end, there was never a single thing that we disclosed that someone else took over and started working on."

Dr. Marchionni saw serious problems in two aspects of the review process:

the composition of the panel, which is largely made up of academics, and the panel's interest in the proposed budget. Academic reviewers often employ criteria that apply to RO1 proposals in reviewing SBIRs, which is not appropriate. Second, many of them fail to fully appreciate the entire process of (bio)technology development, even though many serve as consultants to industry. Many have expressed concerns that were completely out of sequence with standard industry practices. For example, some issues that might come into play in the design of clinical studies simply do not need to be addressed until you begin to get there by overcoming many other, more immediate, hurdles along the way. To use a baseball analogy, a first-base coach does not have to be concerned about whether a runner should slide into third on a triple. As the runner approaches third base, he will get instructions from the third-base coach.

Moreover, some academics have a bias towards SBIR proposals that are written by fellow academics and try to "create the rules so that the funding goes to academics." Dr. Marchionni recommends including more individuals from industry in the review process. While he recognizes that this could create a potential conflict of interest, in today's world it is reasonably common for academic reviewers to have ties to industry and thereby they, too, could have a conflict of interest. He noted that there is not that big a divide between what it takes to be a successful scientist in a company and a successful scientist in academe.

A second concern that Dr. Marchionni expressed relates to the fact that it is not uncommon to get comments back from review relating to the appropriate size of the submitted budget. He argues that this is inappropriate. Just as only the direct portion of the budget is subject to review for RO1 grants, only the direct portion of costs should be considered for review in the SBIR proposal. The other costs should be made "invisible," just as indirect costs are made invisible on RO1 proposals.

Dr. Marchionni had two specific recommendations for changes to the SBIR program. The first concerns the recent VC rule which states that a company that is majority-owned by VC is not eligible for SBIR awards. This rule excludes firms that, in his view, are strong candidates for SBIR awards, not needing to rely exclusively on SBIR awards for funding, but using the SBIR awards to enhance and diversify the research of the company. Both activities result in the creation of new jobs. Dr. Marchionni points out that if the VC rule had been in effect at the time that Cambridge NeuroScience was founded, the company would never have received an SBIR award. (See comments by Dr. Marchionni at *<http://www.zyn.com/sbir/articles/vc/vc-8.htm>* and reprinted in the annex.) Neither would it have created the number of research jobs that it did during the late 1980s and early to mid 1990s.

He also questions the ruling that the company was ineligible for SBIR funding once it had been bought by a British firm, noting that the proposed research was to be done in the United States. In his view, SBIR funding contributes to the employment of life scientists who, especially in recent times, have found it difficult to find employment in research environments.

CROSS-CUTTING RESEARCH QUESTIONS

Dr. Marchionni sees the SBIR program as providing a "signal" to external players, such as firms that the company may be working with to form an alliance. He consistently made the case that the SBIR program affects firm performance through the enhancement of its front-burner programs and the ability to diversity its research portfolio. Moreover, it allows the firm to hire new scientists, thereby creating jobs and providing training consistent with the firm's research agenda. He also noted that certain research areas within the firm were not appropriate for SBIR funding, moving at too fast a pace to write a grant for the research.

CAMBRIDGE NERUOSCIENCE—ANNEX

Statement by Mark Marchionni, Ph.D.

I wish to submit the following comments on the SBIR eligibility requirements pertaining to company ownership by individuals.

The SBIR program has made possible the growth and survival of emerging companies for decades. In helping to create new jobs and advance innovative technology this program has been an essential part of growing and maintaining a vibrant and competitive economy in the United States.

The National Institutes of Health has administered an SBIR program for decades, and this source of funding has been instrumental to emerging companies in a growing biotechnology industry. Many of these companies have been owned in part or were started by venture capital firms. From my personal experience, I was one of the first scientific staff members to join Cambridge NeuroScience, Inc. (CNSI) in 1987, shortly after the company was started by the venture capital firm Warburg Pincus. Prior to the initial public offering, Warburg retained majority ownership of CNSI and we submitted and were funded for more than 10 NIH SBIR grants (five Phase I and five Phase II). I was the Principal Investigator on two such grants. These funds were critical in creating the jobs to grow the company from 18 (when I joined) to more than 60 at the time of the IPO. By virtue of this SBIR support, several product candidates advanced into clinical trails or were partnered with major pharmaceutical companies as part of Phase III commercialization. Thus, not only has it been accepted practice for the NIH SBIR program to support the growth of emerging biotechnology companies, but it has helped to accomplish the mission of the NIH and create new jobs as well.

At the current time, I submit that the Small Business Administration and the NIH need to support the growth of new jobs in this essential economic sector—biotechnology. Many have lost their jobs in the recent economic decline and tax revenues need to be spent wisely to stimulate job growth. Biotechnology represents a sector of the economy where the U.S. has a clear competitive advantage, and provides an important opportunity to revitalize economic growth. Further, a very sizeable percentage of all drugs produced in the past decade have been discovered at small companies. Since companies that are controlled by venture capital firms often represent some of the more innovative and competitive start-ups, policies that support collaboration between these companies and government agencies would represent prudent and productive use of tax revenues. It is these very companies that have the greatest chance of advancing their technology through to commercialization. Therefore, I offer my very strong support for producing a wording of the eligibility requirements that would enable the NIH to include in the SBIR program emerging companies that are majority owned by venture capital firms.

CryoLife[10]

Paula Stephan
Georgia State University

DESCRIPTION OF THE FIRM

CryoLife was founded in 1984 by Steven G. Anderson and Robert McNally. Prior to founding the company, Anderson worked at Intermedics, which is now Guidant. CryoLife was founded to commercialize the cryopreservation of human allograft heart valves. The initial technology was "out there," coming from the University of Alabama Birmingham, which had a cryopreservation lab for allograft heart valves, but was not trying to distribute valves or make the technology widely available. The company has had revenue almost from its inception and had a profit beginning around 1986. Anderson serves as Chairman, President and Chief Executive Officer of the company; McNally is no longer with the company.

CryoLife is located in Kennesaw, Georgia. It is situated in a dedicated building, the second phase of which was completed approximately three years ago. The firm, though incorporated in Florida, has only operated in Georgia. A major factor in the decision to locate in the Atlanta area was the need to be near a large, busy airport, given that the firm deals with human tissue.

CryoLife employs approximately 325 individuals, 10 of whom have a Ph.D. Sales for the year ending 2003 were approximately $60 million; sales had been approximately $88 million for the year ending 2001 (see discussion below).

The firm has developed proprietary processes for the preservation of human heart valves, veins and connective tissue. The firm has developed a tissue engineered heart valve and vascular graft replacement called SynerGraft® Valve and Synergraft® Vascular; the firm also has developed a bioadhesive product, BioGlue® Surgical Adhesive (hereafter referred to as "BioGlue").

The firm faces some competition in the human tissue business, from both the profit and the not-for-profit sector. BioGlue® is approved for sales in approximately 50 countries. "In August of 2002 the firm received an order from the Atlanta district office of the FDA regarding the nonvalved cardiac, vascular, and

---

[10]The interview was conducted February 16, 2005, by Paula Stephan at the CryoLife Corporate Headquarters in Kennesaw, Georgia. Dr. Albert E. Heacox, Senior Vice President, Research & Development invited five other scientists from CryoLife to participate in the interview. Two of the five work for AuraZyme Pharmaceuticals, Inc., a subsidiary of CryoLife. The five were: Dr. Steven Goldstein, Director, Tissue Technologies, Dr. K. Ümit Yüksel, Director BioGlue Technologies, Ms. Patti E. Dawson, Director of Allograft Tissues Research and Development, Dr. Eleanor B. McGowan, Director Research & Development, AuraZyme Pharmaceuticals, Inc. and Dr. Carl W. Gilbert, Director Manufacturing, AuraZyme Pharmaceuticals, Inc.

orthopedic tissue processed by the company since October 3, 2001."[11] Nonvalved cardiac and vascular tissues processed after September 5, 2002 were not subject to the FDA order. Revenue from tissue declined subsequent to this order.

INPUTS

The firm has received an undetermined number of SBIR awards. One database (which counts awards since 1993) says 13; another indicates 12 and excludes awards after 1997. The company reports that they have received several awards with dates subsequent to 1997. Of the 12 in the NAS database, seven were Phase I and five were Phase II. The last SBIR award that the company received was in 2002; the company currently has a Phase I award application pending. Two of the awards have been to the company's subsidiary, AuraZyme. All SBIR awards have been from NIH although not all awards have come from the same institute at NIH.

The SBIR program did not play a role in the founding of the company. The company was six years old and had 50 to 70 employees at the time that it received its first SBIR award. The current Director of Tissue Technologies, Dr. Steven Goldstein, was hired in 1991 to work on the first SBIR award. It is unclear how the company initially found out about the SBIR program, although it is thought that the Director of Research at that time knew about the program. He is no longer with the company.

The company has received funding from several sources in addition to the SBIR program. Specifically, prior to going public, the company had private investors. It has also received a contract from the Office of Naval Research and NIST funding through the ATP program. The company has received no support from the State of Georgia.

The firm went public in 1993, raising approximately $15,500,000 at that time. The company currently trades on the New York Stock Exchange. SBIR funding contributed to the success of the public offering in the sense that it helped the company initiate development of a diversified product mix by the time the company went public. For example, SBIR funding helped in the development of the vascular product that the company had at the time it went public.

KEY OUTCOMES FROM SBIR

The company reports that the SBIR program has proven key to funding the research and development behind almost all new products of the company, and the company continues to see SBIR funding as an important source of research funding. To quote one of the scientists, "SBIR funding has basically been our external source of funding when we want to build a new product."

---

[11]CryoLife annual report, available on Company's Web site, page 4.

The company has used the SBIR program to develop proprietary property that it has purchased. BioGlue® is a case in point. The BioGlue® patent was purchased but CryoLife used SBIR funds to develop concepts from the patent. AuraZyme provides another example. While the patents for AuraZyme technologies were filed before the SBIR grants were awarded, the grants have provided resources to solidify the technology and provide in vivo testing results for proof of principle.

The company holds a number of patents. While some were received subsequent to receipt of SBIR funding, in other instances SBIR funding provided the resources to "solidify the technology" and verify the facts, as noted above. The company currently has at least one pending patent application related to an SBIR award.

Company scientists have a history of publishing. Examples of articles and presentations that have been published as a result of SBIR funding include:

- Paper: Gilbert CW, McGowan EB, Seery GB, Black KS, Pegram MD. Targeted prodrug treatment of HER-2-positive breast tumor cells using trastuzumab and paclitaxel linked by A-Z-CINN Linker. *J. Exp. Ther Oncol.* 2003, Jan-Feb: 3(1), 27-35; supported by NCI grant 1R43CA95937-01.
- Poster: from 1R43CA95937-01; Proceeding of the American Association for Cancer Research 43, #2061, pg. 414, 2002, Gilbert CW, McGowan EB, Black KS, Pegram MD, Seery GB. "Efficacy testing of targeted drug delivery using A-Z-CINN Linker in vivo: a model for a single treatment cancer cure; supported by NCI grant 1R43CA95937-01.
- Poster, American Heart Association 4th Annual Conference on Arteriosclerosis, Thrombosis, and Vascular Biology, Washington, DC; May 8-10, 2003, Poster P112. McGowan EB, Gilbert CW, Black KS. AZ-Plasmin, a Novel Thrombolytic Agent for Treatment of Vascular Occlusions.

A list of published scientific papers and abstracts presented related to SynerGraft SBIR grants are given in an annex at the end of this paper.

The company holds four trademarks (BioGlue®, Synergraft®, AZ-CINN™, AuraZyme™). In three instances the trademarks are associated with products for which SBIR funding has been received. In no instance was SBIR funding used for the initial research. The fourth trademark is the name of a wholly owned subsidiary of CryoLife.

IMPACT OF SBIR ON THE FIRM

The company estimates that SBIR funding has provided 20 percent to 25 percent of all research revenue from external sources. Alternative sources for research include revenue from company sales and NIST ATP awards. The company sees SBIR awards as a continuing source of research and development support.

SBIR PROGRAM ADMINISTRATION

The company has not always chosen to follow up a Phase I with a Phase II application. In the case of BioGlue®, for example, it was decided not to apply for Phase II because the application might have required the company to disclose too much. In another instance, the research objective changed in the middle of the Phase I award. In still other instances, the aggressive timeline of the company requires that they get the project finished before a Phase II grant would become available. As was noted several times in the discussion, the company has a revenue stream from sales which helps to support research and development. It was also noted that there have been instances where the company has submitted a Phase II grant that has not been funded.

Research scientists expressed reasonable satisfaction with the way in which the SBIR program is currently structured, which now allows for larger and longer Phase Is. They see these changes as a significant improvement. A concern was expressed, however, by one scientist that Phase I awards involving animal studies are difficult to accomplish in a short period of time given the amount of paper work that is required to conduct animal studies. Another scientist stated that the duration of Phase I awards, while generally not a problem for a company the size of CryoLife, was really too short for a one or two-person company. AuraZyme is a two-person company that is fortunate to be "nested" within CryoLife's corporate structure, but still suffers from these restraints. The short time frame "pushes them to the edge." Small companies simply don't have the time to get the space and the equipment and perform the research in the traditional Phase I time frame of six months. Another scientist expressed frustration over the amount of time required between submission and the time of the award. To quote the scientist, "You can have a baby in the amount of time it takes."

One of the PIs indicated that she had benefited from the advice, and Web site, of Dr. Gregory Milman, Director, Office for Innovation and Special Programs, National Institute of Allergy and Infectious Diseases, NIH. Dr. Milman has developed a video entitled "Advice on SBIR and STTR Applications" that one can access at: <*http://www.niaid.nih.gov/ncn/sbir/advice/*>. The video takes about an hour to listen to. He also provides annotated examples of outstanding Phase I and Phase II applications at: <*http://www.niaid.nih.gov/ncn/sbir/app*>.

At least one of the PIs has had RO1 experience in the past and commented that the SBIR program would be enhanced if one could receive extension funds like one could with RO1s.

The interviewees viewed the award process as fair; on the other hand they are aware that one is at a disadvantage in writing grants that relate to proprietary information in the sense that the grant application must be vague. Reviewers pick up on this and the scores reflect this. This can be especially a problem in moving forward from a Phase I award to a Phase II application.

Finally, some frustration was expressed that the SBIR program is becoming increasingly competitive as the number of applications from university faculty members' labs increases. The hypothesis expressed was that faculty are increasingly creating companies to apply for SBIR funding to support the research of postdocs and research scientists working directly with the faculty on research. Applications from these "garage-model firms" more closely resemble RO1 grant applications and receive higher scores than do SBIR applications from established companies that are working to develop proprietary information. While it was recognized that some successful companies start out as "garage-model firms," the argument was that in many instances the SBIR program is providing funding for the faculty member's lab, rather than for a viable commercial enterprise.

CROSS-CUTTING RESEARCH ISSUES

Issues related to disclosure can inhibit a firm's participation in the SBIR program, or can affect the prospects of a favorable review. The firm also recognizes that at times the SBIR program is too slow to accommodate the aggressive timeline of the firm. As noted above, the SBIR program has proven key to funding the research and development behind almost all new products of the company and the company continues to see SBIR funding as an important source of research funding.

CRYOLIFE, INC.—ANNEX

PUBLISHED SCIENTIFIC PAPERS
SYNERGRAFT

IN PREPARATION

10 Year Results with the CryoLife O'Brien Valve. *The Journal of Heart Valve Disease.* Prof. Hvass.

2004

A Cautionary Case: The Synergraft Vascular Prosthesis. *Eur J Vasc Endovasc Surg.* 27:42-44. M.A. Sharp, D. Phillips, I. Roberts, L. Hands.

Cellular Remodeling of Depopulated Bovine Ureter Used as an Arteriovenous Graft in the Canine Model. *Journal of the American College of Surgeons.* 198(5):778-783. J. Matsuura, K. Black, C. Davenport, C. Goodman, N. Pagelsen, A. Levitt, D. Rosenthal, E. Wellons, M. Fallon, J. Ollerenshaw

2003

Congenital Abdominal Aortic Aneurysm: A Case Report. *Journal of Vascular Surgery.* 38(1):190-193. P. Bell, C. Mantor, M. Jacocks. [Article featuring use of a SynerGraft allograft.]

Decellularized Pulmonary Homograft (SynerGraft) for Reconstruction of the Right Ventricular Outflow Tract: First Clinical Experience. *Z Kardiologie.* 92(1):53-59. H. Sievers, U. Stierle, C. Schmidtke, M. Bechtel.

Early Failure of the Tissue Engineered Porcine Heart Valve SYNERGRAFT in Pediatric Patients. *Eur J Cardiothorac Surg.* 23(6):1002-1006. P. Simon, M. Kasimir, G. Seebacher, G. Weigel, R. Ullrich, U. Salzer-Muhar, E. Wolner.

Evaluation of the Decellularized Pulmonary Valve Homograft SynerGraft™. *The Journal of Heart Valve Disease.* 12:734-740. J. Bechtel, M. Muller-Steinhardt, C. Schmidtke, A. Brunswik, U. Stierle, H.H. Sievers.

Immunogenicity of Decellularized Cryopreserved Allografts in Pediatric Cardiac Surgery: Comparison with Standard Cryopreserved Allografts. *J Thorac Cardiovasc Surg.* 126(1):247-252. J. Hawkins, N. Hillman, L. Lambert, J. Jones, G. Di Russo, T. Profaize, T. Fuller, L. Minich, R. Williams, R. Shaddy.

Mechanical Heart Valve Prosthess: Identification and Evaluation (Erratum). *Cardiovasc Pathol.* 12(6):322-344. J. Butany, M. Ahluwalia, C. Monroe, C. Fayet, C. Ahn, P. Blit, C. Kepron, R. Cusimano, R. Leask.

2002

Decellularized Pulmonary Homograft (SynerGraft) For Reconstruction of the Right Ventricle Outflow Tract: First Clinical Experience. *Z Kardiol.* 92:53-59. H. Sievers, U. Stierle, C. Schmidtke, M. Bechtel.

Decellularized Cadaver Vein Allografts Used for Hemodialysis Access Do Not Cause Allosensitization or Preclude Kidney Transplantation. *American Journal of Kidney Diseases.* 40(6):1240-1243. R. Madden, G. Lipkowitz, B. Benedetto, A. Kurbanov, M. Miller, L. Bow.

Tissue Restoration and Repair—Application of Cutting-Edge Technologies to Surgical Products. *Business Briefing: Medical Device Manufacturing & Technology (Reference Section)* [CD-Rom]; K. Black.

2001

Cardiac Tissue Engineering: New Life for Ailing Hearts. *The Cardiovascular Watch.* 1(8). Remarks by K. Black. March 16.

Decellularized Human Valve Allografts. *Ann Thorac Surg.* 71:S428-432. R. Elkins, P. Dawson, S. Goldstein, S. Walsh, K. Black.

Humoral Immune Response to Allograft Valve Tissue Pretreated with an Antigen Reduction Process. *Seminars in Thoracic and Cardiovascular Surgery.* 13(4—Suppl 1):82-86. R. Elkins, M. Lane, S. Capps, C. McCue, P. Dawson.

**Recellularization of Heart Valve Grafts (SynerGraft) by a Process of Adap-

tive Remodeling. *Seminars in Thoracic and Cardiovascular Surgery.* 13(4—Suppl 1):87-92. R. Elkins, S. Goldstein, C. Hewitt, S. Walsh, P. Dawson, J. Oller.

2000

Advances in Heart Valve Surgery. *Cardiac Surgery.* November-December. [Article featuring SynerGraft Heart Valve.] Source: John E. Mayer, Jr., M.D.

Tissue Engineered Heart Valves. *The Journal of Biolaw & Business.* 3(2). K. Black.

Transpecies Heart Valve Transplant: Advanced Studies of a Bioengineered Xeno-Autograft. *Ann Thorac Surg.* 70:1962-1969. S. Goldstein, D. Clarke, S. Walsh, K. Black, M. O'Brien.

1999

The SynerGraft Valve: A New Acellular (Nonglutaraldehyde-Fixed) Tissue Heart Valve for Autologous Recellularization First Experimental Studies Before Clinical Implantation. *Seminars in Thoracic and Cardiovascular Surgery.* 11(4—Suppl 1):194-200. M. O'Brien, S. Goldstein, S. Walsh, K. Black, R. Elkins, D. Clarke.

1998

Tissue Modifications. *Transplantation Proceedings.* 30:2729-2731. K. Black, S. Goldstein, J. Ollerenshaw.

1997

Acellular Porcine Pulmonary and Aortic Heart Valve Bioprostheses (book chapter). Pp. 225-233 in *Stentless Bioprostheses Second Edition, Isis Medical Media.* D. Ross, J. Hamby, S. Goldstein, K. Black.

1994

Tissue-Based Heart Valve Grafts—New Developments. *Cardiac Chronicle.* 8(3). S. Goldstein, S. Harris.

ABSTRACT PRESENTATIONS
SYNERGRAFT

2005

Superior Durability of Synergraft Decellularized Pulmonary Allografts Compared to Standard Cryopreserved Allografts (poster presentation). *Soci-*

ety of Thoracic Surgeons. Z. Tavakkol; S. Gelehrter; C. S. Goldbert; E. L. Bove; E. J. Devaney; R. G. Ohye.

2004

Aortic Root Replacement with a Novel Decellularized Cryopreserved Aortic Homograft: Postoperative Immunoreactivity and Early Results [poster presentation]. *Advances in Tissue Engineering and Biology of Heart Valves.* K. Zehr, M. Yagubyan, H. Connolly, S. Nelson, H. Schaff.

Biomechanical Properties of SynerGraft Treated Human Heart Valves and Vascular Grafts [poster presentation]. *Advances in Tissue Engineering and Biology of Heart Valves.* S. Walsh, P. Dawson, K. Black.

Clinical Outcomes of a Depopulated Bovine Ureter (SynerGraft Vascular Graft Model 100) Used as an Arteriovenous Access Graft [poster presentation]. *Advances In Tissue Engineering and Biology of Heart Valves.* C. Darby, A. Cornall.

Decellularized Allograft Heart Valves (CryoValve SG)—Early Clinical Results from a Multicenter Registry. *Advances in Tissue Engineering and Biology of Heart Valves.* D. Clarke, R. Elkins, D. Doty, J. Tweddell.

In Vivo **Remodeling and Tissue Engineering of a Novel Decellularized Bovine Ureter Following Implantation as an Arteriovenous Graft** [poster presentation]. *Advances in Tissue Engineering and Biology of Heart Valves.* C. Hewitt, S. Marra, A. DelRossi.

Mid-Term Findings on Echocardiography and Computed Tomography After RVOT-Reconstruction: Comparison of Decellularized (SynerGraft) and Conventional Homografts. *Third EACTS/ESTS Joint Meeting.* J. M. Bechtel, J. Gellisen, A. W. Erasmi, M. Petersen, U. Stierle, H. H. Sievers.

Multicenter Clinical Outcomes with a Decellularized Porcine Pulmonary Heart Valve (SynerGraft Heart Valve, Model 700) for Reconstruction of the Right Ventricular Outflow Tract. *Advances in Tissue Engineering and Biology of Heart Valves.* R. Chard, G. Gargiulo, U. Hvass, H. Lindberg, I. Mattila, M. Redmond, L. Segadal, P. Simon.

Performance of Decellularized Bovine Ureter as a Peripheral Vascular Graft in the Dog [poster presentation]. *Advances in Tissue Engineering and Biology of Heart Valves.* S. Goldstein, J. Matsuura, C. Ponce, K. Sylvester, D. Fronk, K. Black.

2003

A Xenograft for Vascular Access: A New Start to An Old Idea? [poster presentation]. *3rd International Congress, Vascular Access Society.* A. Cornall, C. Darby.

An Underestimated Resource for Difficult Patients. The Lower Limb A-V Fistulas. *3rd International Congress Vascular Access Society.* L. Berardinelli, C. Beretta, M. Carini.

Bovine Ureter Grafts: Our Intial Experience. *The European Society for Cardiovascular Surgery 52nd Congress.* G. Esposito, T. Castrucci, A. Nicoletti, G. Canu, M. Fusari.

CryoLife O'Brien Stentless Aortic Porcine Valve at 10 Years. *Society for Heart Valve Disease 2nd Biennial Meeting.* U. Hvass, F. Baron, A. Elsebaey, D. Nguyen, E. Flecher.

Early Performance of CryoValve SG Pulmonary Heart Valve Used for the Ross Procedure. *Society for Heart Valve Disease 2nd Biennial Meeting.* J. Oury, P. Wojewski, D. Doty, D. Oswalt, R. Elkins.

Up to 8 Years Experience with the Pulmonary Autograft in Subcoronary and Root Inclusion Technique. *Society for Heart Valve Disease 2nd Biennial Meeting.* H. Sievers, U. Stierle, G. Dahmen, C. Schmidtke.

2002

Cellular Remodeling of Depopulated Bovine Ureter Used as an Arteriovenous Graft in the Canine Model. *American College of Surgeons 88th Annual Clinical Congress/Surgical Forum.* J. Matsuura, E. Wellons, K. Black, J. Ollerenshaw.

CryoVein® SG & CryoArtery® SG: Tissue Engineered Vascular Allografts for AV Access [poster presentation]. *American Association of Tissue Banks 26th Annual Meeting.* K. Sylvester, S. Capps, D. Fronk.

Depopulated Femoral Vein Allograft as an Arteriovenous Graft in High Risk Patients for Infection. *European Society for Cardio-Vascular Surgery.* J. Matsuura.

Depopulated Venacaval Homograft: A New Venous Conduit. *American Association of Thoracic Surgeons.* M. Malas, C. Baker, S. Quardt, M. Barr, W. Wells.

Early Clinical Experience with SynerGraft® for Hemodialysis Access and Peripheral Vascular Disease. *European Society for Cardio-Vascular Surgery.* B. Yoffe, E. Harah, Y. Leonty.

Immunogenicity of Decellularized Cryopreserved Allografts in Pediatric Cardiac Surgery: Comparison with Standard Cryopreserved Allografts. *American Association of Thoracic Surgeons.* J. Hawkins, J. Jones, N. Hillman, L. Lamberg, G. DiRusso, T. Profaizer, T. Fuller, R. Shaddy.

In Vivo **Resistance to Calcification of SynerGraft Tissue Engineered Heart Valve Grafts.** *American Association of Thoracic Surgeons.* R. Elkins, S. Goldstein, S. Walsh, K. Black.

Use of a Bioengineered Vascular Tissue Graft for Use in Battlefield Injuries. *23rd Army Science Conference.* K. Black, J. Matsuura, C. Davenport, C. Goodman, N. Pagelsen, J. Ollerenshaw.

2001

A Tissue Engineered Heart Valve: *In Vitro* **Performance of the SynerGraft Heart Valve.** *The Society for Heart Valve Disease First Biennial Meeting.* S. Walsh, S. Goldstein, C. Bair, K. Black.

Humoral Immune Response to Allograft Valve Tissue Pretreated with an Antigen Reduction Process. *Stentless Bioprostheses 4th Annual Symposium.* R. Elkins, M. Lane, S. Capps, C. McCue, P. Dawson.

In Vivo **Performance of an Unfixed Composite Aortic Xenograft [SynerGraft] as Aortic Root Replacement in the Sheep** [poster presentation]. *The Society for Heart Valve Disease.* R. Elkins, S. Goldstein, S. Walsh, K. Black.

New Technologies in Heart Valve Surgery. *Scandinavian Association for Thoracic Surgery 59th Annual Meeting.* S. Goldstein.

Recellularization of Heart Valve Grafts [SynerGraft] by a Process of Adaptive Remodeling. *Stentless Bioprostheses 4th Annual Symposium.* R. Elkins, S. Goldstein, S. Walsh, J. Ollerenshaw, C. Hewitt, K. Black, D. Clarke, M. O'Brien.

Results of 13-Year Follow-Up Study on Aortic Heart Valve Replacements Utilizing the Ross Procedure. *Western Thoracic Surgical Association.* R. C. Elkins, M. F. O'Brien.

SynerGraft: The First Successful Reconstruction and Regeneration Tissue Products. *Tissue Engineering for Heart Valve Substitutes Symposium.* K. Black.

SynerGraft Vascular Conduit as a Hemodialysis Access Graft in the Canine Model. *2nd International Congress Vascular Access Society.* J. Matsuura, K. Black, E. Wellons, C. Davenport, C. Goodman, K. Greene, J. Ollerenshaw.

SynerGraft Vascular Conduit as a Hemodialysis Access Graft in the Canine Model. *Eastern Vascular Society 15th Annual Meeting.* J. Matsuura, K. Black, E. Wellons, C. Davenport, C. Goodman, K. Greene, J. Ollerenshaw.

SynerGraft Vascular Conduit as a Hemodialysis Access Graft in the Canine Model [poster presentation]. *SVS/AAVS Joint Meeting.* J. Matsuura, K. Black, E. Wellons, C. Davenport, C. Goodman, K. Greene, J. Ollerenshaw.

2000

Advances in Tissue Processing. *24th Annual Meeting American Association of Tissue Banks.* K. Black.

Decellularized Human Valve Allografts. *VIII International Symposium Cardiac Bioprosthesis.* R. Elkins, K. Black, P. Dawson, S. Goldstein, S. Walsh.

In Vivo **Arterialization of SynerGraft Processed Non-Vascular Xenogeneric Conduit.** *2nd International Meeting of the Onassis Cardiac Surgery Center.* R. Hanley, R. Lust, Y. Sin, K. Black, J. Ollerenshaw.

In Vivo **Arterialization of SynerGraft® Processed Non-Vascular Xenogeneic Conduit.** *American Heart Association Scientific Sessions.*

Successful Use of Natural Revitalizing XenoGraft Connective Tissue Matrices in Animal and Human Heart Valve Replacement [poster presentation]. *International Society for Applied Cardiovascular Biology.* S. Goldstein, S. Walsh, K. Black, M. O'Brien.

SynerGraft Heart Valve: Reconstitution of an Unfixed Acellular Xenograft *In Vivo* [received AHA Citation]. American Heart Association. R. Elkins, K. Black, M. O'Brien, S. Goldstein, S. Walsh, D. Clarke, S. Bode, J. Hamby.

SynerGraft Tissue Conduit is Adopted by Host in Aortic Reconstruction. *VIII International Symposium Cardiac Bioprosthesis.* D. Clarke, R. Lust, Y. Sun, K. Black, J. Ollerenshaw.

SynerGraft® Treatment of Valve Allografts [poster presentation]. *24th Annual Meeting American Association of Tissue Banks.* P. Dawson, S. Goldstein, S. Walsh, K. Black.

SynerGraft Vascular Tissue Conduit is Rapidly Recellularized Following Peripheral Bypass. *European Association for Cardio-Thoracic Surgery 14th Annual Meeting.* D. Clarke, M. Tillson, K. Black, J. Ollerenshaw.

Tissue Heart Valve Engineering: Experience with an Autologous Engineered Xenograft. *Peripheral Vascular Surgical Society.* M. O'Brien.

Transpecies Heart Valve Transplant: Advanced Studies of a Bioengineered Xeno-Autograft. *The Society of Thoracic Surgeons 36th Annual Meeting.* S. Goldstein, D. Clarke, S. Walsh, K. Black, M. O'Brien.

Use of Decellularized Cadaver Allograft (SYN) Does not Cause Allosensitization in Hemodialysis Patients and is Safe for Use in Potential Transplant Recipients. *2001 ASN/ISN World Congress of Nephrology.* G. Lipkowitz, B. Benedetto, R. Madden, A. Kurbanov, L. Bow, M. Miller, J. Matsuura.

1999

A New Era in Health Care. *Cambridge Health Care Institute Meeting on Tissue Engineering.* S. Goldstein.

A Simple Tissue Implant Model to Study Xenogeneic and Allogeneic Rejection [poster presentation]. *Association for Academic Surgery.* H. Tran, M. Puc, N. Patel, S. Goldstein, J. Ollerenshaw, K. Black, A. DelRossi, C. Hewitt.

Acellular Porcine Heart Valve Leaflets Do Not Mineralize in the Ovine RVOT; *Stentless Bioprosthesis Third International Symposium.* S. Goldstein, K. Black, D. Clarke, E.C. Orton, M. O'Brien.

Advanced Tissue / Cellular Engineering. *European Medical & Biological Engineering Conference.* K. Black.

Advances in Tissue Processing. *24th Annual Meeting American Association of Tissue Banks.* K. Black.

Decellularized Human Valve Allografts. *VIII International Symposium Cardiac Bioprosthesis.* R. Elkins, K. Black, P. Dawson, S. Goldstein, S. Walsh.

***In Vivo* Arterialization of SynerGraft Processed Non-Vascular Xenogeneric Conduit.** *2nd International Meeting of the Onassis Cardiac Surgery Center.* R. Hanley, R. Lust, Y. Sin, K. Black, J. Ollerenshaw.

***In Vivo* Arterialization of SynerGraft® Processed Non-Vascular Xenogeneic Conduit.** *American Heart Association Scientific Sessions.*

Inflammatory Responses to Uncrosslinked Xenogeneic Heart Valve Matrix. *World Heart Valve Disease Symposium.* S. Goldstein, K. Black, D. Clarke, E.C. Orton, M.F. O'Brien.

Performance of an Acellular, Composite Porcine Heart Valve Bioprosthesis in the Ovine RVOT. *34th Congress of the European Society for Surgical Research.* S. Goldstein, S. Walsh, K. Black, E.C. Orton, D. Clarke.

Successful Trans-Species Implant of a Tissue Engineered Heart Valve. *First Satellite Symposium on Tissue Engineering for Heart Valve Bioprostheses.* K. Black.

Transpecies Heart Valve Transplant: Advanced Studies of a Bioengineered Autograft. *The European Association for Cardio-Thoracic Surgery.* S. Goldstein, S. Walsh, K. Black, D. Clarke, E.C. Orton, M.F. O'Brien.

1998

Meniscal Transplantation to Tissue Engineering: CryoLife's Vision of Orthopedics. *Osteochondral Autograft Transfer System 1998 Meniscus and Cartilage Transplantation Study Group on Meniscus Reconstruction.* K. Black.

1997

Acellular Porcine Pulmonary and Aortic Heart Valve Bioprostheses. *2nd Intl Stentless Bioprostheses Symposium.* D. Ross, J. Hamby, S. Goldstein, K. Black.

Effects of Cryopreservation on Biomechanical Properties of Tissue Engineering Matrices. *Workshop on Biomaterials and Tissue Engineering.* S. Goldstein, J. Hamby.

The Ross Operation in Children: 10-Year Experience. *33rd Annual Meetingof the Society of Thoracic Surgeons.* R. Elkins, C. Knott-Craig, K. Ward, M. Lane.

1995

Age-Dependent Alterations in Collagen Cross-Links in Porcine Aortic Heart Valve Leaflets. *American Society for Biochemistry and Molecular Biology.* S. Goldstein, M. Yamauchi.

Modulation of Human Dermal Fibroblast Remodeling of Porcine Heart Valve Leaflet Matrix. *American Society for Artificial Internal Organs.* S. Goldstein, D. Fronk.

1994

Development of a Chimeric Heart Valve: Effects of Cell Removal upon Leaflet Mechanics and Immune Responses in a Xenogeneic Model. *VI International Symposium Cardiac Bioprostheses.* S. Goldstein, K. Brockbank.

Localization of Epitopes Involved in Hyperacute Rejection in Porcine Heart Valve Xenografts. *Strategies for Xenotransplantation.* S. Goldstein.

1993

Effects of Cell Removal Upon Heart Valve Leaflet Mechanics and Immune Responses in a Xenogeneic Model. *2nd Intl Congress on Xenotransplantation.* S. Goldstein, K. Brockbank.

Illumina[12]

Robin Gaster
North Atlantic Research

EXECUTIVE SUMMARY

Illumina is a successful venture funded biomedical company selling tools for researchers in genomics. Located on a new campus in the San Diego biomedical cluster, the company was expected to reach break-even in 2005.

Illumina is a classic example of a venture-funded biomedical company, one that has gone public and appears poised to reach profitability in 2006, while providing cutting edge technology in an area of critical importance for the large-scale analysis of genetic variation and function. Because Illumina received venture funding during its first year in existence, and subsequently received further rounds before a successful IPO, SBIR was never the primary source of research funding.

However, according to its founder and one of its key initial researchers, Dr. Mark Chee, SBIR did provide funding for projects that would not have been funded in the normal course of company business—and these projects turned out to be of critical importance for the development of core Illumina product lines.

The Illumina case therefore shows that even where companies are well funded, SBIR can have an important impact in funding alternative or complementary lines of business that might not fit within a company's top research priorities, or might not meet projected internal hurdle rates.

Comments from Dr. Chee about the SBIR program focused on the extended funding cycle and time lags, and selection and review procedures.

COMPANY HISTORY AND OBJECTIVES

Illumina was founded in April 1998 by David Walt, Ph.D., CW Group (Larry Bock), John Stuelpnagel, D.V.M., Anthony Czarnik, Ph.D. and Mark Chee, Ph.D., based on core technology developed at—and then exclusively licensed from—Tufts University. The first substantial venture capital funding (about $8.6 million) came in November 1998.

Illumina completed a $28 million Series C financing in December 1999, and an IPO at the end of July 2000, raising just over $100 million.

Illumina's mission is to develop tools for the large-scale analysis of genetic

---

[12]Based on the following interviews: Dr. Jian-Bing Fan, February 25, 2005 (at Illumina); Dr. Mark Chee (founder and former Vice. President of Genomics at Illumina) (now CEO of Prognosys) (by phone) December 14, 2006.

DNA
SNP genotyping is a method of determining
variation in genetic sequences.

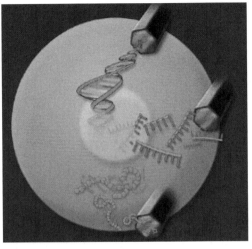

RNA
Gene expression
profiling is the
analysis of which
genes are active—
in particular a cell
or group of cells.

Proteins
Proteomics is the process of determining which
proteins are present in cells and how they interact.

FIGURE App-D-4

SOURCE: Illumina.

variation and function, which in turn will support an understanding of variation
and function at the cellular level, critical for achieving the broad social goal of
personalized medicine.

Illumina's tools convert data generated from human genome sequencing
into medically relevant information, linking genetic variation and genetic function to specific diseases, improving the community's ability to discover drugs,
and permitting diseases to be detected earlier, and with greater accuracy and
specificity.

Massive quantities of raw genetic data have flowed from the successful sequencing of the human genome. This has driven demand for tools that can assist
researchers in processing the billions of tests necessary to convert raw data into
medically valuable information. Such tools must perform functional analysis of
highly complex biological systems. Illumina's technology platform has been
developed into a line of products that can address the scale of experimentation
and the breadth of functional analysis required.

ILLUMINA TECHNOLOGY

BeadArray Technology

Illumina has developed a proprietary array technology that enables the large-scale analysis of genetic variation and function. BeadArray technology combines microscopic beads and a substrate in a simple proprietary manufacturing process to produce arrays that can perform many assays simultaneously.

This approach provides a combination of high throughput, cost effectiveness, and flexibility:

- *High throughput* is achieved by using a high density of test sites per array, and by formatting arrays in either a pattern arranged to match the wells of standard microtiter plates or in various configurations in the format of standard microscope slides, allowing throughput levels of up to 150,000 unique assays per plate. Laboratory robotics are also used to speed processing time.
- *Cost effectiveness* is maximized by reducing consumption of expensive reagents and valuable samples, and through low-cost manufacturing processes that exploit cost reductions generated by advances in fiber optics,

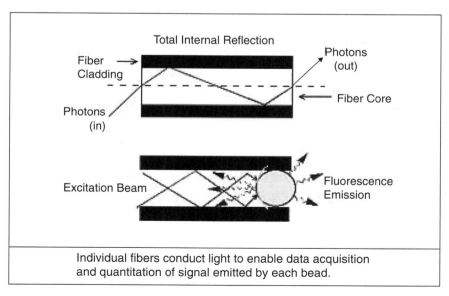

FIGURE App-D-5 Bead-based technology on a chemically etched fiber optic strand.

SOURCE: Illumina.

digital imaging, and bead chemistry technologies. Per-sample running costs, including labeling, will typically range between $80 and $200, comparable in cost to just the sample labeling steps using other microarray platforms.

- The *flexibility* needed to address multiple markets segments is provided by varying the size, shape, and format of the well patterns, and creating specific bead pools or sensors for different applications.

BeadArray technology is deployed by Illumina in two different array formats, the Array Matrix and the BeadChip. Illumina's first bead-based product was the Array Matrix, which incorporates fiber optic bundles, manufactured to Illumina specifications, cut into lengths of less than one inch. Each bundle contains approximately 50,000 individual fibers.

Ninety-six bundles are placed into an aluminum plate which forms an Array Matrix. BeadChips are fabricated in microscope slide-shaped sizes with varying

---

BOX App-D-1
Genetic Variation and Function

Every person inherits two copies of each gene, one from each parent. The two copies of each gene may be identical, or they may be different. These differences are referred to as genetic variation. Examples of the physical consequences of genetic variation include differences in eye and hair color. Genetic variation can also have important medical consequences, including predisposition to disease and differential response to drugs. Genetic variation affects diseases, including cancer, diabetes, cardiovascular disease and Alzheimer's disease. In addition, genetic variation may cause people to respond differently to the same drug. Some people may respond well, others may not respond at all, and still others may experience adverse side effects. The most common form of genetic variation is a Single Nucleotide Polymorphism, or SNP. A SNP is a variation in a single position in a DNA sequence. It is estimated that the human genome contains between three and six million SNPs.

While in some cases a single SNP will be responsible for medically important effects, it is now believed that the genetic component of most major diseases is the result of the interaction of many SNPs. Therefore, it is important to investigate many SNPs together in order to discover medically valuable information.

Current efforts to understand genetic variation and function have primarily centered around SNP genotyping and gene expression profiling.

SNP Genotyping
SNP genotyping is the process of determining which SNPs are present in each of the two copies of a gene, or other portion of DNA sequence, within an individual or other organism. The use of SNP genotyping to obtain meaningful statistics on the effect of an individual SNP or a collection of SNPs, and to apply that information to clinical trials and diagnostic testing, requires the analysis of millions of SNP genotypes and the testing of large populations for each disease. For example, a single large clinical trial could

numbers of sample sites per slide. Both formats are chemically etched, to create tens of thousands of wells for each sample site.

In a separate process, Illumina create sensors by affixing a specific type of molecule to each of the billions of microscopic beads in a batch. Different batches of beads are coated different specific types of molecule. The particular molecules on a bead define that bead's function as a sensor. For example, Illumina creates a batch of SNP sensors by attaching a particular DNA sequence to each bead in the batch. Batches of coated beads are combined to form a pool specific to the type of array. A bead pool one milliliter in volume contains sufficient beads to produce thousands of arrays. This technology permits the creation of universal arrays for SNP genotyping. By varying the reagent kit, users can still use the array to test for any combination of SNPs.

To form an array, a pool of coated beads is brought into contact with the array surface where they are randomly drawn into wells, one bead per well. The tens of thousands of beads in the wells comprise individual arrays. Because the

involve genotyping 200,000 SNPs per patient in 1,000 patients, thus requiring 200 million assays. Using previously available technologies, this scale of SNP genotyping was both impractical and prohibitively expensive. Large-scale SNP genotyping will be used for a variety of applications, including genomics-based drug development, clinical trial analysis, disease predisposition testing, and disease diagnosis. SNP genotyping can also be used outside of healthcare, for example in the development of plants and animals with desirable commercial characteristics. These markets will require billions of SNP genotyping assays annually.

Gene Expression Profiling
Gene expression profiling is the process of determining which genes are active in a specific cell or group of cells and is accomplished by measuring mRNA, the intermediary between genes and proteins. Variation in gene expression can cause disease, or act as an important indicator of disease or predisposition to disease. By comparing gene expression patterns between cells from different environments, such as normal tissue compared to diseased tissue or in the presence or absence of a drug, specific genes or groups of genes that play a role in these processes can be identified. Studies of this type, used in drug discovery, require monitoring thousands, and preferably tens of thousands, of mRNAs in large numbers of samples. Once a smaller set of genes of interest has been identified, researchers can then examine how these genes are expressed or suppressed across numerous samples, for example, within a clinical trial. The high cost of current gene expression methods has limited the development of the gene expression market.

As gene expression patterns are correlated to specific diseases, gene expression profiling is becoming an increasingly important diagnostic tool. Diagnostic use of expression profiling tools is anticipated to grow rapidly with the combination of the sequencing of various genomes and the availability of more cost-effective technologies.

SOURCE: Illumina.

beads assemble randomly into the wells, a final procedure called decoding is used in order to determine which bead type occupies which well in the array. Decoding also validates each bead in the array—a further quality control test. By using multiple copies of each bead type, the reliability and accuracy of the resulting data is improved via statistical processing of results from identical beads.

An experiment is performed on the Array Matrix by preparing a sample, such as DNA from a patient, and introducing it to the array. The Matrix is dipped into a solution containing the sample, and molecules in the sample bind to matching molecules on the coated bead. The BeadArray Reader detects the matched molecules by shining a laser on the fiber optic bundle. Measuring the number of molecules bound to each coated bead, results in a quantitative analysis of the sample.

Oligator Technology

Genomic applications require many different short pieces of DNA that can be made synthetically, called oligonucleotides (single-stranded DNA). For example, SNP genotyping typically requires three to four different oligonucleotides per assay. An SNP genotyping experiment analyzing 10,000 SNPs may therefore require 30,000 to 40,000 different oligonucleotides, contributing significantly to the expense of the experiment.

Illumina's Oligator technology is designed for the parallel synthesis of many different oligonucleotides. Each synthesizer can produce up to 3,072 oligos in parallel, using very small amounts of material.

PRODUCT ROLLOUT AND COMMERCIAL RESULTS

In 2001, Illumina launched its commercial genotyping service product line, combining BeadArray technology with an automated process controlled by a laboratory information management system to provide high throughput identification of the most common form of genetic variation, known as single nucleotide polymorphisms, or SNPs.

In 2002, Illumina launched BeadLab, an integrated turnkey system built around BeadArray technology. The BeadLab can routinely produce up to 1.4 million genotypes per day.

In 2003, Illumina launch several new products, including 1) a new array format, the Sentrix BeadChip; 2) a gene expression product line on both the Sentrix Array Matrix and the Sentrix BeadChip that allows researchers to analyze a focused set of genes across 8 to 96 samples on a single array; and 3) a benchtop SNP genotyping and gene expression system, the BeadStation, for performing moderate-scale genotyping and gene expression using our technology.

As of end-2004, nine BeadLabs were in use, along with 42 BeadStations.

TABLE App-D-4 Revenue and Expenses Trends at Illumina, 2000-2004

	Year				
	2004	2003	2002	2001	2000
Total Revenue ($)	50,583	28,035	10,040	2,486	1,309
Total Costs and Expenses ($)	56,096	54,657	51,895	32,805	24,544
Net Loss ($)	(6,225)	(27,063)	(40,331)	(24,823)	(18,606)

SOURCE: Illumina.

In 2005, Illumina bought CyVera, whose digital-microbead platform is highly complementary to Illumina products and services.

The systematic rollout of these technologies has provided a firm base for rapidly expanding revenues at Illumina. The 2005 annual report shows that revenues have increased dramatically in 2003 and 2004, and that the company seemed poised for profitability in 2005.

This strongly positive trend is reflected especially in revenue growth, as shown in Figure App-D-6.

THE ROLE OF SBIR

Dr. Chee is the source for SBIR related activities at Illumina, as he was the principal investigator on most early SBIR awards, and has been a strong champion of the program within the company.

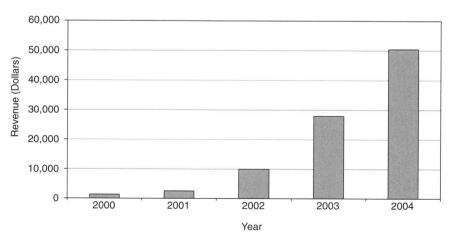

FIGURE App-D-6 Illumina revenue growth 2000-2004.

SOURCE: Illumina.

According to Dr. Chee, SBIR was a very positive experience for working with VCs, as it provided a significant technical validation of the technology. While the initial round of funding came before the first SBIR award, the latter was very important for the second round of financing a year later. However, it would not be accurate to say even here that SBIR made the difference between VC funding or not.

The real importance of SBIR at Illumina was that it provided flexibility in pursuing projects outside the mainstream of immediate research objectives. As such, it provided a key counter-balance to the tendency to over-focus, which is perhaps inevitable in a small company. At Illumina, SBIR definitely promoted a more diverse approach to R&D. And of course, additional funding is always useful to researchers.

The additional opportunities provided by SBIR almost all paid off in commercially successful, revenue-generating products.

(1) Genotyping system.

There was a lack of demand from customers at that time—mainly because they were genotyping at a very much smaller scale (by a couple of orders of magnitude or thereabouts). Although this lack of immediate demand would probably not have blocked the project receipt of SBIR funding allowed this research to progress more rapidly.

The genotyping project became the technical foundations for a critical product line, which in turn became the base for Illumina's work with The International HapMap Project, a highly prestigious international genotyping project. The Illumina technology has been enormously important for cutting end-user costs, which is especially important for some of the developing nations participating in the project—for example China, which is using the technology to meet its commitments to the HATMAP project.

SBIR funding was used to start work on the foundation research used to determine the best way to implement design of the array-based system. Positive results during the SBIR-funded research quickly led to substantial subsequent company investment. According to Dr. Chee, SBIR funding for this project lies somewhere between necessary and really useful.

(2) The pyro-sequencing project.

The pyro-sequencing project was another nonmainstream project that would not have been attempted without SBIR support. In this case, while the technical results from the research were good, the project was eventually abandoned for business reasons.

(3) Gene expression profiling.

This project constitutes an important SBIR success story, leading to significant Illumina products. The profiling project was a lower priority for Illumina, partly because this appeared to be an effort that would compete directly with the much better funded and established Affymetrix.

SBIR allowed the company to project its thinking into the next wave of technology, and this worked out very well. It took about 3 years to develop this tech to the marketplace. That technology created the base for Illumina's product lines covering whole genome expression arrays. These arrays generate superb data, and are a highly successful commercial product. They would not have been possible without SBIR.

Overall, it is clear that SBIR was important at Illumina because it allowed Dr. Chee to successfully champion projects that would not normally have been funded by the company, projects with high-risk/high-return characteristics. SBIR funding was not needed for core company projects and research, but within the company—like any small company, even a well funded start-up, there was limited funding for projects outside the immediate research stream. As a result, the projects funded by SBIR would not have been funded in the normal course of business.

SBIR at Illumina should not be understood as focusing on *peripheral* research; instead, it allowed a focus on *higher-risk* research that was positioned further from the market projects that resulted in dramatic improvements in the core technology.

There has been a very big pay-off from SBIR projects at Illumina:

- Genotyping.
- Parallel arrays.
- Gene expression profiling.

The first two are integral parts of Illumina's main product lines. They have returned many times the original investment in revenues for the company.

It is also worth noting that Illumina's experience in some ways confirms the very short product cycle identified in many Phase II Recipient Survey responses. For example, Illumina's parallel array processor, which addresses multiple arrays at the same time, was originally seen as developed of a technology platform. Research went very fast, and Illumina was able to start selling a commercial system before the end of Phase II.

RECOMMENDATIONS FOR CHANGES TO THE SBIR PROGRAM

The following comments are from the interview with Dr. Chee, as he has been the primary SBIR contact at Illumina.

Dr. Chee said that in general, he believed the SBIR program was highly successful. He had some observations about areas of possible improvement.

Selection/Review

Overacademic Reviews

Like a number of other applicants in his experience, Dr. Chee said that he often received comments that reflected lack of understanding of the differences between commercial and academic R&D. One example of the differences could be found in his work on gene expression profiling

Because the company was building work in this area without a preexisting base, technology was at every stage of the project extremely crude. Initial results of the research were as a result "awful." However, that research also suggested that the theory on which the research was based was valid, and there were some very preliminary indications that the technology would in the end work well.

Reviewer comments seemed to indicate expectations that by end of Phase I, there would be a system in place that was performing well. In Dr. Chee's view, this indicated a very naïve understanding of how new technology gets invented in a company, as opposed to in an academic setting, where preliminary successful research results are usually required before substantial grants are awarded.

This misunderstanding of company-based research led to problems with overacademic reviews. Dr. Chee pointed out that RO1 applications typically came from universities with existing labs already in place for preliminary experiments, a system of research, and staff and grad students. The result was usually lots of good preliminary data.

In contrast, Illumina started with three people at a conference table. Work was literally conducted sitting on the floor, writing on notepads. Everything was built from scratch: the Illumina team wrote their own software, and mixed their own reagents. *Nothing* worked well during the initial period, and even later, good results often took significant amounts of time. As a result, Illumina's applications "looked pretty sketchy" in the initial period. Reviewers of these early applications wanted an R01 type approach, and clearly did not understand the corporate research environment of a start-up.

Commercial Review

Dr. Chee observed that academic review for commercial potential was likely to be a futile exercise for Phase I, and that pressure to develop a complete commercialization plan at this stage was likely to be more trouble than help to the company. He also noted that presenting more material on commercialization did not always work to the company's advantage in review, as it presented more material for reviewers to criticize. Instead of the focus on commercialization planning, Dr. Chee suggested that NIH focus on commercialization potential—commercialization track record and other funding.

In general, Dr. Chee thought that the concept of a commercialization consultant attached to the review section might be worth exploring, but he did not think that a separate two-stage review which separated commercialization and technical review would be a good idea, as it could add additional delays.

Delays/Funding Cycle

Dr. Chee emphatically noted that the primary problem with the SBIR program was the funding cycle and the long delays between application and funding. In his recent experience, applications were likely to be "in limbo for a year even if the application was eventually successful."

Anything that could reduce the length of the funding cycle was worth exploring, and Dr. Chee was very positive about the possibility of offering applicants a chance to provide a short written "rebuttal" to the comments of the lead reviewer during the first. This in his view fit well with the new system implemented at CRS about 2 years ago whereby lead reviewers prepared comments before the meeting of the study section.

Similarly, Dr. Chee strongly favored "instant scoring"—the notion that the actual score should be developed very quickly after the meeting of the review panel, and also that scores once assigned should be released electronically to applicants immediately, well before final pink sheet comments could be available. He also favored any methods for accelerating pink sheet distribution itself.

Conflict of Interest

Dr. Chee said that in the real world it is difficult to avoid conflicts completely and still get good reviewers. However, his approach was not to worry too much about potential conflicts, trusting to the system to sort that out. He has tried to point out direct competitors (e.g., for Illumina applications, Affymetrix) who should not act as reviewers.

However, Dr. Chee also recognized that the review process is intrinsically difficult, and that the SBIR program does this work reasonably well, in comparison with other NIH selection panels with which he has been involved.

Award Size

In general, Dr. Chee noted that there was room for much better cost-benefit analysis by NIH, and that one size (award) did not and should not fit all applicants. He believed that if different size awards became the norm, it would be especially important for NIH to developed procedures for assessing the relative costs and likely benefits of applications.

Specially, he saw merit in testing approaches that would weight applications inversely for the funding required, so that applications that were especially expensive would have to provide correspondingly greater benefits.

He observed that the very small size of Phase I actually worked against startups that had no other resources on which to draw and possibly no infrastructure, while being required to present feasible and exciting projects. However, he believed that the current size of the Phase I award at NIH was appropriate and should not be increased.

Inhibitex[13]

Paula Stephan
Georgia State University

DESCRIPTION OF THE FIRM

Inhibitex was founded in 1994 at Texas A&M where the co-founders, Dr. Joseph M. Patti and Dr. Magnus Höök, were on the faculty. The firm moved to Atlanta in 1998 when it received funding from Alliance Technology Ventures, which has a mandate to build biotech in Atlanta. Inhibitex hired its first employee soon after arriving in Atlanta. During its early years in Atlanta the firm worked out of lab space at Georgia State University; the firm got dedicated space after hiring its fifth employee. It currently has approximately 75 employees and will relocate to new space, financed in part by the state of Georgia, in 2005. Dr. Patti has been full-time with the firm since 1998. He currently serves as Vice President, Preclinical Development and Chief Scientific Officer. He is also a director of the firm. Dr. Höök remains on the faculty of Texas A&M and serves on the Scientific Advisory Board of the company.

The scientific platform for the company is MCSRAMM; the research underpinning this platform came out of Texas A&M. The company has two products in clinical trails: Veronate and Aurexis. Vernoate is in Phase III clinical trials as an anti-infectious drug to prevent hospital-associated infections in very low birth weight infants (VLBW infants). There are approximately 60,000 VLBW infants born each year in the United States and studies indicate that 30 to 50 percent develop at least one hospital-associated infection while in the neonatal intensive care unit, resulting in significant mortality and morbidity. Veronate has been awarded Fast Track and Orphan Drug status by the FDA. Clinical trials started in 2002. Aurexis is designed to combat *S. aureus* blood stream infections in hospitalized patients (staph). It is a leading cause of hospital-associated infections and related mortality. It is estimated that there were approximately one million cases of hospital-associated *S. aureus* infections worldwide in 2002. Aurexis is designed to be used in tandem with standard antibiotic treatments. Aurexis has recently completed a 60-patient Phase II clinical trial.[14] The company has three additional product candidates in preclinical development. The company has a collaborative agreement with Wyeth for global development of vaccines targeting staphylococcal infections. The company also has a co-collaboration agreement with Dyax, a company based in Cambridge, MA, for the development of human monoclonal antibodies targeting enterococci.

---

[13]Based on an interview with Dr. Joseph Patti, February 17, 2005, at Inhibitex, Alpharetta, GA.

[14]See company prospectus, dated June 3, 2004, pages 1-2, for discussion of products.

The company has raised approximately $174 million since 1998: $85 from private funding, which includes venture capital from New Enterprise Associates and Alliance Technology Ventures, and $39 million from its IPO in June of 2004 and $50 million from a PIPE financing in November 2004. The company's stock is traded on NASDQ under the ticker "INHX."

Since inception the company has not generated any revenue from the sale of products and does not expect to until it receives regulatory approval for commercialization of products. Its current revenue (approximately $1 million in 2003) comes from the amortization of an up-front license fee, quarterly research and development support payments received in connection with a license and collaboration agreement with Wyeth and a grant received from the FDA's Office of Orphan Products Development.

INPUTS

The company has had one SBIR, Phase I. It was awarded 9/15/1999, for $99,350; Joseph Patti was the PI. The company, according to Dr. Patti, applied for three other SBIR Phase Is which were unfunded. The funded study, according to Dr. Patti, was designed to look at the potential of a multicomponent *S. aureus* vaccine. The company had fewer than five employees at the time the SBIR award was received.

The co-founders of the firm were both affiliated with Texas A&M at the time the company was founded. Joseph M. Patti was assistant professor at Texas A&M Institute of Biosciences and Technology (1994-1998) and co-founder Magnus Höök was Regents Professor and Director for the Center of Extracellular Matrix Biology. Prior to his appointment as an assistant professor, Dr. Patti was a postdoc in the lab of Dr. Höök. The SBIR award played no role in the creation of the firm.

In addition to SBIR funds, the firm received government funds from the FDA as a result of Veronate being awarded orphan drug status. The company has a collaborative agreement, noted above, for the global development of vaccines against staphylococcal infections. The company received venture funding and had an initial public offering in 2004 (see discussion above).

The company does not see the SBIR program as playing a role in the decision of external partners to provide funding. Indeed, Dr. Patti expressed the view that SBIR funding could potentially be a drawback to the receipt of VC money because VCs might look unfavorably on the reporting obligations associated with an SBIR award. He also noted that disclosure can be an issue: "The IP people get kind of antsy when you start talking about some of these grants, whether they are considered confidential or not confidential. Who's reviewing it? Do they have an alliance? Are they competitors?"

KEY OUTCOMES FROM SBIR

According to Dr. Patti, although preliminary, the data generated from the SBIR award were "interesting and prompted us to continue the work, albeit with a slightly different focus." Dr. Patti went on to say that "one could argue that it [SBIR] helped get our Wyeth deal, although even in the absence of this funding, we would have pursued the vaccine approach." The collaboration agreement with Wyeth was executed prior to the company going public and was described in the IPO prospectus.

The company has a number of patents, and Inhibitex scientists have published papers in scientific journals. None of the patents or publications relate directly to the SBIR award.

IMPACT OF SBIR ON THE FIRM

The company does not see the SBIR award as playing a key role in the company's strategy and currently does not anticipate applying for further SBIR awards. "I have struggled with the applicability of this program to a growing firm; useful if you have a side project—useful to look at noncore areas—but how could you survive if you were asking SBIR funding for the core?" According to Dr. Patti, even the larger grants of several million dollars that are currently being awarded are too small to run a small Phase II clinical trial. The Inhibitex 60-person Phase II clinical trial for Aurexis cost approximately $5 million. SBIRs can be helpful in funding exploratory science, but "you can't grow your organization based off of them."

Dr. Patti expressed the opinion that SBIRs were not compatible with the life span of an early stage biotech company. He estimates the length of time between writing the proposal and receiving SBIR money to be approximately a year, while the life span of an early stage biotech company is at most two years. Moreover, the amount of money (at least when Phase Is were limited to $100,000) was insufficient to support research unless the firm is in an academic lab and does not have to pay overhead, etc. If the firm is in dedicated space, the amount of funding is insufficient and is "not compatible with the expectations of the investors that you are going after."

SBIR ADMINISTRATION

The company sees the SBIR program as being well publicized. Other than venture capital and participation in the ATP program, "this is it."

The company has not participated in any business/commercialization support activities sponsored by (a) the funding SBIR agencies, (b) the states related to SBIR opportunities.

The company clearly sees the size and duration of the SBIR awards (at least as they existed in the late 1990s and early 2000s) to be insufficient for a dedicated biotech firm. The greatest drawback, from the company's point of view, is the speed of the SBIR process. The 18 months that elapse between writing the proposal and receipt of the money is too long. Increasing turnaround takes precedence over increased funding: "Raising the amount is good but turning it around faster is really important."

The company sees the award selection process as biased toward the NIH mentality of "show me the data and then I'll fund it." While this is a workable model for established PIs at universities, it is inconsistent with an early stage biotech company that is "pushing the envelope to find out what the data is." Companies need the money before the study has been conducted; not after it has already been done. Yet the review process (and resulting priority score) is biased towards proposals that provide data and "have all the answers."

Dr. Patti suggests that NIH have representatives from biotech companies as part of the review process. People who work in the biotech world on a daily basis understand the issues faced by a biotech company while university-based scientists appear to have less appreciation for such issues. NIH priority scores often reflect this lack of understanding.

There was a considerable amount of paperwork associated with the grants and at the time the company had the SBIR there was a fairly obtuse financial reporting system that required one to report via a dial in system. The reporting period extended for two-to-three years even though that grant was for one year.

CROSS-CUTTING RESEARCH ISSUES

The company expressed the view that two factors inhibit participation in the SBIR program: VC-related issues and the slowness of turnaround. The turnaround issue was discussed above. The VC-related issues cut both ways. First, as noted above, venture capitalists may see the reporting requirements associated with the SBIR award as detracting from the desirability of investing in the firm. Second, a company is now excluded if VC has more than a 50 percent stake in the company. This means that the most attractive companies with the most potential are now excluded from SBIR eligibility. If this rule had existed when Inhibitex applied, they would not have been able to get an SBIR award. "If you are successful, VC owns lots of your company."

SBIRs provide peer recognition of quality science. SBIR awards are noted in BioWorld as well as trade magazines. The receipt of an award sends a signal comparable to that of having a paper published. In some cases one may be able to leverage this recognition and associated funding with an early investor. But there are also downsides, as noted above.

JP Laboratories

Andrew Toole
Rutgers University

COMPANY AND FOUNDER BACKGROUND

JP Laboratories, Inc. is a privately-held research and development (R&D) company located in Middlesex, NJ. The company was founded by Dr. Gordhan N. Patel in 1983 to invent products based on his research experience and emerging knowledge in the fields of chemistry, physics, and biology. Dr. Patel founded the company using an initial $100,000 investment of personal funds. Over the last twenty-two years, he has implemented a successful business strategy that has allowed JP Laboratories (Labs) to remain a small—never more than five employees—but innovative firm. His business strategy focuses on inventing products and licensing them to larger firms for commercialization. According to Dr. Patel, his company has invented twenty products and successfully licensed ten of these for commercialization, SBIR funds, other governmental funds, and the royalty streams from these licenses have allowed the company to expand its R&D capabilities and continue the invention process.

In 2003, following the invention of their Self-indicating Instant Radiation Alert Dosimeter (SIRAD), Dr. Patel decided to expand the company's business strategy to include manufacturing and sales for their SIRAD product. This product, which is described more fully in the next section, evolved from decades of Dr. Patel's research and discovery activities related to various types of indicator devices. SBIR awards from DoD and NIH supported part of the research underlying SIRAD. JP Labs and Dr. Patel received several major awards and recognitions for developing SIRAD: (1) in 2003, Dr. Patel was invited to testify to a congressional subcommittee about SIRAD's use in counterterrorism, *<http://reform.house.gov/UploadedFiles/Patel%20Testimony.pdf>*; (2) in 2004, JP Labs received the Frost & Sullivan Excellence in Technology Award in the field of homeland security for this product; and (3) in 2005, JP Labs received R&D-100 award (see the photo at the end of this document). Dr. Patel's decision to expand the business strategy of JP Labs is likely to dramatically change their corporate profile going forward. The next several years will be a critical transition period involving additional private investment, facilities expansion, new hiring, and internal corporate restructuring.

Dr. Patel is a good example of a "scientist-inventor-entrepreneur." Born and raised in Manund (Gujarat state), a small village in India, Dr. Patel developed a strong scientific background as a university student and research scientist. He earned undergraduate and graduate degrees in chemistry and physics from Sardar Patel University in Vidyanagar. In 1970, having just completed his Ph.D. in physics on the crystallization of polymers, he joined the research lab of Dr. Andrew Keller at the University of Bristol, UK. After three years at Bristol, he spent a short time in a research position at Baylor University in Waco, Texas. As a scientist, Dr. Patel published over 65 research papers in peer-review journals. From there, he joined Allied Corp. and worked as a bench-level scientist for nearly ten years. While at Allied, Dr. Patel was an inventor and co-inventor on numerous patents on polymers, crystals, and time-temperature indicators. According to the U.S. Patent and Trademark Office, Dr. Patel is an inventor on 38 different U.S. patents since 1975. In 1983, when he lost his job due to downsizing at Allied, Dr. Patel became an entrepreneur by founding JP Laboratories.

SBIR AND INVENTION AT JP LABS

The SBIR Program provided vital financial support for product innovation at JP Labs from the very beginning of the company. Their first SBIR award, granted by the U.S. Army, was received in the year the company was founded, 1983. Since that time Dr. Patel has successfully won seventeen SBIR awards from a variety of agencies including the DoD, NSF, NIH, and EPA. Twelve of these awards were for Phase I feasibility studies and five were for Phase II product development. The awards total over $2.6 million (in nominal dollars) through 2005 (their last SBIR award was in 2001). Figure App-D-7 shows the time profile of awards to JP Labs.

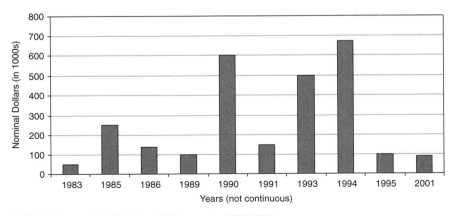

FIGURE App-D-7 JP Labs SBIR awards, 1983-2004.

SOURCES: U.S. Small Business Administration, Department of Defense, National Institutes of Health, and National Science Foundation.

These SBIR awards have supported the R&D activities at JP Laboratories in four major research areas (SBIR funding agencies in parentheses):

(1) Color changing indicators for perishable goods and sterilization (Army, NIH, NSF, USDA).
(2) Radiation sensing devices (DARPA, Navy, NIH).
(3) Synthetic lipids and blood (NIH).
(4) Etching and metallization of plastics (EPA).

JP Labs has successfully licensed many of the products discovered in these four research areas. In the area of color changing indicators, Dr. Patel developed a sticker that is a time-temperature indicator for use with perishable items like foods and medicines. He licensed this technology to Rexam PLC, a UK-based firm focused on beverage and plastic packaging. In another example, the EPA supported research into a system for etching plastics for plating. The new method Dr. Patel developed is less expensive and environmentally safer than the prevailing technology, Chromic acid. Four of JP Labs' U.S. patents are related to this technology and it was successfully licensed to Enthone, Inc., a specialty chemical firm in New Haven Connecticut. Six of the products, indicators for monitoring sterilization of medical supplies are licensed to NAMSA, Northwood, OH.

Research funded through the SBIR program also played an important role in the development of their Self-indicating Instant Radiation Alert Dosimeter (SIRAD) product. The last funding of almost a million dollars for development SIRAD for first responders was provided by Technical Support Working Group, Arlington, VA (funded in part by the Department of Homeland Security, the Department of State, the Department of Defense, and the Department of Justice). As shown in Figure App-D-8, SIRAD is a credit card sized badge that detects radiation levels instantaneously and indicates the radiation level using a color changing strip. It can be used as an inexpensive but accurate dosimeter in situations where radiation exposure is likely.

While the discovery of SIRAD draws on decades Dr. Patel's research and experience, the SBIR program helped finance some the practical research underlying its invention. In 1985, DARPA funded a Phase I study into "monitoring radiation with conductive polymers." At the end of 1988, the NIH funded a Phase I, and subsequently a Phase II, study into the use of a radiographic film dosimeter to examine the dose, dose rate, and energy of neutrons. Finally, the Navy funded a Phase II development study for a radiation dosimeter that evolved directly into the SIRAD product.

KEY SBIR HIGHLIGHTS

Dr. Patel was very positive about the role and contribution of the SBIR Program to the success of JP Laboratories. The key outcomes from SBIR participa-

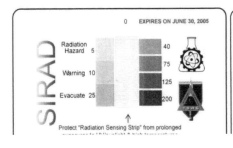 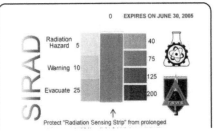

FIGURE App-D-8 SIRAD.

NOTE: Photos of SIRAD badges before (left) and after (right) irradiation with 100 rads of 100 KVP X-ray. Dose is estimated by comparing the color of the sensing strip with the color reference bar printed on each side of the strip. The closest match indicates the dose in rad.

SOURCE: JP Laboratories Web site, <*http://www.jplabs.com/html/what_is_sirad.html# SIRAD*>.

tion are numerous new patents, several new products, and royalty payments from licensing agreements. Dr. Patel said SBIR is a "great program" and that JP Labs "would not have survived without SBIR." Two of the most important benefits for JP Laboratories from participation in the SBIR Program were:

(1) Key Source of Early-stage Financing

Dr. Patel used his personal savings to start JP Labs. At that time, he also worked as a consultant for his old employer Allied Corp. and this provided some cash flow. Nevertheless, additional investment was needed to keep the company going. Venture capital sources were either not interested or demanded control of the company, an option Dr. Patel did not find attractive. As he searched for funding sources, he learned about the newly started SBIR program and decided to apply. The Army funded his first SBIR feasibility study in 1983. The SBIR program became a critical source of R&D funding over the next twelve years and enabled a significant portion of the inventive activity that led to licensing revenue for JP Labs.

(2) Helpful for Business Deals

Having invented a potential product, Dr. Patel indicates that the SBIR Program serves as a recognizable source of credibility. Dr. Patel is more of a scientist

Presented to

JP LABORATORIES, INC.
TECHNICAL SUPPORT WORKING GROUP

for the Development of

SIRAD™
(Self-Indicating Instant Radiation Alert Dosimeter)

Selected by *R&D Magazine* as One of the 100 Most
Technologically Significant New Products of the Year

2005

Tim Studt
Chairman of R&D 100 Awards

FIGURE App-D-9

SOURCE: JP Laboratories.

than a businessman. When searching for licensees, Dr. Patel found it helpful to let people know that the research was backed by a particular U.S. government agencies through the SBIR Program. This would draw the attention of firms and facilitate the licensing process.

ISSUES WITH THE CURRENT SBIR PROGRAM

Dr. Patel does not see any significant problems with the SBIR Program. His experience has been very positive. He noted two points. First, the funding gap between Phase I and Phase II awards was not a major problem. He was able to adjust largely because the licensing revenues from prior inventions began to flow. Second, his research focuses mainly on the chemistry and physics of materials, which is relatively less expensive and involves shorter research lags than product innovation related to biopharmaceutical medicines.

Nanoprobes

Andrew Toole
Rutgers University

COMPANY BACKGROUND

Nanoprobes, Inc., is a privately-held research and development company located in Yaphank, NY. The company was founded in 1990 by James Hainfeld and Frederic Furuya to commercialize research products based on a new method for labeling biological molecules. The founders had discovered a new way to design gold labels to increase the label's effectiveness as a molecular detection tool. Even before the company was fully operational, it had identified its first commercial product based on this technology, which was later introduced as Nanogold.

Over the past fifteen years, the managers of Nanoprobes have successfully grown the company. In 1990, Nanoprobes started with one full-time scientist and a business manager working in the basement of the life sciences building at the State University of New York at Stony Brook. In 1992, the company became one of the first occupants in a new facility constructed for the Long Island High Technology Incubator (LIHTI) at Stony Brook. Over the next few years, Nanoprobes expanded to eight full-time employees. In 2000, the company "graduated" from LIHTI and moved to its current research facility in Yaphank, NY. Today, Nanoprobes has 15 employees, 13 of them full-time, engaged primarily in research and product fulfillment activities.

The business evolution of Nanoprobes reflects its scientific orientation. Nanoprobes' product innovation and improvement depend heavily on successful laboratory-style research. An explicit part of its business strategy is to expand its scientific capabilities in order to broaden and deepen its scientific knowledge surrounding gold labeling technologies. This requires expertise in fields like chemistry, biophysics, biology, and microscopy. With over half its employees engaged in research activities, Dr. Powell notes that Nanoprobes has an "academic culture" that supports discovery, publication, and involvement in professional societies such as the Microscopy Society of America. In fact, it is commonplace for Nanoprobes to subcontract research with academic scientists at universities and other research institutions. This strategy has worked well. Nanoprobes currently offers numerous product variations within about ten separate product categories. Some of these categories are Nanogold conjugates, Nanogold labeling reagents, FluoroNanogold, Ni-NTA-Nanogold, Undecagold reagents, negative stains, silver and gold enhancers, etc.

Building an outstanding scientific reputation is also critical for marketing and sales at Nanoprobes. Researchers are the primary buyers of Nanoprobes' products. Reaching these customers requires active participation in research networks and professional organizations. For example, Nanoprobes will be displaying and discussing staining procedures in a poster session at the United States and Canadian Academy of Pathology meetings in February 2006. Also, publishing in top journals is necessary for illustrating the value of their products and building credibility among researchers. Dr. Powell notes that their Nanogold probes have been cited in over 150 publications.

NANOPROBES' TECHNOLOGY, SBIR, AND INNOVATION

Gold labeling is the core technology at Nanoprobes. The company scientifically investigates, innovates, and markets a variety of forms of this labeling technology which have a number of potential uses. Most of the company's revenue stream is produced by various forms and enhancements of its primary product, Nanogold.

Nanogold is a larger gold cluster compound (1.4 nm in diameter) that is an uncharged separate molecule in solution that does not interfere with antibody binding. When coupled with Fab' fragments, it is the smallest gold-antibody probe commercially available. It offers improved labeling density and greater staining of hard-to-reach antigens. Visualization is further intensified when combined with visual enhancing methods such as immunogold silver staining. The pictures in Figure App-D-10 are pictures of Nanogold-Fab' conjugate using a scanning transmission electron microscope (STEM) and a transmission electron microscope (TEM).

The SBIR program provides vital financial support for product improvement and innovation at Nanoprobes. Soon after the company was founded, it received its first SBIR award, granted by the U.S. Department of Energy in 1991. Since that time the scientists at Nanoprobes have successfully won thirty-five SBIR project awards, mostly from the U.S. National Institutes of Health (NIH). Twenty-five these awards were for Phase I feasibility studies, seven were for product development in Phase II, and three were Fast Track (combined Phase I and II). The awards total over $9 million (in nominal dollars) through 2005. Table App-D-5 lists the date, agency, topic, and phase for Nanoprobes' SBIR awards (Fast Track awards are identified in the topic field).

According to Dr. Powell, the management at Nanoprobes decided to maintain a steady stream of SBIR grants. This steady stream has been valuable to the company's success in a variety of ways. The grants have contributed to the firm's patenting activity. Nanoprobes currently has ten patents and several patent applications pending approval from the U.S. Patent and Trademark Office. The grants have contributed to the maintenance of their academic research ties through subcontracting. They have contributed to the firm's publication and conference activity, which is vital for establishing credibility with potential partners and investors, marketing, and sales. And, they have contributed to the firm's

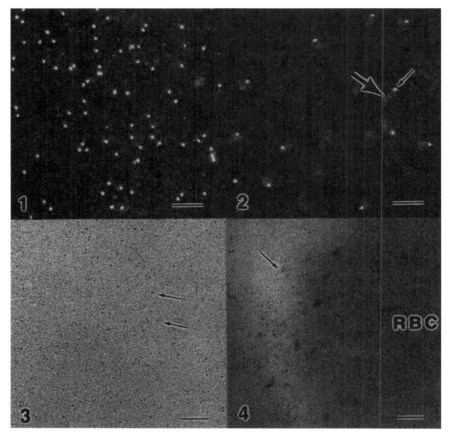

FIGURE App-D-10 (1) STEM darkfield micrograph of 1.4 nm gold particles (bright dots). Full width 128 nm, 500,000 X mag. Bar=20 nm.; (2) STEM darkfield micrograph of Fab's (thick arrow) with 1.4 nm gold particles (thin arrow) attached. Full width 128 nm, 500,000 X mag. Bar=20 nm.; (3) TEM brightfield micrograph of 1.4 nm gold particles (arrows). 300,000 X. Bar=30 nm.; (4) TEM micrograph of human red blood cell (RBC) with Fab'-1.4 nm gold particles attached (arrow). Magnification=300,000 X. Bar=30 nm.

SOURCE: Nanoprobes Web site, <*http://www.nanoprobes.com/MSA92ng.html*>.

internal research capabilities by allowing Nanoprobes to broaden and accelerate its R&D process.

These SBIR contributions have impacted the company's product offerings. Dr. Powell notes that SBIR funds supported modifications and reformulations of the company's core product, Nanogold. SBIR funds supported part of its work on Undecagold reagents. For this product category, the company actually responded to an SBIR solicitation. For their FluoroNanogold product, Dr. Powell notes

TABLE App-D-5 Nanoprobes' SBIR Awards

Nanoprobes, Inc., SBIR Award History

Year	Funding Agency	Topic	Phase I	Phase II
1991	DoE	Ultrasensitive detection system for DNA sequencing	YES	NO
1991	NIH	Amplification of silver-gold immuno and DNA probes	YES	NO
1992	NIH	Combined fluorescent and gold immunoprobes	YES	YES
1993	NIH	Gold coupled nucleotides for ultra sensitive probes	YES	NO
1993	NIH	Large metal cluster and cluster-polymer immunoprobes	YES	YES
1993	NIH	Caged metal cluster and colloid immunoprobes	YES	YES
1994	NSF	Molecular cytoskeletal microscopy probes and biological wires	YES	NO
1995	NSF	Gold & combined fluorescent/gold labels for solid-phase DNA synthesis	YES	NO
1996	NSF	Metal cluster and combined fluorescent/gold labels for peptide synthesis	YES	NO
1996	NIH	Nickel chelating polyhistidine specific molecular probe	YES	NO
1997	NIH	Heavy atom clusters for membrane protein derivatization	YES	NO
1997	NIH	Fluorescent and large metal cluster combination probes (FAST TRACK)	YES	YES
1998	NIH	Signal amplification by catalytic nanogold deposition	YES	NO
1999	NSF	Autometallographic fabrication of copper interconnects	YES	NO
1999	NIH	Luminescent lanthanide chelates labels for immunoassays	YES	NO
1999	NIH	Large covalent gold labels and probes (FAST TRACK)	YES	YES
2000	NIH	Nonradioactive southern blotting detection system	YES	NO
2000	NIH	Metal enhanced radiation therapy	YES	NO
2000	NIH	Gold quenched molecular beacons	YES	YES
2001	NIH	Enzymatic metallography for biological detection	YES	YES
2001	NIH	Durable emitters for nanospray mass spectrometry	YES	YES
2002	NIH	Live cell correlative imaging probes	YES	NO
2002	NIH	Improved MRI contrast agents	YES	YES
2003	NIH	Reiterative signal amplification by gold deposition	YES	NO
2004	NIH	Correlative chromogenic gene and protein assessment	YES	NO
2004	NIH	Gold enhanced angiography (FAST TRACK)	YES	YES
2005	NIH	Nanogold enhanced head & neck cancer radiotherapy	YES	N/A
2005	NIH	Correlative Enzymatic and gold probes	YES	N/A

SOURCES: U.S. Small Business Administration; National Institutes of Health; and National Science Foundation.

that it "owes the most to the SBIR program." In 1992, the NIH funded Phase I research into combined fluorescent and gold immunoprobes. This research was further supported by a Phase II grant with financing in 1994 and 1995.

SBIR has also enabled Nanoprobes to investigate the applications of its core technologies to improved public health. One such technology, "Enzyme Metallography" has proven to be a better detection method for pathological assessment of human biopsy material. In collaboration with the Cleveland Clinic, this has been used to develop a test for Her-2/neu breast cancer, an important marker for aggressive malignant behavior that occurs in about 30 percent of breast cancer cases. Due to the improved detection, this technology was recently licensed to Ventana Medical Systems, Inc., which makes automated tissue staining equipment used in many hospitals and testing labs worldwide. This test is expected to be released in the next 1-2 years worldwide and should result in improved detection and management of this condition by helping to identify patients suitable for treatment with the very promising therapeutic, Herceptin. The SBIR program played a key part by enabling the research to be carried out for this development.

Nanoprobes is also investigating the use of gold nanoparticles in vivo (at the animal stage) as X-ray contrast agents for better visualizing coronary disease and tumors. Gold absorbs X-rays more strongly than current iodine agents and stays in the blood longer, allowing better images to be obtained and potentially enabling the noninvasive assessment of coronary arteries and earlier detection of tumors. In addition, Nanoprobes is investigating the use of gold nanoparticles to enhance radiotherapy. Since gold absorbs x-rays, its presence in tumors can increase the specific dose. This approach has yielded promising results in mice, where one study resulted in 86 percent long term (>1 year) survival, vs. 20 percent without gold. The SBIR program has been absolutely necessary to provide capital for these high-risk, early-stage studies.

KEY SBIR HIGHLIGHTS

Dr. Powell is very positive about the role and contribution of the SBIR Program to the success of Nanoprobes. He says it is a "tremendously good program" that has allowed the company to become "more sophisticated and successful." Dr. Powell highlights the following three benefits of the SBIR program for Nanoprobes:

(1) Key Source for Building Research Capabilities

In the beginning, personal funds were used to get Nanoprobes started. However, for small science-intensive companies, it is important to achieve a critical mass of research personnel and equipment to sustain the firm going forward. SBIR funds are an important source of capital for this process, especially Fast Track awards. Fast Track awards decrease the uncertainty the firm faces by providing a longer period of continuous funding. They know in advance how much money will be coming and for how long and can plan more ambitious and longer-term projects.

(2) SBIR Financing Provides Flexibility

SBIR funds are more flexible than other sources. Dr. Powell notes that these grants provide a "degree of creative freedom that one does not normally get with venture capital funding." This is important since opportunities change as research moves forward. Another aspect of this flexibility is that SBIR projects allow the firm to explore higher risk research avenues.

(3) An Important Source of Credibility

Small firms, especially start-ups, must somehow achieve credibility in the marketplace. Potential buyers and business partners are skeptical about the capabilities of small firms and the claims they make about their products. Over time, the business strategy at Nanoprobes is shifting toward licensing and partnering. SBIR helps overcome the credibility barrier.

ISSUES WITH THE CURRENT SBIR PROGRAM

Dr. Powell does not see any significant problems with the SBIR Program. He made the following point about the current program.

Onerous Application Process

Over time, the SBIR application process has become more complicated. A complete Phase I application can now be up to sixty-four pages long. It requires too much time to prepare. Further, the recent movement to the "grants.gov" on-line application system was very demanding.

Neurocrine, Inc.[15]

Robin Gaster
North Atlantic Research

EXECUTIVE SUMMARY

Background

Neurocrine is a drug development/biotech company, and one of the largest companies in the study. Publicly traded on the NASDAQ, and with a market cap of about $1.5 billion, Neurocrine has approximately 450 employees, has been heavily backed by venture capital from inception in 1992, and recently moved into a brand new purpose-designed campus. It has no products on the market but several in the pipeline, two very close to market.

Neurocrine is located in an R&D hub (San Diego), is not woman or minority owned, was venture backed, and while a multiple winner generates only a small percentage or R&D funds from SBIR.

SBIR History and Status

Since 1992, Neurocrine has received 22 Phase I awards and 14 Phase II awards, generating a total of just over $10 million in SBIR funding. (See annex.) In 2004, Neurocrine became ineligible for further SBIR funding in light of the recent interpretation of ownership regulations, as Neurocrine is 89 percent owned by institutional investors.

Key Utilization of SBIR

Neurocrine does not use SBIR for projects within the company's critical development path. SBIR is however seen as an important source of discretionary funding that allows for more speculative or longer-term research, or research on alternative mechanisms for achieving critical path results. This research has in some cases subsequently led to internally funded research and to integration into Neurocrine's primary product pipeline.

---

[15]Based on interview conducted at Neurocrine in San Diego, February 24, 2005. Dr. Paul Conlon, director of Research and Development; Rich Maki. Dr. Conlon has been at Neurocrine since 1993 and wrote several of the original applications. Rich Maki is a senior researcher with extensive SBIR experience.

Key Issue/Concern

The new interpretation of SBA guidelines, which Neurocrine continues to vigorously contest (Neurocrine in fact continues to apply for SBIR awards). Also, Neurocrine is concerned about a perceived shift in the Phase II application process, where it believes reviewers now require much more detailed technical information, which constitutes a risk to critical company intellectual property (IP).

Outcomes:

- Products: none yet.
- Commercial pipeline: products on the way, some with significant input from SBIR.
- Knowledge: at least 30 papers and several patents based directly on SBIR.
- Employment: dramatic expansion, not directly related to SBIR.

Recommendations/Comments:

- Eliminate commercialization plans from review process from both Phase I and Phase II.
- Reconsider VC and large drug company participation on panels.
- Reconsider demands for increasingly detailed IP during Phase II application process.
- Increase size/duration of Phase I awards to $300,000, for one year.
- Increase size of Phase II to $1 million per year.
- Reduce duration of Phase II awards: two year maximum, with second year entirely conditional on meeting year one milestones.

Additional Lessons Learned

Neurocrine's story appears to indicate that there is considerable confusion at NIH and in reviewer panels about the focus of SBIR within the product development cycle. On the one hand, increased focus on commercialization inevitably means pressure to move downstream toward products; commercialization plans based on very early basic research are not defensible. On the other, Neurocrine had an otherwise high scoring proposal rejected as being too close to the market.

BACKGROUND

Neurocrine started with $5 million in venture funding in 1992-1993, raised $50 million though a series B offering in 1993-1994, made three major partnership deals amounting to well over $100 million in funding in 1994-1995, and completed an IPO in 1996. Today, Neurocrine is a public company more than 89 percent owned by institutional investors, with a market capitalization of approximately $1.5 billion on the NASDAQ, and cash reserves of more than $300 million, against debt of approximately $66 million. The company recently moved to a large new custom built campus in San Diego.

Employment

Neurocrine has always been a much larger company than the norm for SBIR. By 1993, a year after founding, it already employed 25-30 people, reaching 100 in 1996. Currently, Neurocrine employs about 450 people, and will soon be above the SBIR limit of 500 for small companies.

Products

To date, the company has no products that have reached market, although a recent filing problem with FDA is being resolved and Indiplon, a new drug for insomnia, is expected to reach the market in approximately 14 months. Indiplon has two formulations that have completed all clinical trials with apparent success.

The outcomes described above help to define both the importance of SBIR to Neurocrine and some limitations. Of the 11 programs now making their way toward or through clinical trials, 5 received significant support from SBIR. As Conlon noted, the primary pipeline is not dependent on SBIR, but at the same time SBIR has opened the door to research that is clearly now part of that primary pipeline.

TABLE App-D-6 Research and Development Pipeline

Program	Targeted Indication	Status	SBIR
Indiplon IR	Insomnia	Filed	NO
Indiplon MR	Insomnia	Filed	NO
GnRH Antagonist—56418	Endometriosis	Phase I	YES
Altered Peptide Ligand	Multiple Sclerosis	Phase II	YES
Altered Peptide Ligand	Type 1 Diabetes	Phase II	NO
D2 Receptor Agonist	Sexual Dysfunction	Phase II	NO
Urocortin II	Cardiovascular/Endocrine	Phase I	NO
CRF R1 Antagonist	Anxiety	Phase I	YES
Melanocortin Receptor Agonist/Antagonist	Obesity/Cachexia	Preclinical	YES
Melanin Concentrating Hormone Antagonist	Obesity	Preclinical	YES
Sleep Program	Insomnia	Preclinical	NO

SOURCE: Neurocrine, Inc.

PHASES IN NEUROCRINE'S USE OF SBIR

SBIR never had a real financial impact at Neurocrine, where financial resources were substantial even in the mid-1990s.

Neurocrine's use of SBIR changed over time, falling into three distinct time periods:

- **Stage 1. Initial Use, 1993-1996.** During this period, Neurocrine was at least initially still a relatively small company (about 25 employees in 1994), and was still seeking its intellectual way. While focused on the intersection of biology and chemistry at the cellular level, it had not yet tightened its focus on small molecule bioscience.

Neurocrine received ten Phase I awards between 1994 and 1997, of which eventually became Phase II awards. Paul Conlon, now VP research and development and Principal Investigator on the first awards, noted that they played a key role in allowing Neurocrine to test ideas and explore possible directions for the company.

They also gave Neurocrine valuable credibility when exploring partnerships with other much bigger and more established companies: "Validation was important—being able to say that our work was being funded by the National Institutes of Health, and that it had been approved by a peer review panel, provided tremendous credibility."

That credibility may have a made a key difference for Neurocrine. In 1994-1995, Neurocrine made deals with three major pharmaceutical companies, including a $70 million agreement with Ciba Geigy. These deals in turn provided the evidence of progress on which to base Neurocrine's public offering in 1996.

- **Stage 2. Consistent Success.** During 1997-2002, Neurocrine "figured out the SBIR application process." They were now consistently putting together good applications, and their success rate rose substantially. They understood what review panels wanted to see, and they also found that their cutting edge work on small molecules was being well received. The result was a string of Phase I and successor Phase II awards.

These awards now filled a somewhat different function at Neurocrine. With the new focus on small molecules, the earlier search for scientific identity was largely concluded; now SBIR was being used much more directly to explore promising offshoots of core research. For example, awards for work on the GNRH receptor.

- **Stage 3. Difficulties, 2003-2005.** In 2003, Neurocrine suddenly found its long string of SBIR successes under pressure, from two directions.

  - First, the application process changed. Neurocrine found that work programs and descriptions that had been sufficiently detailed for success during stage 2 were now challenged by panel members seeking much more specific detail. Neurocrine strongly believes that this change accompanied changes in the composition of review panels, with the introduction of members from major pharmaceutical companies. Essentially, Neurocrine believes that this change puts its crown jewels of intellectual property at risk. (see box App-D-2). After substantial negotiations, a compromise was reached in the case of one application, but it became moot for reasons described immediately below.
  - Second, Neurocrine was ruled ineligible for further SBIR awards as a public company that was more than 50 percent owned by institutional investors (Neurocrine is in currently owned 89 percent by institutions). This resulted from the new interpretation of existing SBA statutes and regulations, implemented at NIH in 2003.

The changing application process is taken very seriously at Neurocrine, for which protection of its IP is central. Neurocrine notes that it makes no sense to jeopardize major potential partnership agreements for $50 million or more, to pursue a $1 million SBIR award. Neurocrine would walk away from SBIR altogether before risking its IP.

Clearly, Neurocrine is different from the majority of small and poorly funded SBIR companies, who may not have any alternatives to SBIR funding. However, even if smaller companies have little alternative to accepting these new demands from reviewers, they may still be unfair, and they may still pose long-term problems for the NIH SBIR program. Other interviews may help to determine whether this is an unusual case, or whether there has been a real change in requirements from review panels. Neurocrine is arguing from two cases in 2003.

STRATEGIC ROLE OF SBIR AT NEUROCRINE

Neurocrine has clearly used SBIR to its advantage. However, Neurocrine sees SBIR as filling a very specific function, and one that is *not* at the very core of Neurocrine's research program. SBIR provides discretionary funding to allow research into promising offshoots and alternatives that likely would not otherwise get done.

But this funding is and must be unprogrammed within the company precisely because it is high risk. According to Conlon, no company can afford to place SBIR at the heart of its research program because the money is unpredictable (although of course many smaller companies do precisely that). So for companies that do have other resources—and Neurocrine has many, not least $300 million in cash in the bank—SBIR is useful funding that allows interesting and sometimes important work (at least in retrospect) , but does not fund core research of critical strategic importance to the company.

Example 1: CRF research funding through SBIR allowed exploration of a range of possible applications, for anxiety, IBS, depression. And further SBIR funding allowed the company to explore R1 and R2 receptors, identifying ways to separate out the different R1 and R2 receptor sites. This exploration would not have been possible without SBIR.

Example 2. Neurocrine's core MC4 program was focused on researching agonists for use in treating endometriosis. SBIR funding allows the company to explore an alternative mechanism—antagonists—which resulted in a new program focused on treating a related disease, Cachexia. Again, SBIR funding supported new research directions.

Example 3. SBIR funding paid for upgrading Neurocrine's proprietary library of GPCR molecules. While core work is based on that library, its function is to provide better leads for small molecule efforts. However, even with better leads, it takes two years to turn a lead into a drug candidate.

In several cases, Neurocrine has made significant use of SBIR even when not receiving Phase II awards (and in one case, Neurocrine withdrew its application for Phase II funding after being given an award, as the company changed strategy), (e.g., Nonpeptide Antagonists of CCR-7 for Immunosuppression. The Phase II was funded, but not included in the list of awards, as Neurocrine declined the monies.

Legitimation Effects.   Conlon believes not only that SBIR lent important legitimacy when talking to potential partners, but that receiving their first Phase II —after a number of awards ended at Phase I—provided important internal validation that the company was on the right track (and given the early involvement of venture capital, probably helped validate the company with funders as well).

COMMERCIALIZATION AND THE REVIEW PROCESS

Neurocrine is deeply dissatisfied with the actual impact of efforts to improve commercialization review at NIH, and in fact now recommends that on balance commercialization review should be eliminated, and NIH should return to simply funding the best science.

Neurocrine offered a number of objections to the current approach:

- **Timing.** Even Phase II is too early for an effective commercialization plan. For Neurocrine, Phase II is about narrowing down candidates for further drug development. At Phase II application, the market may be 8-12 years of further development away—a period which will require a partnership with a major drug company and many millions of dollars. It is hard to see what value any commercialization plan might have at this stage.

- **Reviewer Capacity and Conflicts of Interest.** Few academic reviewers have any effective capacity to review commercial plans. And Neurocrine was clearly very perturbed about potential conflicts of interest stemming from the addition of reviewers from major drug companies.
- **Less Focus on the Quality of Science.** The new emphasis on commercialization means by definition that less attention is paid to the quality of the science.
- **Phasing.** If commercialization plans for Phase II are impractical, Phase I plans are even less realistic. Neurocrine thought they should be eliminated.

Example

Neurocrine's Phase I award to research development of TCR antagonists to MBP-reactive t-cells generated an extraordinarily high score of 131 at Phase I. After successfully completing Phase I, Neurocrine submitted a Phase II application. This was rejected primarily (according to Neurocrine) on commercialization grounds—reviewers claimed that the project was too development oriented, too close to market. However, within months, Neurocrine had reached agreement to develop the project further through a $ partnership worth approximately $70 million with Ciba Geigy. Neurocrine sees this as evidence that the commercialization assessment process is fatally flawed. However, it is also possible to argue that the reviewer was correct—and that the Ciba Geigy deal provides that the product was indeed sufficiently developed to receive fully commercial funding.

Neurocrine has also noted considerable confusion at NIH about the relationship of SBIR awards to product development cycles. In the example above, the product is still years away from the market, so it is hard to see how it could be too commercial.

OTHER ISSUES

In discussing SBIR with Conlon and Maki, it became apparent both that the company is very experience with and sophisticated about SBIR. Maki has served on NIH study sections, but only for RO1s, not SBIR. However, understanding of the new competing continuation awards was still limited.

Neurocrine saw no significant differences between the ICs from the perspective of applicants; had generally had excellent experiences with program staff at all ICs.

A final note. Neurocrine's experience confirms once again that resubmission is a normal part of the NIH SBIR process; Conlon expects to resubmit on a regular basis.

RECOMMENDATIONS

Neurocrine had a number of recommendations for improving the SBIR program:

- **Commercialization plan.** Even though Neurocrine scores reasonably well on commercialization plans, as it has a large business development unit that writes these plans, it believes that they in the end simply randomize panel review results. The difficulties of getting good commercialization reviews in their view much more than outweigh the benefits, and Neurocrine believes that panels should go back to simply reviewing the science. This would also resolve important difficulties with the make-up of panels.
- **Study sections.** If commercialization reviews are not eliminated, Neurocrine believes changes should be made in the composition of review panels, and that venture capitalists should be excluded (and possibly representatives from large drug companies).
- **Track record.** Neurocrine rejected possible mechanisms for ensuring that past track record would be taken into account during panel discussions.
- **Size/Duration (Phase I).** Neurocrine would be willing to trade off significantly larger Phase I awards for significantly fewer of them: $300K awards, with 1/3 as many would be appropriate.
- **Size/Duration (Phase II).** Neurocrine is not at all convinced about the need for longer awards, but believes both in larger awards and in more pay for performance. It suggested that awards should be $1 million per year for two years, with the second year being completely conditional on achieving specific research milestones.
- **Direct access to Phase II.** Neurocrine strongly agreed that companies should be allowed to apply directly for Phase II awards, without going through Phase I first. They pointed out that by definition this would bring better quality projects now excluded into the program.

BOX App-D-2
Patenting

Neurocrine has filed numerous patents during the past ten years. It regards protection of IP very seriously. However, the new demands for specific structural details of proposed molecules to be developed during Phase II projects are difficult to handle because the company is not yet in a position to file patent applications for several reasons:

- The company files patents only on the final molecule that has emerged as a strong candidate for drug development. Prior to Phase II, the identity of the final molecule is simply known: Phase II work is largely the process of sorting and testing different promising candidates, with the view of emerging at the end of Phase II with one or possibly more final candidates.
- If patents applications were filed pre-Phase II, they would have to be filed for a considerable number of candidate molecules. Given that patents cost $25,000 each, this would be a huge potential burden, essentially undercutting the $1 million in support available under Phase II.
- Patents become harder to defend the further upstream from actual drug deployment they are filed. Forcing early filing makes the IP less defensible, and hence less valuable.

From Neurocrine's perspective, these new demands would mandate withdrawal from the SBIR program were they to be fully implemented (i.e., were detailed structures to become a necessary part of SBIR filing). Of course, Neurocrine has other resources. Many smaller companies do not.

NEUROCRINE—ANNEX

TABLE App-D-7 Neurocrine SBIR Awards and Outcomes

Year	Fiscal Year	Phase Type	Award Size ($)	Project Title	Funding Institute-Center	Notes/Outcomes
Phase II						
1	1994	Phase I	77,250	COMBINATORIAL ORGANIC SYNTHESIS OF A CRF-R ANTAGONIST	NS	Research assisted in the development of a CRF-1 Receptor antagonist that showed efficacy in depressed patients in the clinic. However, the compound was discontinued due to toxicological findings. Additional CRF-1 antagonists are presently being evaluated in the clinic.
2	1996	Phase II	384,159	COMBINATORIAL ORGANIC SYNTHESIS OF A CRF-R ANTAGONIST	NS	
3	1997	Phase II	365,804	COMBINATORIAL ORGANIC SYNTHESIS OF A CRF-R ANTAGONIST	NS	
1	1994	Phase I	77,250	CRF BINDING PROTEIN AND DEMENTIA	NS	Effort discontinued in preclinical studies due to lack of progress.
02A1	1997	Phase II	375,000	CRF BINDING PROTEIN AND DEMENTIA	NS	
3	1998	Phase II	375,000	CRF BINDING PROTEIN AND DEMENTIA	NS	
1	1995	Phase I	100,000	NOVEL CRF—RECEPTOR	NS	CRF-2 was identified, characterized, and is still an area of active interest in our company. To date however, no small molecule compounds have progressed into clinical trials.
2	1998	Phase II	398,300	NOVEL CRF RECEPTOR	NS	
3	1999	Phase II	351,700	NOVEL CRF RECEPTOR	NS	

continued

TABLE App-D-7 Continued

Year	Fiscal Year	Phase Type	Award Size ($)	Project Title	Funding Institute-Center	Notes/Outcomes
1	1996	Phase I	100,000	CRF BINDING PROTEIN AND OBESITY	DK	Effort discontinued in preclinical studies due to lack of progress.
2	1998	Phase II	595,586	CRH BINDING PROTEIN AND OBESITY	DK	
3	1999	Phase II	570,105	CRH BINDING PROTEIN AND OBESITY	DK	
01A1	1997	Phase I	100,000	CRF RECEPTOR ANTAGONISTS IN NEUROPROTECTION	NS	Research assisted in the development of a CRF-1 Receptor antagonist that showed efficacy in depressed patients in the clinic. However, the compound was discontinued due to toxicological findings. Additional CRF-1 antagonists are presently being evaluated in the clinic.
2	2000	Phase II	635,649	CRF RECEPTOR ANTAGONISTS IN NEUROPROTECTION	NS	
3	2001	Phase II	666,805	CRF RECEPTOR ANTAGONISTS IN NEUROPROTECTION	NS	
1	2000	Phase I	240,527	A SCREENING LIBRARY FOR PEPTIDE-ACTIVATED GPCRs	GM	An acitve area of research within the company to improve the quality and quantity of small molecule leads.
2	2002	Phase II	889,087	A Screening Library for Peptide-Activated GPCRs	GM	
3	2003	Phase II	638,359	A Screening Library for Peptide-Activated GPCRs	GM	

TABLE App-D-7 Continued

Year	Fiscal Year	Phase Type	Award Size ($)	Project Title	Funding Institute-Center	Notes/Outcomes
1	2000	Phase I	99,931	NOVEL NON-PEPTIDE ANTAGONIST OF THE GNRH RECEPTOR	HD	Research assisted in the development of a GnRH Receptor antagonist that has recently shown efficacy in Phase I studies.
2	2001	Phase II	471,580	Novel Non-Peptide Antagonists of the GnRH Receptor	HD	
3	2002	Phase II	471,580	Novel Non-Peptide Antagonists of the GnRH Receptor	HD	
1	2001	Phase I	138,733	NOVEL NON-PEPTIDE ANTAGONIST OF THE MCH RECEPTOR	DK	Research assisted in the development of a MCH Receptor antagonist that is still being evaluated preclinically.
2	2003	Phase II	466,776	Novel non-peptide antagonist of the MCH receptor	DK	
		8	8,589,181			
Phase I Only						
1	1994	Phase I	75,000	DEVELOPMENT OF TCR ANTAGONISTS TO MBP-REACTIVE T-CELLS	NS	Research assisted in the development of an APL that is in Phase II clinical studies.
01A1	1995	Phase I	95,025	CRF RECEPTOR ANTAGONISTS FOR TREATMENT OF COCAINE ABUSE	DA	Research assisted in the basic understanding of CRF-1 Receptor antagonist.
01A1	1995	Phase I	100,000	ANTI INFLAMMATORY/ EFFECTS OF CRF BINDING PROTEIN	AR	Effort discontinued in preclinical studies due to lack of progress.
1	1996	Phase I	99,634	NOVEL TYPE III INTERLEUKIN 1 RECEPTOR	AI	Effort discontinued in preclinical studies due to lack of progress.

continued

TABLE App-D-7 Continued

Year	Fiscal Year	Phase Type	Award Size ($)	Project Title	Funding Institute-Center	Notes/Outcomes
1	1997	Phase I	100,000	IGFBP 3 LIGAND INHIBITORS FOR THE TREATMENT OF DIABETES	DK	Effort discontinued in preclinical studies due to lack of progress.
01A1	2000	Phase I	133,067	DEVELOPMENT OF A TOXIC FUSION PROTEIN FOR BRAIN TUMORS	CA	Research assisted in the development of a Fusion Toxin that has been through Phase II clinical studies.
01A1	2000	Phase I	99,390	DEVELOPMENT OF NEUROPROTECTIVE DRUGS FOR RETINAL DISEASE	EY	Effort discontinued in preclinical studies due to lack of progress.
1	2001	Phase I	207,018	Melanocortin-4 Receptor Antagonists	CA	Research assisted in the development of a MC-4 Receptor antagonist that is still being evaluated preclinically.
1	2002	Phase I	133,540	CRF-Type-2 Receptor Antagonists for Anxiety	MH	Research assisted in the basic understanding of CRF-2 Receptor antagonist. It is still an active area of interest, however no small molecule antagonists have been identified to date.
01A1	2002	Phase I	137,959	Nonpeptide Antagonists of CCR-7 for Immunosuppression	NS	Effort discontinued in preclinical studies due to lack of progress.
01A1	2002	Phase I	130,048	Novel CRF antagonists for inflammation and pain	AR	Research assisted in the basic understanding of CRF-1 Receptor antagonist.

TABLE App-D-7 Continued

Year	Fiscal Year	Phase Type	Award Size ($)	Project Title	Funding Institute-Center	Notes/Outcomes
1	2002	Phase I	139,336	Novel non-peptide agonists of the melanocortin receptor	DK	Research assisted in the development of a MC-4 Receptor agonist that is still being evaluated preclinically.
1	2002	Phase I	139,497	Selective CRF Antagonists for Bowel Disorders	DK	Research assisted in the basic understanding of CRF-1 Receptor antagonist.
1	2002	Phase I	170,479	Corticotropin Releasing Factor Receptor Function in CNS	MH	Research assisted in the basic understanding of CRF-1 Receptor antagonist.
		14	1,759,993			

NOTE: For a list of codes for National Institutes of Health institutes and centers, see Box App-A-1.

SOURCE: Neurocrine, Inc.

Optiva, Inc.[16]

Robin Gaster
North Atlantic Research

COMPANY HISTORY

Optiva was founded in 1988 by David Giuliani, an entrepreneur formerly in management at Hewlett Packard,[17] and two faculty members at the University of Washington in Seattle, Drs. David Engel and Roy Martin.

The company (originally called GEMTech) was founded to pursue the idea of a dental hygiene device using a piezoelectric multimorph transducer that worked on sonic principles, using sound waves to dislodge plaque. Such an approach could have significant advantages, for example, in addressing plaque below the gum line.

Early financing came from the founders, and was used to develop the original technical ideas. However, three years of effort and prototypes convinced the team that the original technology could not be made into a commercial product. Instead, they adopted an alternative technology based on activating water in the mouth by use of sonic technology somewhat akin to a tuning fork. This approach was compatible with another objective—putting all the moving parts in the head of the toothbrush, so that the more expensive body could be sealed against water leakage.

After considerable experimentation, the team determined that when tuned to 520 vibrations per second, the vibrating brush head generated fluid dynamics that would erode plaque beyond the reach of the brush itself.

To commercialize the product, Giuliani raised $500,000 from 25 private investors, and also benefited from an NIH SBIR award which effectively doubled the size of the investment.

The company faced both very substantial opportunities, and significant challenges. Almost all Americans suffer from periodontal disease at some point, so the *potential* market was very large. Even the *existing* market was substantial—12 percent of Americans used electric toothbrushes, generating a total market of $125 million annually.

---

[16]Aside from the interview with David Giuliani, the major sources for this case study were the long profile of Optiva published in *Inc Magazine* in 1997 (David H.Friedman, "Sonic Boom," *Inc Magazine*, October 1997), and the Sonicare Web site which provides considerable documentation for the product. Other sources are cited individually.

[17]According to Friedman, "Sonic Boom," op.cit., Giuliani had spent 12 years at HP before seeking more entrepreneurial work. He developed a hand held ultrasound device that could measure bladder volume without using a catheter, as part of a company called International Biomedics (eventually acquired by Abbott).

However, that existing market was dominated by very strong brands owned by large companies—Braun, Bausch and Lomb, Teledyne. Also, the Sonicare® toothbrush was more expensive to make, and would have to sell for considerably more than standard electric toothbrushes ($129, vs. $50-70 for other brands).

Optiva therefore decided to focus on dentists as the critical intermediary between the company and consumers.

The Sonicare® toothbrush was launched at a periodontal convention in Florida in 1992. According to the company history, the first dentist to visit the booth bought 36, and the company sold 70 altogether. Optiva hired a sales manager.

This approach proved very successful. Using studies (some funded by SBIR) that demonstrated the benefits of Sonicare® technology, dentists proved interested; 98 percent of those who tried the product recommended it to their patients,

FIGURE App-D-11
SOURCE: Optiva.

and some even signed on a resellers. Optiva began advertising in dental journals, and formed a small sales force to reach out to dentists.

This was a taste of the very rapid success to come. Again according to the company, by 1995 more than one-third of dentists in the U.S. were recommending Sonicare®. Consumer-direct marketing was also expanding rapidly, as Shaper Image featured Sonicare® products in their catalog and in their stores. GEMTech changed its name to Optiva.

The company's sales strategy focused on dentists, partly mandated by its situation as small under-capitalized company without distribution agreements. For Giuliani, leverage was key: he developed the company's core strategic thrust of "borrowing other assets and using them to build the company's reputation. The reputation of dentists was what we leveraged. This approach took advantage of the dentists' patient contacts and patients' trust in their dentists." Giuliani went on to note that "Borrowed reputation had to be returned with interest—through a product that dentists were confident in and proud to be related to."

Starting in 1992, the company sought to move beyond its dentist-based sales strategy, into direct consumer marketing. Initial efforts at direct mail failed resoundingly (the product was too complex to explain in a page), but in 1993 Sharper Image ordered 4,000 units and then 16,000 more a few months later—a huge sale at the time, which stretched the company's manufacturing and fulfillment operation to the limit.

Following the Sharper Image sale, consumer word of mouth purchases expanded rapidly, bolstered by a strong endorsement from Oprah Winfrey on her TV show.

In 1994, Optiva received a patent for «high-performance acoustical cleaning apparatus for teeth.»[18] This effectively blocked competitors such as Teledyne (which had to settle a subsequent patent infringement case).

Optiva also boosted both manufacturing capacity—including redesigns which cut manufacturing costs by 60 percent—and sales capacity,. It added 50 manufacturer reps working on commission to pursue sales at Costco and other high volume outlets, as well as specialty stores like Brookstone. Approximately 25,000 retail stores were stocking the product. Optiva also continued to focus on dentists, claiming that by the end of 1995, one-third of all U.S. dentists were recommending Sonicare®.

By 1996, Optiva and Sonicare® were being recognized as a major U.S. success story: Giuliani was invited to breakfast at the White House in May 1996, where the company was cited for its exemplary employment practices and employee benefits. Giuliani and was named SBA's Small Business Person of the Year in 1996.

In 1999, Optiva Corporation relocated its headquarters and manufacturing operations from Bellevue, Washington, to a new 176,000-square-foot, state-of-

---

[18]Patent no. 5,378,153.

the-art facility in Snoqualmie, Washington. The company also launched its television ad campaign featuring a decidedly unconventional Tooth Fairy, and it ended the year with more than 600 employees.

In October 2000, Philips Domestic Appliances and Personal Care (DAP), a division of Royal Philips Electronics acquired Optiva Corporation. In January 2001, Optiva Corporation changed its name to Philips Oral Healthcare, Inc. With the combined resources of the former Optiva Corporation and its new owners, Philips DAP, the company set forth to leverage its tremendous research and development capabilities to create the next generation of power toothbrushes. By the end of 2001, the company produced its 10 millionth Sonicare® and became the #1 rechargeable power toothbrush in the United States.

According to Giuliani, the sale to Philips substantially benefited the former owners, who were able to cash out, the new owners who gained a proved and market leading technology which has become a center piece for their dental product line, and the asset itself, where the company's limited ability to maximize commercial value was dramatically improved by Philips which brought to the table major international marketing capabilities. For founders and company employers, this also provided significant emotional return on the original investment. Overall, some years after the sale, Giuliani continues to see this as an excellent outcome for the company. As he observed, "You want your child to marry well."

OUTCOMES

Sonicare® has been a huge commercial success. It currently serves about one-third of the U.S. market for electric toothbrushes. By 1996, it was selling 1 million brushes annually, and generating revenues of more than $70 million.

In October 1997, Optiva was named the fastest growing private company in the country by Inc Magazine, topping the Inc 500 list. By January 2001, shortly after the sale to Philips in 2000, the company was generating approximately $200 million annually in revenues, and employed more than 600 people in Snoqualmie, Washington, where the company's headquarters and manufacturing facilities are located.[19] Although the size of the acquisition transaction was not publicly revealed, industry sources estimated that it was worth more than $1 billion.

Philips has strongly backed the product line, continuing to introduce new products, and has retained the Snoqualmie operation as world headquarters for the division. Philips has also leveraged its international capabilities—noting for example that after successful launches, more than 65 percent of UK dentists recommend Sonicare®, and that the product had captured 21 percent of the Dutch market four months after launch in the Netherlands.[20]

---

[19]"Sonic toothbrush maker takes acquirer's name," *Puget Sound Business Journal*, January 8, 2001.

[20]*<http://www.homeandbody.philips.com/sonicare/gb_en/03d-story.asp>*.

The product has also clearly had a significant impact on public health. It is in use in a very substantial number of homes, and studies show that it provides better results than a manual toothbrush (e.g., it removes about 40 percent more plaque[21]).

SONICARE® TECHNOLOGY

The technology behind Sonicare® has been validated in extensive academic studies. According to Philips Sonicare® Division, there have been 85 published studies by 119 researchers at 40 universities.[22] These studies have covered a wide range of topics related to Sonicare®, including:

- Plaque removal.
- Gum health.
- Biofilm removal.
- Dental hypersensitivity.
- Stain removal.
- Dry mouth.

According to Sonicare®, the technology achieves its bristle velocity through a combination of high frequency and high amplitude bristle motion. This velocity generates dynamic fluid action, which is gentler on dentin than a manual or an oscillating toothbrush. The cleaning power of dynamic fluid action, coupled with the specially designed bristle orientation, results in deep penetration of interproximal spaces, where the "shear force" of the fluids help dislodge the biofilm.[23]

Thus like other electric toothbrushes, the primary mode of cleaning produced by a sonic toothbrush is created by the scrubbing action of the brush's bristles on the surfaces of teeth. However, Sonicare® also produces a secondary cleaning action founded in the intense speed at which the bristles of the sonic toothbrush vibrate. This vibratory motion imparts energy to the oral fluids that surround teeth (such as saliva). The motion of these agitated fluids can dislodge dental plaque, even beyond where the bristles of the toothbrush actually touch.

The brush head of the toothbrush is designed to vibrate at over 30,000 brush strokes per minute. This high speed brushing motion creates movements in the fluids that surround the teeth, creating fluid pressure and shear forces. These fluid dynamics can dislodge dental plaque in hard-to-reach areas between teeth and below the gum line. The cleaning effect of these fluid forces has been measured to occur at distances of up to 4 millimeters (slightly more than one-eighth of an inch) beyond where the bristles of the sonic toothbrush actually touch.

---

[21]K. Moritis, M. Delaurenti, M. R. Johnson, J. Berg, and A. A. Boghosian, "Philips Oral Health-care," American Journal of Dentistry, 15 (Special Issue): 23B-25B, 2002.

[22]*<http://www.sonicare.com/why/proven.asp>*.

[23]*<http://www.sonicare.com/dp/why_reco/why_reco_superior.asp>*.

It is worth noting that this secondary action is considerably less important overall than the actual brush impact (which is shared with all toothbrushes, although nonsonic brushes typically generates less than 10,000 brush strokes per minute, or less than one-third the level of Sonicare®). Also there are no studies about long term impacts in terms of plaque removal. However, the unique technology of the Sonicare® product has clearly been the primary differentiator, and is the basis for the product's very substantial commercialization success.

The Sonicare® technology is still apparently regarded as the gold standard for technology in this area, and has not been successfully replicated by other companies (partly because it is patent-protected).

THE IMPACT OF SBIR FUNDING

Even though Optiva received only one SBIR award (for both Phase I and Phase II), from two applications, Optiva's need to create a product matched well with the commercialization focus of the SBIR program. (Giuliani noted that the rejected application was focused on methods, not products, and was therefore correctly rejected for being insufficiently commercial).

Phase I SBIR funding had some immediate and significant validation effects for the company, in particular in relation to third-party investors. Giuliani noted that "Once we got notice of our award, it provided important validation of scientific merit for potential investors. Even the Phase I award helped us to close investment deals. And our Investor presentations afterwards included the pink comment sheet from NIH."

Phase II funding came at a critical time for the company. Just as the company was moving toward productization, significant clinical research and validation was required. SBIR funding helped specifically meet that need. Without SBIR funding, Giuliani believed that the company "Would have had less success; SBIR money and the SBIR process played a significant role in the company's early success."

The SBIR provided substantial additional benefits to the project, beyond the funding. It focused the research team onto being able to present a cogent explanation of its product and the latter's potential unique benefits to society. According to Giuliani, the SBIR application "forced us to take our existing, less coherent approach and mold it into a better project."

The SBIR award also played in important role in helping the company to gain the trust of its core market—the dentists. Giuliani had researched the dental care industry, and noted that "no new tech had ever succeeded in home dental care without backing of dental community." Dentists became the critical intermediary between the company and the marketplace, and the NIH award was once again an important step in validating the product.

In addition, the SBIR application process generated very important advice and contacts. Most unusually, the study section visited both the company and the University of Washington. This brought the company into contact with very experienced people with "phenomenal capabilities." The company prepared very extensively for the site visit, and found the feedback from the visitors hugely helpful.

Giuliani believes that this input as extremely important: "We could not have paid people to do what the visitors did, especially as we had—at the time—no money and no useful contacts. They provided insights that money literally could not buy."

The visitors brought both genuine interest and also a certain degree of healthy skepticism to the visit. According to Giuliani, the visit had been arranged because the study section was impressed by the potential of the project, and believed that they could help tune company plans through a visit.

EXPERIENCE WITH SBIR

Giuliani had had experiences with SBIR from several perspectives.

Although the impact of SBIR has been highly positive at Optiva, the company subsequently had two or three "bad experiences" with the application process, where applications were poorly understood, or, on the opinion of the applicants, rejected on unsubstantiated or inappropriate grounds. As a result, Optiva ceased applying for SBIR funding (and is of course on longer eligible after its purchase by Philips).

Giuliani has continued to pursue SBIR funding for his new company, Pacific Bioscience, which seeks to adapt sonic technology to skin cleaning. These applications have also been rejected, again on grounds that Giuliani found unconvincing. As a result, he is a strong supporter of the notion that better communication between applicants and reviewers might generate better results and consequently more enthusiasm for the program.

Giuliani also has experience working as a reviewer on study sections. Overall, he was quite impressed. He found the materials were well prepared, that proposers were in general doing a good job of presenting their projects effectively, and that the study section was well organized and reviewers took their work seriously. He noted that there was a considerable amount of group discussion, which was in general not dominated by one person. However, he also noted that there would always be some degree of arbitrariness in the process. There was also a range of quality among the reviews, and that it was clear to him that at least one person—and maybe more—had not read any of the materials in preparation for the study section meeting.

RECOMMENDED IMPROVEMENTS IN SBIR

Most of Giuliani's comments focused on the selection process, where he was especially concerned by the overall quality of the review, and also by problems with cycle time partly caused by slow responses to review from NIH.

Review for Commercialization

Giuliani noted that there were considerable difficulties in reviewing applications for commercial merit. Specifically, academic reviewers were poorly suited for this function. However, he believed that adding additional layers of review—on the NSF model—while it might improve commercial review would also provide additional burdens for applicants, and might also therefore tend to slow the application process still further.

Instead, he suggested that NIH consider hiring a "genuine business person" as a consultant to the study section. These consultants could read each application for commercial context, and would provide a report to the study section. Such a consultant could provide further support by ranking applications, or even triaging them from a potential investor's point of view. Such an approach would be educational for the study section, and could increase the comparability of reviews across the entire set of SBIR applications.

Cycle Time

Giuliani was concerned about the slow cycle time, and in particular the delays in receiving reviews. He believed that two improvements might be especially useful:

- **Rebuttal**, whereby applicants would be provided with an electronic copy of the lead reviewer's written summary some period before the meeting of the study section, and would be able to submit a short rebuttal. Giuliani suggested that this approach might help to clarify issues concerning technical aspects of the application, as well as providing some significant additional incentives for reviewers to improve the quality of their reviews.
- **Shortened Response Time.** Giuliani was especially concerned that current response times (the time required between the study section meeting and delivery of the pink sheet with comments and scoring to applicants) was still much too long, and that NIH had taken insufficiently advantage of electronic communications tools to shorten the cycle time. He offered two suggestions:
  - **Electronic communications replacing written responses.**
  - **Immediate communication of raw scores to applicants,** which would improve company planning around applications.

Award size

Giuliani did not favor any increase in funding, either for Phase I or Phase II, nor did he favor the option of allowing the company to apply directly for Phase II skipping Phase I).

Role of Venture Capital in SBIR

Optiva did not receive significant VC funding. According to Giuliani, some had been received rather late in the company's independent life. An IPO had been planned but the company chose not to pursue it. Overall, venture funders owned less than 10 percent of the company at the time of its sale to Philips. In general, Giuliani supported the current SBA interpretation which excludes companies more than 50 percent owned by institutional investors.

OSI Pharmacueticals, Inc.

Andrew Toole
Rutgers University

COMPANY BACKGROUND

OSI Pharmaceuticals (henceforth: OSI) is a biotechnology company focused on the discovery, development, and commercialization of pharmaceutical products intended to extend life or improve the quality-of-life for cancer and diabetes patients.

OSI was originally founded in 1983 under the name "Oncogene Science, Inc." to focus on cancer therapeutics. At that time the company was backed by venture capital investors. Oncogene Science went public in 1985 and eventually changed its name to OSI Pharmaceuticals to reflect the expansion of its drug discovery and development activities beyond cancer indications. As of December 2004, OSI employs 452 people. Over 80 percent of these employees are located within the United States and about 34 percent are primarily engaged in research activities.

OSI owns facilities in the United States and the United Kingdom. Its U.S. operations are divided between Colorado and New York. Colorado is home to its drug development group while New York hosts its corporate headquarters and its drug discovery operations. Somewhat uniquely, its drug discovery facility is part of the Broad Hollow Bioscience Park on the campus of Farmingdale State University. This park is a collaborative effort between Cold Spring Harbor Laboratory and Farmingdale State University intended to grow the bioscience industry on Long Island. OSI's early collaborations with Cold Spring Harbor made it a good candidate for the science park. As for their overseas facility, OSI owns a subsidiary called Prosidion that focuses on diabetes and obesity research. Prosidion was spun out from OSI in 2003.

OSI has been successful at discovering and developing novel pharmaceutical agents through its research activities supported by large pharmaceutical partners, the SBIR program, and other sources of private investment. They now have three FDA-approved drugs on the market. Tarceva is their flagship product and is the first drug they took from concept to market. To date, it is the only epidermal growth factor receptor (EGFR) inhibitor to have demonstrated the ability to improve survival from non-small cell lung cancer and it may have therapeutic activity against certain forms of pancreatic cancer. OSI also markets Novantrone (mitoxantrone concentrate for injection) for approved oncology indications and Gelclair for the relief of pain associated with oral mucositis.

Over time, OSI evolved from a contract research firm to a fully integrated pharmaceutical company. Their contract research experience helped them build strong capabilities in High-throughput Screening (HTS), chemical libraries, medicinal and combinatorial chemistry, and automated drug profiling technology platforms. To this research base OSI added sales and marketing capabilities as well as a regulatory affairs group. According to Hoovers corporate information, OSI had $42.1 million in sales in 2004, a 32.1 percent increase over 2003 sales.

OSI's CEO is Colin Goddard, Ph.D., who was appointed in October 1998. Before joining the company, Dr. Goddard spent four years at the National Cancer Institute in Bethesda, Maryland. He was trained as a cancer pharmacologist in Birmingham, U.K., and received his Ph.D. from the University of Aston in September 1985.

TECHNOLOGIES AND INTELLECTUAL PROPERTY

One of OSI's core research strengths is High-throughput Screening (HTS). HTS works by testing hundreds of thousands of compounds to identify hits against a biological target. These hits are subsequently developed into drug leads through medicinal chemistry. A state-of-the-art HTS operation is located at their Farmingdale, NY, research facility. OSI has compiled a library of more than 300,000 compounds which are formatted in plates in a manner that allows them to be accessed and mobilized quickly for screening. Assays are developed using advanced technologies, mostly involving fluorescence-based readouts that are easily miniaturized for processing in a 384-well plate. The screens themselves are run at very high throughput on robotic screening systems that are capable of running in unattended mode for extended periods. The screening operation can test an entire library against a biological target in about two weeks. A companywide data management system stores all the data generated throughout the discovery cascade. A series of querying and data mining tools enables OSI researchers to examine the data and decide which compounds to synthesize and which compounds should be taken to preclinical and clinical testing.

Intellectual Property

OSI owns approximately 90 U.S. patents and about 150 foreign patents. They have about 100 U.S. patent applications and 200 foreign patent applications pending. Pfizer is a long-standing collaborative partner of OSI Pharmaceuticals in the area of cancer research. This relationship led to co-ownership of about 600 U.S. and foreign granted patents and patent applications in about 50 patent families. Two of these co-owned patent families cover the method of preparation for OSI's leading product, Tarceva. Moreover, OSI jointly owns some patents and patent applications with North Carolina State University.

IMPORTANCE OF SBIR

OSI received 51 SBIR grants between 1983 and 2002. Of these, 17 proposals progressed to Phase II awards. The Phase I grants totaled US$2.92 million (nominal dollars) while the Phase II grants totaled US$8.96 million (nominal dollars).

Dr. Haley was very positive about the role and contribution of the SBIR Program to the success of OSI. However, he stressed the difficulty of assessing the SBIR program using commercialization outcomes and readily quantifiable measures. Pharmaceutical innovation takes an average of 12-15 years from concept to market and most SBIR awards are facilitating concept and preclinical research. Even with SBIR awards contributing to research success, there are numerous factors and other inputs that cloud any clear picture of SBIR's individual contribution. So, while OSI has no commercialization outcomes resulting directly from SBIR grants, Dr. Haley said the program added value to the firm and its technology development in a variety of other ways. He says, "We developed a lot of technology using SBIR grants and generated a lot of hits." Four of the most important benefits for OSI from participation in the SBIR Program were:

(1) Key Source of Early-stage Financing

OSI actively pursued SBIR financing in its early years. The company submitted between one and three project proposals per quarter. Most of these proposals were intended to explore highly uncertain and risky research possibilities. Even though many of these SBIR projects created knowledge, patents, and academic publications, private investors would never have backed these projects because they were "too innovative," exploring too far beyond the known technological frontier. Thus, Dr. Haley believes that several of OSI's research achievements might not have happened without SBIR funding. He stated many of the SBIR-backed research projects supported some of the most creative work that has ever been undertaken at OSI. More broadly, Dr. Haley noted "This program does have a big impact on early-stage biotechnology companies. They all rely on it for funding early-on."

(2) Helpful for Obtaining Follow-on Funding and Business Deals

Specific technological and scientific knowledge generated with SBIR funding laid the foundation for new contract research projects for big pharmaceutical firms that provided an important source of additional funds for OSI. The SBIR knowledge base helped to create opportunities for additional projects with companies like Pfizer, Aventis, Novartis and others. Thus, SBIR funding did not lead directly to products but to valuable research services and business deals. In another example, OSI was exploring two alternative approaches to the discovery of its flagship product, Tarceva. One approach, which turned out not to be successful, was supported by an SBIR award from the National Cancer Institute. The other approach was not initially funded and OSI was having a difficult time convincing Pfizer, a long-term research partner, to finance that research. Dr. Haley recalled that government funding for the project made Pfizer nervous and leveraged them to go forward. The Pfizer-backed approach led to the discovery of Tarceva.

(3) Facilitated Hiring of Employees

Dr. Haley emphasizes the importance of the SBIR awards for making it possible to hire additional scientists. He suggests "it was the development of a critical mass of experienced technologically savvy people all in one spot that made lots of things happen." Even hiring two employees can have a big impact by allowing the firm to reach a critical mass of people involved in knowledge creation. The SBIR Program contributed to a more stable employment environment that ended up having multiple benefits for OSI.

(4) Facilitated Cross-fertilization Between Research Programs

Cross-fertilization is one of the most significant benefits resulting from strategic hiring and sustained research on multiple projects. OSI had scientists working on different projects, some SBIR supported and some supported through private financing sources. Dr. Haley was clear to point out that the synergies flowing from this multiple project model "had a big impact on us." And this relates back to his comments on the difficulty of measuring the impact of SBIR awards. Dr. Haley says that SBIR research frequently resulted in "tangential impacts" that happen through cross-fertilization mechanisms. It could be as simple as two scientists discussing research problems and solutions. These effects can be seen as spillovers between research projects that result in economies of scope in scientific discovery. In Dr. Haley's opinion tangential effects are not exceptions occurring through SBIR but frequent events that push research forward.

ISSUES WITH THE CURRENT SBIR PROGRAM

Dr. Haley highlighted a few issues of concern about the SBIR Program at NIH. His perspective is based on many years of experience with the program beginning in the late 1980s. He also serves on NIH review panels that evaluate new SBIR proposals for funding.

Before mentioning the issues of concern, Dr. Haley noted that the SBIR Program has evolved in a positive and useful direction since its inception. Prior to the 1991 reauthorization of the SBIR Program, the funding levels and the intellectual property rights (IPR) provisions were inadequate. Quite simply, the $50,000 for Phase I and $500,000 for Phase II did not provide enough money to get anything done. However, the current funding levels are sufficient. He also stated that it is also important that private firms be allowed to retain exclusive rights to their discoveries made possible by SBIR funds. While this is currently the case, government's emergency march-in rights are still a concern for private firms. Because of this, firms do not want to link their discoveries directly to government funding. By blurring this link, firms can protect their IPR from government march-in.

With regard to concerns about the program, Dr. Haley made the following three points:

(1) "Funding Gap" Between Phase I & Phase II Awards

The funding gap between Phase I completion and Phase II approval, which can be six months or more, creates an unstable employment environment. Stable employment of scientific staff is critical for small firms, especially in the early-years when there are very few sources of capital. The funding gap can induce key scientific personnel to leave the firm and force the firm to abandon that line of research. This is an unintended but significant problem with the current SBIR approval and funding system.

(2) Funds Tied Too Tightly to the Project's Specific Aims

Biomedical research is notoriously unpredictable. SBIR funding is not nimble enough to change as research opportunities evolve within a project. SBIR funds are tied immutably to the specific aims laid out in the initial proposal. There should be a mechanism in place that allows some flexibility in the SBIR award to address new and unanticipated avenues that are not explicitly part of the grant's specific aims. This is particularly important in highly competitive research areas where the extra time and effort required to initiate a new SBIR cycle is overly burdensome.

(3) Proposal Reviewers Tend to Favor Less Risky Projects

While Dr. Haley is generally pleased with the NIH SBIR review system, he still sees a tendency for reviewers to favor less risky projects that might have a good chance of attracting private investment. Sometimes the most innovative and risky projects are also the ones with the highest social returns. Private investors may not see these as attractive opportunities and the SBIR Program, in principal, can fund these higher risk projects. This problem is intertwined with the SBIR Program's focus on commercialization. High risk projects, by definition, have a lower chance of commercialization but may produce other intermediate outcomes of value via patenting, publication, etc. Greater weight could be placed on these intermediate outcomes in the proposal review and approval process.

Retractable Technologies Inc.—RTI

Robin Gaster
North Atlantic Research

OVERVIEW

The case of Retractable Technology, Inc., (RTI) illustrates the difficulties that can confront companies even if their research and development effort is entirely successful. Especially in health care, where the final consumers are rarely the purchasers of medical goods and services, and where skewed incentives often generate negative outcomes. The existence of extensive middleman interventions has in some cases—as RTI demonstrates—meant that even products that are technically successful, effectively manufactured, and backed by a strong entrepreneurial team can still face substantial difficulties in the marketplace.

COMPANY HISTORY

RTI was founded by Thomas Shaw, an engineer, in 1994, after he was inspired by a TV program about a doctor who had contracted HIV from an accidental needlestick injury. The company focused on building a retractable needle that would essentially eliminate needlestick injuries altogether.

The objective was and remains important. As of 2005, CDC estimates that 800,000 needlesticks occurred annually in the United States, with about 385,000 in hospitals[24] (where the true number was likely much higher, according to another CDC study.[25]) About 1,000 nurses and health care workers contracted hepatitis C from needlesticks—a potentially deadly disease—and about 35 contracted HIV.

Needle safety is not just a U.S. problem. In developing countries, retractable needles could be a critical tool in the fight against deadly diseases such as AIDS, as these needles prevent the re-use of needles by multiple patients, a procedure that can have a devastating accelerating impact on the propagation of diseases.

In January 1992, Shaw received a $50,000 Phase I SBIR grant from NIH to develop one of his syringe prototypes. In October 1993, he received a $600,000

---

[24]Statement for the record by Linda Rosenstock, MD, MPH, director, National Institute for Occupational Safety and Health, Centers for Disease Control and Prevention, before the Subcommittee on Workforce Protections Committee on Education and the Workforce, U.S. House of Representatives, June 22, 2000.

[25]CDC study estimated that 54 percent of needlesticks were reported through the hospital needlestick surveillance mechanism is 1996. CDC: better needlestick reporting required. Centers for Disease Control and Prevention, (3):30-31, March 12, 1997.

Phase II award to further develop and commercialize one of his designs, and to produce 10,000 samples for clinical trials.

The current company, RTI, was founded in 1994, and in 1997 RTI completed work on a state of the art manufacturing facility in Little Elm, Texas, capable of manufacturing approximately 50 million retractable syringes annually.

Initial trials were greeted very enthusiastically by local doctors and hospitals; a number of doctors became investors in the company, as it raised $42 million in investment funding: "I was immediately impressed when I saw the prototype," recalls Dr. Lawrence Mills, former chief of thoracic surgery at Presbyterian Hospital in Dallas. "I thought the biggest problem was the company wouldn't be able to make them fast enough to keep up with the demand." (Now retired, Dr. Mills is a shareholder in Retractable.)[26]

THE ROLE OF SBIR AT RTI

RTI is in one sense a classic SBIR story: one Phase I, one Phase II, a product in clinical trials at the end of Phase II, and a successful product in the marketplace within two years thereafter.

It is also clear that SBIR had a significant role to play in the evolution of RTI; while the company was not founded specifically on the basis of the SBIR funding, its primary product (sole product for some years) was derived directly from the SBIR-funded project. In an interview, the founder, Tom Shaw, said that the funding from SBIR had come at a critical time in the development of the project.

However, Shaw noted that product development is one thing; being able to break into the medical market in the U.S. is something else entirely, and here the SBIR award had minimal impact.

BACKGROUND ON MEDICAL PURCHASING
IN THE UNITED STATES

Most U.S. hospitals and large clinics belong to group purchasing organizations (GPOs). These organizations were originally designed to work on behalf of hospitals, to aggregate demand for medical products, and hence to position hospitals to get a better deal from the large manufacturers who largely dominate the health care market place.

These GPOs have developed a range of business practices designed to ensure on the one hand that hospitals purchase medical services and other medical services only through the GPO, and on the other that manufacturers supplying the GPO do not sell directly to hospitals, thus by-passing the GPO.

These arrangements include:[27]

---

[26]Patricia B.Gray, "Stick it To 'Em," *Fortune Small Business*, March 1, 2005.

[27]This section is based on Prof. Einer Elhague, "The Exclusion of Competition for Hospital Sales through Group Purchasing Organizations," unpublished manuscript, 2002.

- Exclusive dealing agreements, whereby the hospital agrees to purchase all medical devices through the GPO, except where the GPO does not offer the goods or services needed.
- Near-exclusive agreements, where the hospital agrees to purchase to fixed amount—90-95 percent perhaps—of a given commodity through the GPO. Because standardization is now the rule with a given hospital—it is documented to save lives—a commitment at this level is effectively a commitment to purchase 100 percent through the GPO.
- Bundling agreements, whereby a number of products are bundled into a single contract. As in DoD, bundling has the effect of excluding potential competitors for a single good or service.
- Rebates or discounts, where GPOs do not mandate an exclusive agreement, but instead offer substantial rebates for reaching targets—e.g., 90-95 percent of product purchases through the GPO. Rebates may be even more effective than exclusionary deals, as they may be easier to enforce. Sometimes all past rebates remain conditioned on continued meeting of the objective—creating golden handcuffs.
- Nonvolume payments from the GPO, which can include stock options or other investment opportunities in favored manufacturers for senior hospital staff including purchasing managers. Other fixed fees appear to be used as a means of encouraging loyalty to the GPO program.

Together, these incentives and arrangements have allowed GPOs to dominate the market for sales to hospitals in the United States. And in turn, GPOs have developed close relationships with the manufacturers whose products they carry. While GPOs charge hospitals a membership fee, they also have developed arrangements whereby manufacturers pay large sums of money in "fees" of various kinds—for appearing at GPO-sponsored events, for example. While in most areas of the economy these arrangements would be illegal, in this case they are not:

"Most troubling is that some GPOs are funded by suppliers rather than solely by hospitals. The fees that suppliers pay, which would normally be considered illegal kickbacks, are allowed by the 1986 amendment to the Social Security Act. Thus, buying groups may serve the interests of the suppliers that provide their funding, not providers, thereby undermining value-based competition. While the extent of this bias is contested, the potential for conflict of interest is indisputable.

To enable value-based competition, every buying-group practice should be consistent with open and fair competition. There is no valid reason for buying groups to accept financing or any payments from suppliers: if a buying group adds value, the customers (hospitals) should voluntarily pay for it."[28]

---

[28]Michael E. Porter and Elisabeth Olmsted Teisberg, "Redesigning Healthcare: Creating Value-Based Competition and Results," Harvard University Press, Cambridge MA. 2006, pp. 361-362.

RTI AND THE GPOS

RTI has been forced to confront the GPO issue head-on since very soon after its foundation. Early positive reviews from local oppositely staff did not result in sales, or even the opportunity to make sales. At the hospital where the syringe was initially tested, Presbyterian Hospital of Dallas, Dr. Edward L. Goodman, then Presbyterian's director of infection control, wrote that the new syringe was "essential to the safety and health of our employees, staff, and patients" and urged management to buy it. (Goodman was an early RTI shareholder, too.) Mr. Shaw claims that hospital officials told him that the institution could not purchase from RTI because of an exclusionary agreement with Premier, one of the two largest U.S. GPOs.[29]

RTI notes on its Web site that "Often our salespeople are not even allowed to show our products to hospital materials managers because restrictive GPO contracts effectively preclude their purchase of our products, regardless of their effectiveness in preventing needlestick injuries."[30]

QUALITY AND COST

RTI has always claimed that its technology was fundamentally better than that of its competitors. It provided numerous documents to bolster its case, and there has been some analysis in the industry. In October 1999, a nonprofit agency which publishes the medical industry's widely circulated counterpart to *Consumer Reports*, published its thorough evaluation of safety needle devices. RTI's VanishPoint syringe was the only syringe to receive the agency's highest possible rating.[31] The editors of *Health Devices* went to some pains to object to this characterization of their research, but it appears accurate. Cost concerns have also been an issue. RTI's retractable technology was initially somewhat more expensive than competing products (although Shaw noted that these competing products were considered ineffective by users and hence could not be viewed as strictly comparable).

In June 2002, RTI signed a long-term contract with Double Dove Co., Ltd., one of the largest syringe makers in China. Double Dove supplies syringes to RTI at an average unit cost of 8.5 cents, fully packaged, sterilized, and ready for use; this allows RTI to price its needles below competitors even in markets in developing countries.

---

[29]Patricia B.Gray, "Stick it To 'Em," op.cit.
[30]*<http://www.vanishpoint.com/simple4.aspx?PageID=130>*.
[31]*Health Devices*, October 1999.

MARKETS AND SALES

Despite the barriers presented by the GPO structure, RTI has had some significant commercial success. In 2002, it delivered its 100 millionth VanishPoint syringe; in July 2002 it recorded its first profitable quarter.

This success, however, was based only on the exploitation of niche markets within the U.S. that were not dominated by the GPO structure. These included federal prisons, Indian reservations, the Mississippi health department, the Veterans' Administration, as well as some foreign markets such as South Africa. RTI also worked with the Service Employees International Union, the largest union of health care workers in the U.S., which had concerns about large numbers needle-stick injuries, particularly in California the center of the AIDS crisis.

RTI has also had important symbolic wins. Its syringe has been selected for use in the U.S. government Global HIV/AIDS Initiative in Africa, winning three successive contracts. The CDC chose the VanishPoint syringe and blood collection tube holder for a large four-year study on the safety and effectiveness of an anthrax vaccine. In August 2005, RTI signed a licensing agreement with BTMD, a Chinese government-designated medical device manufacturer, for distribution of RTI's safety needle devices in China.

Despite these limited successes (RTI is still not making an operating profit as of 2006, although sales reached a record $25 million for the year), the difficulties

BEFORE

AFTER

FIGURE App-D-12 The RTI VanishPoint® syringe.

SOURCE: Retractable Technologies, Inc.

RTI experienced in reaching U.S. markets led RTI to file antitrust lawsuits against both Becton Dickinson, the largest U.S. syringe manufacturer, and large GPOs, including the two biggest, Novation and Premier.

LAWSUITS AND REGULATORY REFORM

Pretrial proceedings lasted six years, during which RTI's Dallas factory operated at 50 percent capacity for much of the time. In 2003, as the trial date neared, Becton Dickinson, Tyco International, and the other defendants offered substantial settlements—but no change in their business practices.

Ultimately the GPOs did agree to change their business practices, although what they promised exactly remains confidential. Premier and Novation, along with syringe manufacturer Tyco, paid about $50 million to settle the case. Finally, under pressure from lawyers and shareholders, RTI accepted a settlement of $100 million from Becton Dickinson.

The issue of obstructive business practices has been taken up elsewhere. The U.S. Senate antitrust subcommittee has held hearings on whether hospitals' buying practices are stifling competition. The U.S. Department of Justice opened a criminal investigation of hospital purchasing last year. The New York State attorney general's office is investigating Becton Dickinson's sales practices.[32]

The settlements have funded further expansion in RTI's marketing, but in the view of RTI, they have not substantially affected the tight relationships between hospitals, GPOs and large manufacturers.

On the regulatory front, the problem of needlestick injuries generated some significant changes. In September 1998, California became the first state to pass a law requiring the use of safety needles. Cal-OSHA enforces this law. Many other states have passed similar laws.

In November 1999, both the Occupational Safety and Health Administration (OSHA) and the National Institute for Occupational Safety and Health (NIOSH), a component of the Centers for Disease Control and Prevention (CDC), issued documents insisting on the use of safety needle devices.

In November 2000, President Clinton signed into law the Needlestick Safety and Prevention Act (Public Law 106-430). As a result, OSHA's revised Occupational Exposure to Bloodborne Pathogens rule became effective in 2001. Employers are required to identify, evaluate, and implement the use of safer medical devices. They must also maintain a sharps injury log and involve frontline health care workers in the evaluation and selection process for safety devices.

RTI was involved in this legislation, but believes that its implementation has been too weak to change business practices among hospitals.

---

[32]Patricia B.Gray, "Stick it To 'Em," op.cit.

CURRENT STATUS

RTI is concerned that the long view for their company is dark. The original patents will expire in 2015, at which point many of RTI's technical advantages over its competitors will disappear. Unless the market opens dramatically before then, RTI will continue to be overwhelmed by the huge resources available to its competitors and a marketplace that is structured in ways that sharply tilt the competitive edge away from outsiders like RTI.

CONCLUSIONS

RTI can be seen in some ways as company with severe Phase III problems, somewhat analogous to companies with successful products trying to negotiate the tortuous twists and turns of the DoD acquisition community. Yet unlike DoD, it does not appear that there are institutional forces anywhere in the health care sector working on behalf of the smaller firms with innovative technologies that are trying to reach the market.

For RTI, the years spent fighting for the right to deploy VanishPoint technology in the marketplace have left scars. In an interview, Tom Shaw expressed considerable concern at what he saw as the lack of support from the Congress once VanishPoint technology moved from the lab into the marketplace. As Shaw put it, "While the SBIR award made it possible to develop the VanishPoint technology and to make RTI potentially successful, it is inexplicable that the Congress would fund good research but simply lack the courage to stand up against large corporations' anticompetitive practices. The current situation does not support the financial and medical interests of America's taxpayers."[33]

---

[33]Interview.

SAM Technologies

Robin Gaster
North Atlantic Research

BACKGROUND AND HISTORY

Dr. Gevins founded SAM in 1986, at about the same time that he founded its sister nonprofit research organization, the San Francisco Brain Research Institute, founded by Gevins in 1980 and previously part of the University of California School of Medicine in San Francisco. Dr. Gevins was focused on a project he had conceived while a freshman at MIT, to build a technology that could measure the intensity of mental work in the brain—reflecting in real time the concentration and attention capacity of the user.

Since 1986, SAM has consistently pursued this single goal, using all its SBIR and other awards to help build a prototype to measure signals in the brain that reflect attention and memory. This is, in short, a case study in how multiple SBIR and other awards can help to support a visionary and very high risk project in long-term biomedical research.

Dr. Gevins had received RO1 grants at UCSF, where he was offered a tenured position in the psychology department. However, RO1 reviewers were not in the mid-1980s friendly to technology-oriented projects, and Dr. Gevins found that SBIR was a better channel for his engineering activities.

Over the past 30 years, Dr. Gevins has received continuous federal support from the Air Force, the Navy, DARPA, NASA, NSF, and seven NIH institutes. These awards have been used to maintain a core staff working on the central project of the company. To fund such a complex and long-term project, Dr. Gevins systematically divided it into essential individual subprojects, and sought funding for them through unsolicited federal basic research and SBIR awards. This minimized overall risk, and SAM's work has been supported by many SBIR awards from many agencies, as the project covers many possible applications of the technology. For example, SAM has received significant support from support from the National Institute on Ageing, because SAM's assessment and analysis technology could have a very large impact on seniors facing performance deficits of many kinds.

SAM has developed both the hardware that measures brain signals and transmits that signal to a processing device (now a PC), as well as the software used to integrate different kinds of brain stimulation signals. One early SBIR was designed to build the meters necessary to capture the EEG signals SAM intended to work with, as these meters were not then available elsewhere.

In 2005-2006, SAM completed the first commercial product in the MM line, the world's first medical test that directly measures brain signals regulating attention and memory, the SAM Test (Sustained Attention & Memory Test). The SAM Test is covered by four U.S. patents and by a number of trade secrets. The test is designed to fill an urgent need for an objective measure of how a patient's cognitive brain functioning is affected by a disease, injury, or treatment in a wide range of areas including head injuries, sleep disorders, mild cognitive impairment of aging, attention deficit hyperactivity disorder, epilepsy, and depression.

The next proposed product, the Online Mental Meter, is designed for wide-spread use beyond medical care as a computer peripheral that provides continuous information about the user's state of alertness and mental overload or underload, by measuring mental activity in real time while people perform everyday tasks at a computer. The Online MM constitutes a substantial technical leap from the SAM Test, which requires that a subject perform a standardized repetitive psychometric test of sustained focused attention and memory. SBIR projects 17 and 18 (see Table App-D-8) have paved the way for this advance.

Continuous real-time measurement of mental effort could become a key enabling technology for a wide variety of advanced adaptive systems that will vary the sharing of tasks between a human and a computer in an optimal manner depending on the user's cognitive state. SAM believes that such systems may well be ubiquitous in the future.

SAM aims to become the gold standard for medical testing in neurology. Currently, the brain measurement component of psychological testing requires a PET or MRI scene, which is inconvenient and very costly ($4,000 or more each). In addition, existing performance-based tests can be misleading, as they fail to measure brain activity directly. For example, early Alzheimer's patients often produce acceptable memory and brain performance, because these patients are able to compensate for their initial problems. Direct brain measurement would reveal what performance analysis obscures—the actual problem at the neuron level.

Federal funding has allowed SAM to reject overtures from venture capital companies. According to Dr. Gevins, venture companies have "a different agenda, timescale, and process." In contrast, SBIR supports a transition from basic research to the next step." Dr. Gevins observed that venture capital companies in general have declining interest in truly innovative work, because such work often takes too long to get to market for venture capital timescales.

SAM is currently working with consultants under the LARTA commercialization support program to develop a strategic alliance with a large corporation in order to make the SAM Test commercially available as a fee-for-service medical test. The partner will need to undertake independent clinical trials, FDA registration, approval for third-party reimbursement and a major marketing and sales campaign, activities that could take at least 3 years and cost in excess of $15 million.

TABLE App-D-8 SAM Technologies SBIR Awards

	Funding Institute- Center or Agency	Year		Description
1	NIH	1988	R44-RR03553	Removal of Distortion from Magnetic Resonance Images
2	NIMH	1989	R44-MH42725	Active Electrode Hat for EEG Imaging of Schizophrenics
3	AFOSR	1989	F49620-89-C-0049	Software Tools for Signal Identification using Neural Networks
4	AFRL	1989	F33615-89-C-0605	Flight Helmet EEG System
5	NIMH	1991	R44-MH43075	EEG ArtifactDetection
6	AFOSR	1991	F49620-92-C-0013	Physiological Indices of Mental Workload
7	NINDS	1992	R44-NS27392	Neurofunctional Research Workstation
8	NIAAA	1993	R44-AA08680	Cognitive Performance Assessment for Alcohol Intoxication
9	NIMH	1993	N44-MH30023	128 Channel Automated EEG Recording System
10	NINDS	1994	R44-NS28623	Multimodality Workstation for Seizure Localization
11	NINDS	1995	R44-NS32241	Functional Brain Imaging of Unconstrained Subjects
12	NASA	1995	NAS9-19333	Spacecrew Testing and Recording System
13	AFRL	1995	F41624-95-C-6000	Decontamination of Physiological Signals of Mental Effort
14	NIMH	1996	N44-MH60027	Neurocognitive Experiment Authoring Tool
15	NIAAA	1997	R44-AA11702	Attention & Alertness Neurometer
16	NINDS	1998	N44-NS-0-2394	Assessment of Alertness in Patients with Sleep Disorders
17	AFOSR	1998	F49620-98-C-0049	Sustained Attention Meter for Monitoring Cognitive Load
18	AFOSR	1998	F49620-96-C-0021	Brain Automatization Monitor
19	AFOSR	1998	F41624-98-C-6007	Operator State Classifier Developer's Toolkit
20	AFRL	1998	F41624-97-C-6030	Rapid Application Cutaneous Electrode (RACE) System
21	AFRL	1999	F41624-99-C-6007	An Ambulatory Neurophysiological Monitoring System
22	NIMH	2001	R44-MH60053	Neurocognitive Assessment Meter For Psychiatric Drugs
23	NIDA	2001	R44-DA12840	Neurocognitive Index of Cannabis Effects
24	NIA	2002	R44-AG17397	Neurocognitive Assessment of the Elderly
25	NICHD	2002	R44-HD37728	
26	NINDS	2003	R44-NS42992	
27	DARPA	2003	DAAH01-03-C-R292	Multicompartment Neuroworkload Monitor
28	NHLBI	2004	R44-HL065265	
29	ONR	2004	N00014-04-C-0431	Multitasking Personnel Selection Test

SOURCE: SAM Technologies.

Staff

SAM has 13 scientists, engineers, and associates, and several outside consultants, covering a range of disciplines. The 8 most senior staff members have been with SAM an average of 11 years. Collaborations with scientists and doctors at universities, medical schools, and government labs are used to leverage internal research efforts, and SAM has made distribution agreements with medical device companies which account for most product sales. SAM is an FDA registered Medical Device Manufacturer.

OUTCOMES

Commercial Products

Six of the SBIR-funded projects (#3, 7, 9-12) have to date resulted in two commercial products.

Image VueTM

Image Vue™ is a software package for visualizing brain function and structure by fusing EEG data with MRI (Magnetic Resonance Images), using patented algorithms to integrate functional and structural information about the brain, in order to localize epileptic seizures in a patient's brain. A wizard-driven software system running under Windows XP, it co-registers EEGs with MRIs, performing patented DEBLURRING™ spatial enhancement and several types of source localization analysis, and provides interactive 3-D graphics visualization. The patented XCALIPER™ hardware and associated software facilitates rapid measurement of EEG electrode positions needed for co-registration with MRIs.

The product is used primarily to visualize and localize the origin and spread of epileptic seizures in the human brain in planning neurosurgical treatment of complex partial seizure disorders that are refractory to treatment with antiepileptic drugs.

Image Vue™ is FDA-registered and is sold by Nicolet Biomedical Inc. (a subsidiary of Viasys Healthcare, Inc), the world's largest supplier to the clinical neurology market. Nicolet has purchased approximately 100 systems from SAM to date, from which they have generated about $2,000,000 in revenues. A number of competing products worldwide have been modeled on Image Vue™.

MANSCAN®

MANSCAN® evolved from basic research completed under prior NIH R01s, which with the aid of SBIR awards has been turned into robust algorithms embodied in a convenient, integrated system to enable research on human brain function that would not otherwise be commercially available.

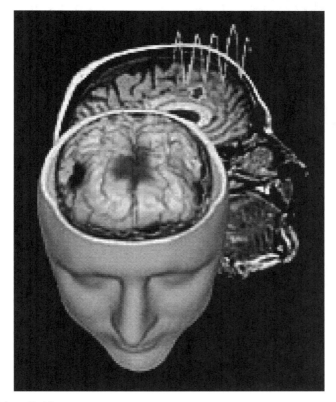

FIGURE App-D-13

SOURCE: SAM Technologies.

MANSCAN® is an integrated software and hardware system for performing brain function research via high-resolution EEG and event-related potential
(ERP) studies, and for integrating the results with magnetic resonance images.
MANSCAN® It was the first system to integrate the high time resolution of EEG
with the high anatomical resolution of MRI, and the first to allow subsecond
measurement of rapidly shifting functional cortical networks. It enabled a new
generation of research, and a number of significant advances in understanding
attention, memory and other basic cognitive brain functions have been made
with it.

Results of these studies provide unique views of structural and functional
neuroanatomy. MANSCAN® analysis and visualization functions quickly and
easily quantify features from EEGs and ERPs, leading neuroscience toward the
goal of uniting brain electrical activity with brain anatomy.

MANSCAN®'s hardware includes quick application electrode caps, an efficient device called XCALIPER™ for measuring electrode positions, and an advanced digital amplifier called MICROAMPS™. MANSCAN® software is fully integrated with the Microsoft NT/W2000/XP operating system running on a PC.

Thirty MANSCAN® systems have been sold to qualified scientists at U.S. universities, medical schools and government labs, where it is helping them to perform advanced research. MANSCAN® has generated approximately $650,000 of revenue. Several competing products worldwide have been modeled on MANSCAN®. MANSCAN® is also specifically designed as a step toward the MM.

Knowledge Effects

SAM aims to produce commercial products—clearly its entire mission is focused on commercial outcomes in the long run. However, there are been significant knowledge effect benefits during the course of this high risk research. Nine of the projects listed below led to over 50 peer-reviewed scientific and engineering publications. Thirteen of the projects led to 18 U.S. patents. More widely, SAM staff have published more than 150 peer-reviewed publications including five papers in *Science*.

SBIR ISSUES AND CONCERNS

The Selection Process

In recent years, SAM has been criticized as being "insufficiently innovative," possibly because the field is catching up with SAM. Dr. Gevins notes that the recent drive for closer attention to commercialization is impacting SBIR reviews at NIH, but also that reviewers from academia have a better understanding of more basic research and are likely to be somewhat biased toward it. He also notes that academic reviewers are not themselves unbiased, in that they tend to focus on whether outputs from the project in question will be useful in their own research. Review quality and outcomes also vary very substantially by study section.

Conflict of Interest

Dr. Gevins is very concerned about potentially *major* conflicts of interest stemming from the use of industry participants on study sections. He procedures for addressing such conflicts are "pathetic": Section members are handed a written conflict of interest description immediately before the panel meets, and are then on the honor system to disqualify themselves. No NDA is signed, and little attention is paid to the process. Dr. Gevins believes that the current approach is just designed to protect NIH from awkward questions, rather than to provide real protections to applicants.

SAM always reviews membership of review panels, and not infrequently requests that the SRA exclude a panel member from reviewing a SAM proposal. This veto process mostly works, according to Dr. Gevins, but is not foolproof. There is clearly some element of risk involved in releasing internal plans to outsiders (Dr. Gevins pointed out that this risk is endemic to the funding process—and that venture capitalists never sign NDAs, so there are also risks involved in working with venture capitalists).

In short, Dr. Gevins argues that while formal protections are in place, no effort is made at NIH to define or verify the absence of conflicts of interest, even though SRAs are in general honest and conscientious.

Recommendation. NIH must take the conflict of interest problem much more seriously. It should implement its own conflict of interest policies more effectively, and should consider mechanisms for auditing reviewer activities, at least on a random basis.

Expertise

Dr. Gevins sees a disconnect between the SRA running the selection process, and the IC which will eventually fund the project, and which has technical expertise in the subject area. This contrasts with funding at DoD, NSF, where a single point of contact essentially determines funding and manages the award.

Recommendation. It should be mandatory that the primary reviewer should have technical competence in the field covered by the proposal.

Commercial Review

Dr. Gevins sees substantial room for improvement in addressing commercializing concerns. He does not support the commercialization index in use at DoD, which he regards as highly oversimplified and biased toward short-cycle projects. He did agree that a commercialization review could be useful, but though there might be helpful ways to separate out technical/scientific review from commercialization, which could be addressed by a separate perhaps permanent panel of experts, and where problems could be addressed within a single funding cycle rather than requiring full resubmission, which means at least one and possibly two funding cycles delay.

Reviewer Evaluation

CSR manages this function and does so fairly effectively. Dr. Gevins believes that the key motivation for reviewer participation is to follow activity near the cutting edge in a particular field.

Overly Random Scoring

Like many interviewees, Dr. Gevins noted the substantial random element in the review process. In particular, he believed that that there are quite substantial differences in scoring tendency between different review panels.

Recommendation. Scores should be normalized across study sections, just as they are for RO1s. Otherwise it is perfectly possible—indeed likely—that one study section will tend to systematically provide higher scores than another. As all scores are integrated into a single priority score list for a given IC, this would inevitably generate a bias toward projects that were reviewed by the higher scoring study section.

Funding Issues

Size of Awards

Dr. Gevins said that in his experience, over-limit applications were always discussed beforehand with the program manager. SAM makes a point of mentioning this in the application, to ensure that reviewers know that the relevant program manager is in the loop. Extra-large awards can sometimes be held up for one or more funding cycles by the program manager, even if they are technically inside the Payline. This is a less formal procedure, but similar to that in place at NIH for RO1 awards.

Recommendation. SAM supports increasing the size of Phase I awards and reducing the number of awards. Some program announcements already call for Phase Is in the vicinity of $500,000. SAM also supports increasing the size of Phase II awards and reducing the number. SAM would recommend a three year Phase II award for $1 million, possibly requiring prior approval from the program manager.

Commercialization Support

SAM has actively participated in the new LARTA-led program. It sees the program as useful, particularly because it forces companies to focus on commercialization. However, at the time of the interview, SAM had received almost no useful time from the consultants, who appeared to have too many companies in their portfolios (20 or more each).

In general, the basic outline of the LARTA support program met SAM's needs, which focused not on preparing for public presentations, but on moving steadily through the steps of developing a good commercialization plan, focused on strategic alliances. SAM had received a steady flow of reminder/check-up calls, pushing the company to focus on the commercial element of the business.

Recommendation. NIH should consider allowing LARTA to focus its resources more tightly on fewer companies, providing them with more resources.

Program Managers

Dr. Gevins wondered what role SBIR program managers or liaisons play at the various institutes. As they did not appear to manage the financial and reporting aspects of individual grants, and did not substantially influence selection, he was unclear as to their role. He saw program managers as providing no value added for recipients, and added that in some cases they could be highly destructive (a point also made by other interviewees). He believed that some managers clearly had their own research agendas, and sought to impose these on the SBIR application process. Dr. Gevins also noted that being a program manager with responsibilities for many SBIR awards was not a plum job at NIH; as a result, it was often handed off to the least senior staff member.

Recommendation. SAM suggested adding a program manager review section to the final report for each project, which would allow NIH to gather better feedback about program manager performance.

Funding Gap Issues

SAM handles the gap by operating with multiple overlapping project. Current work has taken six years on a specific stage of the overall brain measurement project.

Dr. Gevins also noted that ICs do not always fund projects immediately—the latter can be delayed for one or more funding cycle. SAM believes that April applications are likely to be funded fastest, because they show up at the beginning of the fiscal year and require least juggling from the IC. For example, SAM had a project that was approved during a September Council meeting, but had still not been funded by the IC as of the following March.

Other Concerns and Recommendations

SAM offered a range of other concerns and suggestions:

- Believes awards are too short. Six months for Phase I is "a joke," as Phase I research always takes about a year. SAM has never completed a two-year Phase II award in the standard two years.
- Does not support Fast Track, partly because reviewers tend to split Fast Track applications in two anyway, and also because there are too few advantages in this process for companies, in comparison to the additional uncertainty.

- Supports direct access to Phase II, without prior Phase I. SAM believes that the time required to apply for and complete a Phase I is the problem, as this can add a year or more to a project.
- Supports the view that drug development funding could be distorting the overall shape of the SBIR program.
- Supports competing continuation awards.

Sociometrics Corporation

Robin Gaster
North Atlantic Research

EXECUTIVE SUMMARY

Sociometrics is a woman- and minority-owned company of approximately 16 employees that develops commercial products and services based on state-of-the-art behavioral and social research. Located in California's Bay Area R&D hub, it has received funding from diverse government and private sources, generating revenues of more than $28 million since its foundation in 1983. By introducing the concept of packaged replication programs to the practice of social behavior change, Sociometrics has changed the nature of the field.

Sociometrics illustrates well one type of successful SBIR company. Winner of an NIH SBIR award during the first year of the program in 1984, Sociometrics has gone on to continue winning awards with consistency. Between 1992-2002, the company received 19 Phase I awards from NIH (see annex); 18 of these have become Phase II projects, a very high 94 percent conversion rate. Since 2003, six new Phase I projects have been awarded; two additional Phase I applications have received priority scores at the fundable level. Sociometrics will be submitting eight Phase II applications from these projects at the appropriate time.

Consistent with the goals of the SBIR program, Sociometrics has become a product-oriented company. Every Phase II it has conducted has generated a highly marketable, commercial product. Sociometrics has now developed several product lines, secured a distribution agreement with a major electronic publishing house, and had products chosen by the Centers for Disease Control and Prevention (CDC) for its public health initiatives. In 2002, the majority of the firm's profits shifted from contract and grant research fees to product sales, a testament to the firm's SBIR-related success. This success has been due in part to the cumulative and strategic nature of Sociometrics' efforts. New projects and products build on previous ones, adding one or more innovations in the process. The company has also leveraged the ubiquity of the World Wide Web to increase its products' reach and public access, designing most products for download or interactive use on the Internet.

Despite Sociometrics' own success, its market niche—the development of behavioral and social science-based commercial products—remains under-resourced with private funds. In part, this situation obtains because typical customers for such products are nonprofit organizations that appreciate and use the resources but cannot afford to pay very much for them. Responding to this important need, Sociometrics has kept its product pricing at close to production cost, leveraging instead good business practice with a sense of public service.

For these reasons the company continues to rely on SBIR funding to provide the necessary development support to create new products and expand its range of services. While SBIR grants remain an important revenue stream for many small companies in the program, Sociometrices involvement has distinguished itself with: a) a wide set of well-regarded and widely-used products; b) a profitable business-model, distinguishing it from many other behavioral and social science-focused SBIR firms; and c) Phase II funding as sufficient support to bring the company's products to market, in contrast to many biotech- and pharmaceutical-oriented companies.

Motivated once again by both business and public service concerns, Socio-metrics staff members have published extensively in peer-reviewed journals. The company does not develop patentable products.

PRIMARY OUTCOMES

- Four major product lines, with a fifth under development;
- Consistent profitability from inception;
- Industry standard for topically focused social science data and program archives;
- Distribution agreement for data products with world-leading provider of authoritative reference information solutions;
- CDC adoption of program archive products for nationwide distribution;
- Product line with social impact: effective program replication kits have changed the way behavioral practitioners operate in community settings; and
- More than 60 peer-reviewed publications based on SBIR-funded projects.

KEY SBIR ISSUES

Sociometrics has found the SBIR program a very productive platform for its work. The program has allowed Sociometrics to: a) take state-of-the-art social and behavioral research in its topical areas of expertise (reproductive health, HIV/AIDS, drug abuse, and mental health); b) use Scientist Expert Panels to assess the research and identify the best available data, practices, and knowledge; and c) develop commercial products and services based on the panels' selections. Sociometrics' products are aimed at a diverse set of target audiences. For example, its data archives and evaluation instruments are meant for use by researchers, faculty members, and students. Its effective program replication kits, evaluation publications, and program development and evaluation training workshops are intended for use by health practitioners in schools, clinics and community-based organizations. Its forthcoming Web-based behavioral and social science informa-tion resources, summarizing state-of-the-art research knowledge in select topical areas in nonscientist language, will be aimed at both academic and practitioner audiences.

BACKGROUND

Sociometrics Corporation was established in September 1983 as a corporation in the State of California. Within a year of its founding, Sociometrics had applied for its first SBIR project. Dr. Josefina Card, Sociometrics founder and CEO, was encouraged to found the company as a for-profit organization by Dr. Wendy Baldwin, then a program manager at NIH. (Dr. Baldwin later went on to become Deputy Director of NIH for Extramural Research.) Dr. Baldwin suggested that Sociometrics incorporate as a for-profit entity in order to benefit from the newly created SBIR program without precluding the possibility of obtaining basic and applied research grants.

Sociometrics' goals are:

- To conduct applied behavioral and social research to further our understanding of contemporary health and social problems;
- To promote evidence-based policymaking and intervention program development;
- To conduct evaluation research to assess the effectiveness of health-related prevention and treatment programs;
- To facilitate data sharing among social scientists as well as public access to exemplary behavioral and social data; and
- To help nonexperts utilize and benefit from social science and related technologies and tools.

In carrying out its mission, several areas of corporate expertise have been developed:

- The design and operation of machine-readable, topically focused data archives;
- The development of powerful, yet user-friendly, software for search and retrieval of information in health and social science databases;
- The harnessing of state-of-the-art developments in computer hardware and software to facilitate access to, and use of, the best data in a given research area;
- Primary and secondary analysis of computer data bases using a variety of commercially available statistical packages as well as custom-designed software;
- The design, execution, and analysis of program evaluations;
- The design, execution, and analysis of health and social surveys;

- The collection and analysis of social and psychological data using a variety of modes (mail; telephone; focus groups; in-person interviews);
- The collection and dissemination of social intervention programs with demonstrated promise of effectiveness; and
- The provision of training and technical assistance on all the above topics.

Sociometrics currently has 16 employees, 6 of whom have Ph.D.s and 5 of whom have Masters degrees. Expertise of its staff spans a diverse set of behavioral and social science fields, including: sociology, social psychology, clinical psychology, demography, linguistics, education, and public administration. In June 2005, a seventh Ph.D., specializing in political science and international relations, will be joining the firm.

Sociometrics has been the recipient of several awards including:

- A Medsite award for "quality and useful health-related information on the Internet";
- A U.S. Small Business Administration Administrator's Award for Excellence "in recognition of outstanding contribution and service to the nation by a small business in satisfying the needs of the federal procurement system"; and
- A certificate of recognition for Project HOT (Housing Options for Teachers) by the California State Senate, the California State Assembly, and the Palo Alto Council of PTAs "in appreciation for service supporting Palo Alto Unified School District's teachers and staff."

PRODUCTS

Sociometrics currently provides four research-based product lines:

- Data archives and analysis tools;
- Replication kits for effective social and behavioral intervention programs;
- Evaluation research; and
- Training and technical assistance services.

Sociometrics staff members are currently developing a fifth product line, online behavioral and social science-based information resources, to facilitate "distance learning."

Data Archives and Analysis Tools

The Sociometrics data archives are collections of primary research data. Each collection is focused on a topic of central interest to an NIH Institute or Center (IC). The data sets comprising each collection are selected and vetted by high-level scientist advisory boards to ensure that each collection is best-of-breed. Data sets are acquired from their holders, packaged and documented in standard fashion, and then made publicly available both online and on CD-ROM. Each successive data archive has leveraged features of previous archives that promote ease of use and has then developed new features of its own. This cumulative development effort has resulted in a digital library consisting of several hundred topically focused data sets that are easy to use, even by novices such as students and early-career researchers. Data in the Sociometrics' archives are accompanied by standard documentation, SPSS and SAS analytic program statements, and the company's proprietary search and retrieval tools. By providing high-quality data resources, and adding features facilitating appropriate and easy use, Sociometrics has created a niche of standardized, quality data products that complement the larger, though not universally standardized, data resources offered by other major data providers such as the University of Michigan. Currently, Sociometrics publishes nine data archives, with a tenth data archive on childhood problem behaviors under development. The nine collections are disseminated both as single

---

BOX App-D-3
The Sociometrics Data Archives: Topical Foci and Scope

- **AIDS/STD.** Nineteen studies comprising 30 data sets with over 18,000 variables.
- **Adolescent Pregnancy & Pregnancy Prevention.** 156 Studies comprising 260 data sets with over 60,000 variables.
- **Aging.** Three studies comprising 22 data sets with over 19,000 variables.
- **American Family.** Twenty studies comprising 122 data sets with over 70,000 variables.
- **Child Well-being and Poverty.** Eleven studies comprising 35 data sets with over 20,000 variables.
- **Complementary and Alternative Medicine.** Eight studies comprising 17 data sets with over 10,000 variables.
- **Contextual Data Archive.** Thirteen data sets compiled from over 29 sources with over 20,000 variables.
- **Disability.** Nineteen studies comprising 40 data sets with over 23,000 variables.
- **Maternal Drug Abuse.** Seven studies comprising 13 data sets with over 5,000 variables.

data sets (by Sociometrics) as well as via institutional subscriptions to the entire collection known as the *Social Science Electronic Data Library* (SSEDL).

Sociometrics' first data archive on adolescent pregnancy and pregnancy prevention was funded as part of the very first cohort of SBIR awards at NIH. More than 20 years after its initial release, this archive continues to be highly relevant, utilized, and regularly updated by Sociometrics. The archive was originally published on mainframe tapes, and has since been delivered to its customers using ever-changing computer data storage technologies including 5¼ and 3½ inch floppy disks and CD-ROM. It is now available 24/7 on the World Wide Web, where it has been accessible for the past 8 years.

A year ago, Sociometrics entered into a five-year distribution agreement for the *Social Science Electronic Data Library* (SSEDL) with Thomson Gale, a world-leading provider of authoritative reference information solutions. The agreement calls for Thomson Gale to market SSEDL via subscription to its wide range of academic and research library customers, while allowing Sociometrics to continue selling individual data sets from its own Web site. Sociometrics receives a portion of the Thomson Gale subscription sales in royalties. Several thousand SSEDL data sets have been downloaded from the Sociometrics Web site over the last three years, some by pay-as-you-go customers who execute a secure credit card transaction, others by faculty members and students able to download the data sets at no charge because their university is an SSEDL subscriber. Sales of the data archives and associated products have yielded approximately $125,000 in profits over the last three years. This figure does not yet include royalties from the Thomson Gale agreement which came into effect at the close of the 2003-2004 fiscal year, the latest date for which figures are available.

Competition

The Inter-University Consortium for Political and Social Research (ICPSR) data archive at the University of Michigan provides the main competition for Sociometrics' data archives. Despite ICPSR's much larger collection of data sets, Sociometrics has maintained a specialized niche, leveraging organization around selected health-related topics, careful selection of exemplary data by Scientist Expert Panels standardized documentation, ease of use of data sets, and value-added SPSS and SAS data analytic statements into a specialized collection tailored to data novices such as students and early-career researchers, that complements the ICPSR collection.

Replication Kits

Having developed the Data Archives, Sociometrics realized in 1992 that additional public service would occur if it extended and adapted its work beyond selection, packaging, and distribution of exemplary data to selection, packaging, and distribution of practices shown by such data to be effective in changing unhealthy or problem behaviors. Diverse health issues with important behavioral determinants (adolescent pregnancy, STD/HIV/AIDS, and substance abuse) were selected to showcase the new product line. Prior to 1992, information on effective programs was limited to brief descriptions in scientific journals often not read by health practitioners, a serious barrier to their widespread use. Using SBIR funding, Sociometrics sought to overcome this barrier by adapting to this new product line the time-tested methods it used to establish its data archives.

The company again worked with Scientist Expert Panels to identify and select effective programs based on their empirical support, collaborated with developers of selected programs to create replication kits for Panel-selected behavior change interventions, and partnered with networks of health professionals to disseminate these kits to schools, clinics, and community-based organizations. Replication kits were conceptualized as boxes containing all the materials required to reimplement the effective intervention. Typical replication kits contain a user's guide to the program, a teacher's or facilitator's manual, a student or participant workbook, one or more videos, and forms for "homework" assignments or group exercises.

The first effective program collection, the Program Archive on Sexuality, Health & Adolescence (PASHA) now comprises 29 replication kits. The newer HIV/AIDS Prevention Program Archive (HAPPA) and the Youth Substance Abuse Prevention Program Archive (YSAPPA) encompass 11 and 12 replication kits, respectively. Despite considerable initial skepticism from academics and some practitioners, the program-in-a-box approach has been received with considerable enthusiasm. PASHA, HAPPA, and YSAPPA have all proven to be social successes, with their programs being implemented in hundreds of schools, clinics, and communities across the country. They have also proven to be commercial successes, generating profits totaling over half a million dollars in the last three years.

Like the Sociometrics Data Archives, the Sociometrics replication kits (known collectively as the Sociometrics Program Archives) are topically-focused, best-of-class collections, selected using clearly defined effectiveness criteria by Scientist Expert Panels, and sold with free technical support for purchasers. This complimentary technical assistance has been lauded as an extremely valuable service by Sociometrics' customers and the company's reputation follows, in part, from the excellent product support it provides.

The replication kits have been sold individually from the Sociometrics Web site to such customers as schools, community health and service organizations, and medical clinics. They have also been displayed at exhibit booths at annual meetings of health practitioner professional organizations. Their dissemination is further supported by a company newsletter published three times annually and reaching 30,000 recipients. In 2003, CDC became an important customer for several replication kits, providing "train the trainer" workshops in Atlanta for hundreds of practitioners in use of selected kits. These new trainers have in turn returned to their hometowns and home organizations to train more staff, resulting in further sales of the replication kits and dissemination of important prevention programs.

Competition

CDC has provided the impetus for the only real competition to Sociometrics in the field of replication kit development for effective teen pregnancy and HIV/AIDS prevention programs. The CDC initiatives have been based on a decentralized distribution model, with CDC funding program developers to publish their programs themselves or to seek out their own commercial distributors. In contrast, the Sociometrics distribution model is centralized, with Sociometrics' Web site serving as a one-stop-shopping-point for highly effective programs in the areas in which the company operates.

The inevitable delays in implementing a new initiative, plus changes in policy at CDC and substantial budget cuts have put one of the CDC's two development programs on hold (the one on teen pregnancy prevention), leaving the Program Archive on Sexuality, Health, and Adolescence without significant current competition. The other CDC program on HIV/AIDS prevention, a competitor to Sociometrics' HIV/AIDS Prevention Program Archive, is using replication kits developed at Sociometrics for some of its selected programs, boosting sales of the Sociometrics HIV/AIDS Prevention Program Archive by an order of magnitude. In this manner Sociometrics Program Archives have complemented the larger CDC efforts, just as the Sociometrics Data Archives have complemented the larger University of Michigan efforts.

Longer-term Challenges and Opportunities

Over the longer term it is possible that commercial challenges may arise from changes in the academic world, where more and more developers of effective programs are deciding to publish their work themselves, releasing kits or parts of kits through their own Web sites or negotiating other arrangements with commercial publishers. Sociometrics is not overly concerned by these developments as it regards its work as complementary to, and supportive of, developers' efforts to get their effective programs in the public domain. Recent history is supportive. During the initial establishment of Sociometrics' HIV/AIDS Prevention Program Archive (HAPPA), the advisory panel recommended 18 effective programs for inclusion in the archive; of these one was withdrawn as "obsolete" by its original developer, seven developers had previously decided to use a commercial publisher, and ten were made available through HAPPA. Thus with the help of Sociometrics' efforts complementing existing efforts, replication kits for almost all effective programs are now publicly available to community-based organizations striving to prevent HIV. This constitutes an important public service in terms of: (1) packaging the most promising interventions to enhance their usability; (2) facilitating low-cost access to, and widespread awareness of, these interventions; (3) encouraging additional rigorous tests of the interventions' effectiveness in a variety of populations; and (4) demonstrating the value of, and providing a model for, the research-to-practice feedback loop.

Further opportunities for enhanced product dissemination arise from Sociometrics' collaboration with other organizations besides CDC. In particular, large nonprofit networks provide many opportunities for partnership. For example, on the teen pregnancy prevention program archive, Sociometrics has worked with the National Campaign to Prevent Teen Pregnancy, Advocates for Youth, and the National Organization for Adolescent Pregnancy, Prevention, and Parenting Inc. These organizations have become bulk purchasers of replication kits. They have provided other marketing support as well; for example, Advocates for Youth placed a link to Sociometrics on its Web site, and marketed Sociometrics' kits to its constituency from there.

Evaluation Research

Sociometrics has considerable expertise in program evaluation research and technical assistance. Over the last 15 years, the company has conducted many studies and provided technical assistance to many nonprofit organizations to determine whether a particular social intervention program was able to meet its short-term goals and long-term objectives. While most of the company's work developing its data and program archives has been funded by the SBIR program, Sociometrics' evaluation work has been funded primarily by state governments (such as California, Minnesota, and Wisconsin), local governments (such as Santa Clara County) and private sources especially foundations seeking an evaluation of the efforts of their grantees (the Packard Foundation, the Mott Foundation, the Northwest Area Foundation, and the Kaiser Family Foundation) and nonprofits seeking an evaluation of the effectiveness of their work. Sociometrics publishes a number of books and resource materials on program evaluation (e.g., *Data Management: An Introductory Workbook for Teen Pregnancy Program Evaluators*). It also offers at low cost (15 cents per page) evaluation research instruments that have been used in national surveys or in successfully implemented and published evaluation efforts.

Training Services

Sociometrics conducts workshops and courses to familiarize practitioners with the tools and benefits of social science and related technologies. These courses have recently been put online to increase their reach while lowering access costs. Training is offered in a variety of social science areas, particularly in effective program selection, development, and implementation; program evaluation concepts, design, and execution; and data collection, management, and analysis.

Science-Based Information Modules

These new products, still under development, will integrate the research literature in a given topical area, describe what science says in language and format easily understood by nonscientists (eighth grade reading level), and disseminate the information online via the Internet for easy "distance learning" access by all.

Into the Future

Most of Sociometrics' products are available for 24/7 download (with payment by credit card) on its award-winning Web site at *<http://www.socio.com>*. Its data archives, collectively known as *The Social Science Electronic Data Library* (SSEDL), are also available via institutional subscriptions marketed to universities and research libraries by Sociometrics' dissemination partner Thomson Gale. Sociometrics will continue its development of its Web site as a major product platform. In 2003, this Web site received over 1.7 million hits resulting in 29,729 product downloads. The company will also continue to develop additional subscription products. For example, Sociometrics plans to bundle its HIV and teen pregnancy replication kits, evaluation resources, training courses, and information module products and disseminate these bundled products to academics and health practitioners via online subscriptions. Eventually, mental health resources will be added to the Data Library, HIV, and teen pregnancy subscription resources as a fourth subscription line. Two current and two forthcoming SBIR Phase I grants support expansion into this important topical focus of mental health.

Profits and Revenues

Sociometrics' gross annual revenues are approximately $2.3 million with approximately 22 percent of this amount being profit (Table App-D-9). Profits from product sales are now substantially larger than profits from SBIR project fees, a testament to the success of Sociometrics as an SBIR firm. However, the profit stream is still insufficient to replace SBIR as the primary funding engine for future development efforts. Sociometrics does not market price its products, as many of its customers are small, community-based nonprofits that cannot afford products fully priced to market. Rather Sociometrics' products are priced at the cost of production with a small profit mark-up equivalent to a technical assistance retainer. Sociometrics sees this focus on widespread dissemination

Changing the Field

NIH and CDC officials, among others, regard Sociometrics' effective-program replication kits as an important innovation helping to bridge the gap between health-related research and practice. The program-in-a-box opened the door for researchers to generate something more than an article or book as an output from their studies, and many researchers were especially pleased to find a way to connect their work to the improvement of practice.

Publications

Sociometrics' staff members have more than 250 peer-reviewed publications; approximately 60 of these are based on the company's SBIR work.

Training Effects

Sociometrics is poised for expansion now that younger staff are becoming qualified as PIs in their own right, which relieves some of the PI burden from the two senior managers, who were until recently PIs on all projects.

NIH Institutes and Centers (ICs)

Sociometrics has had the longest relationship with—and is closest to—the National Institute on Child Health and Human Development (NICHD). Sociometrics' Founder and CEO, Dr. Josefina Card, has served on several NICHD study sections, and has also been on the NICHD National Advisory Council.

SUPPLEMENTAL FUNDING

Supplemental funding procedures vary substantially by IC. Typically, small supplement requests—up to 25 percent of the annual award amount at NIMH—are available at the discretion of the program officer (depending on funding availability). These are referred to as "noncompeting administrative supplements." Large supplementary funding requests must compete with other similar requests, seeking a "competing supplementary award." Sociometrics has obtained a few noncompeting administrative supplements. It has also obtained three larger supplement awards by expanding the scope of the funded Phase II grant in a way deemed "high priority" by the funding agency or by competing successfully via another funding mechanism (such as an RFA) with the funding agency then deciding, for administrative simplicity reasons, to add monies to the SBIR grant instead of issuing a new grant award. Examples include:

- **Teen pregnancy prevention program replication kits.** Originally funded by NICHD, Sociometrics sought a third year of Phase II support through a supplement to expand the scope of the Program Archive on Sexuality, Health & Adolescence (PASHA) from teen pregnancy prevention alone to teen STD/HIV/AIDS prevention as well. This expansion had been recommended by the PASHA Scientist Expert Panel, in light of the national spotlight on HIV/AIDS and the similar sexual-risk behaviors underlying both unintended pregnancy and STD/HIV/AIDS. The supplement request was forwarded by NICHD to the Deputy Assistant Secretary of Population Affairs who serves simultaneously as Director of the Office of Population Affairs. This political appointee interviewed the Sociometrics PI, and then personally approved the requested additional $750,000 in Phase II funding, transferring the monies to the NICHD grant.

- **Program archive for HIV/AIDS in adults.** Initially funded by the National Institute of Allergy and Infectious Diseases (NIAID), supplementary funding was requested for the HIV/AIDS Prevention Program Archive (HAPPA) to expand the project to include programs targeted directly at minorities. In this case, the request was for approximately $575,000 over three years. However, an end-of-year budget underrun at NIAID resulted in the full requested funding being provided over one year, instead the requested three years.

- **Complementary and alternative medicine data archive.** Sociometrics had Phase II funding from the National Center on Alternative Medicine (NCAM) to establish the Complementary and Alternative Medicine Data Archive (CAMDA) when it responded to an RFA issued by NCAM encouraging research on minorities and CAM. Sociometrics responded to the RFA by proposing to expand CAMDA to include data sets especially focused on minority populations. Its proposal received a high priority score and NCAM decided to fund the project via an administrative supplement to the Phase II project rather than via a new grant award.

RECOMMENDATIONS

Sociometrics believes that the SBIR program provides an essential resource for generating innovative and effective research-based products in efficient fashion. In response to questions about its support for various issues and trends in the program, Sociometrics makes the following recommendations:

- **Normalization of scores.** Scores should be normalized across SBIR study sections.
- **Award size and duration.** Phase I duration should be one year, and additional funding (beyond $100,000) should be available with justification. Phase II size and duration limits could remain as they are ($750,000 over two years); Sociometrics has always found it possible to split larger projects into two or more ideas qualifying for separate SBIR funding. While

TABLE App-D-10 Continued

Fiscal Year	Phase Type	Award Size ($)	Project Title	Funding Institute-Center
1995	Phase I	70,770	ESTABLISHING A CONTEXTUAL DATA ARCHIVE	HD
1996	Phase II	338,926	ESTABLISHING A CONTEXTUAL DATA ARCHIVE	HD
1997	Phase II	409,298	ESTABLISHING A CONTEXTUAL DATA ARCHIVE	HD
1993	Phase I	49,975	ESTABLISHMENT OF A RESEARCH ARCHIVE ON DISABILITY	HD
1994	Phase II	236,145	ESTABLISHMENT OF A RESEARCH ARCHIVE ON DISABILITY	HD
1995	Phase II	230,731	ESTABLISHMENT OF A RESEARCH ARCHIVE ON DISABILITY	HD
1996	Phase II	31,288	ESTABLISHMENT OF A RESEARCH ARCHIVE ON DISABILITY	HD
1997	Phase II	149,709	ESTABLISHMENT OF A RESEARCH ARCHIVE ON DISABILITY	HD
1992	Phase II	220,955	ESTABLISHMENT OF AN AIDS/STD DATA ARCHIVE	HD
1993	Phase II	236,301	ESTABLISHMENT OF AN AIDS/STD DATA ARCHIVE	HD
1994	Phase II	42,070	ESTABLISHMENT OF AN AIDS/STD DATA ARCHIVE	HD
1995	Phase II	10,000	ESTABLISHMENT OF AN AIDS/STD DATA ARCHIVE	HD
1998	Phase II	350,655	HIV/AIDS PREVENTION PROGRAM ARCHIVE	AI
1999	Phase II	397,449	HIV/AIDS PREVENTION PROGRAM ARCHIVE	AI
2000	Phase II	574,670	HIV/AIDS PREVENTION PROGRAM ARCHIVE	AI
1997	Phase I	99,221	INSTITUTE FOR PROGRAM DEVELOPMENT AND EVALUATION	HD
1999	Phase II	337,465	INSTITUTE FOR PROGRAM DEVELOPMENT AND EVALUATION	HD
2000	Phase II	412,385	INSTITUTE FOR PROGRAM DEVELOPMENT AND EVALUATION	HD
2001	Phase II	49,987	INSTITUTE FOR PROGRAM DEVELOPMENT AND EVALUATION	HD
1993	Phase I	49,494	INSTRUMENT ARCHIVE OF SOCIAL RESEARCH ON AGING	AG
1992	Phase II	223,017	MICROCOMPUTER DATA ARCHIVE OF SOCIAL RESEARCH ON AGING	AG
1993	Phase II	19,106	MICROCOMPUTER DATA ARCHIVE OF SOCIAL RESEARCH ON AGING	AG
2001	Phase I	197,562	PROMOTING EVALUATION/TEACHING/RESEARCH ON AIDS (PETRA)	MH

continued

TABLE App-D-10 Continued

Fiscal Year	Phase Type	Award Size ($)	Project Title	Funding Institute-Center
2003	Phase II	420,281	PROMOTING EVALUATION/TEACHING/RESEARCH ON AIDS (PETRA)	MH
2004	Phase II	329,507	PROMOTING EVALUATION/TEACHING/RESEARCH ON AIDS (PETRA)	MH
1994	Phase I	80,991	SOCIONET—ONLINE ACCESS TO SOCIAL SCIENCE DATA	HD
1995	Phase II	446,682	SOCIONET: ONLINE ACCESS TO SOCIAL SCIENCE DATA	HD
1996	Phase II	303,033	SOCIONET: ONLINE ACCESS TO SOCIAL SCIENCE DATA	HD
1998	Phase I	99,340	STATISTICS ON DEMAND: DATA ANALYSIS OVER THE INTERNET	HD
2000	Phase II	381,525	STATISTICS USING MIDAS: DATA ANALYSIS OVER THE INTERNET	HD
1999	Phase II	367,726	STATISTICS USING MIDAS: DATA ANALYSIS OVER THE INTERNET	HD
2002	Phase I	99,011	VIRTUAL PROGRAM EVALUATION CONSULTANT (VPEC)	HD
2003	Phase II	383,357	VIRTUAL PRACTIONER EVALUATION CONSULTANT (VPEC)	HD
2004	Phase II	363,923	VIRTUAL PRACTIONER EVALUATION CONSULTANT (VPEC)	
1993	Phase I	50,000	ARCHIVE OF TEEN PREGNANCY PREVENTION PROGRAMS	HD
1995	Phase II	408,644	PROGRAM ARCHIVE ON SEXUALITY, HEALTH, & ADOLESCENCE	HD
1996	Phase II	987,378	PROGRAM ARCHIVE ON SEXUALITY, HEALTH, & ADOLESCENCE	HD
1997	Phase II	45,000	PROGRAM ARCHIVE ON SEXUALITY, HEALTH, & ADOLESCENCE	OPA
1994	Phase I	79,383	ESTABLISHING A STROKE DATA ARCHIVE	NS
2002	Phase I	199,922	PROMOTING CULTURALLY COMPETENT/ EFFECTIVE HIV/AIDS PREVENTION PROGRAMS	AI
2004	Phase II	361,223	PROMOTING CULTURALLY COMPETENT/ EFFECTIVE HIV/AIDS PREVENTION PROGRAMS	AI

NOTE: For a list of codes for National Institutes of Health institutes and centers, see Box App-A-1.

SOURCE: Sociometrics Corporation.

Both the characteristics and versatility of the polymer-based carrier backbone are important to the broad-based applicability of VectraMed's TADD technology. The polymer is water-soluble and consists of well-defined physiological components. Moreover, with information on disease-specific triggers, a variety of polymer cleavable linking groups can be designed with different circulation times and solubility properties.

VectraMed was pursuing three lines of product development. One of these areas involved cancer drugs through the joint venture with Elan Pharmaceuticals. Prior to Elan's restructuring, their cancer applications looked promising with increased efficacy and reduced toxicity in animal models. VectraMed is currently in discussions with Elan to terminate the joint venture. In addition to cancer, VectraMed had active programs in chronic inflammatory diseases and fibrotic diseases including pulmonary hypertension, pulmonary fibrosis, and surgical adhesion prevention. While most of the preclinical results were positive, the lack of financial backing has suspended further research.

IMPORTANCE OF SBIR

Dr. Pachence was very positive about the role and contribution of the SBIR Program to the success of VectraMed. While there were no commercialization outcomes, he said that the program has "made an impact" in a variety of ways including contributions to multiple patent applications and multiple papers published in prestigious journals. Four of the most important benefits for VectraMed from participation in the SBIR Program were:

(1) Key Source of Early-stage Financing

Dr. Pachence used his personal savings to start VectraMed and to license its initial product candidates from Rutgers University and the University of Medicine and Dentistry of New Jersey in 1997. In that same year, VectraMed won two Phase I awards to perform proof of principle animal studies, one for postsurgical adhesions and the other for pulmonary hypertension. Each of these feasibility studies succeeded and VectraMed went on to win Phase II awards for each of these lines of research. SBIR Phase I funds were received in 1997 ($200,000) and the Phase II awards extended over two years, 1999 ($776,699) and 2000 ($846,783). (The SBIR investment into these lines of research totals $1,823,482.) Dr. Pachence noted that these SBIR awards were "key to getting the first stages of development done" and that SBIR was "important for early money."

(2) Critical for Follow-on Funding and Business Deals

Given the expense and time required to develop new drug therapies, it was incumbent upon VectraMed to obtain additional private financing beyond the initial personal funds that Dr. Pachence had invested. To a significant degree, SBIR funds allowed VectraMed to obtain additional follow-on private investment. The SBIR Program financed a large part of the preclinical animal studies. These initial SBIR awards used academic labs as subcontractors to complete the research since VectraMed did not have its own laboratory facilities. The positive results from these studies helped convince corporate partners and angels to invest in VectraMed. This new scientific evidence was combined with previously collected information on the market opportunity to facilitate the completion of a formal business plan. Dr. Pachence noted that "SBIR was super critical for getting the angel financing but also for getting the deals done with the corporate community."

(3) Facilitated Hiring of Employees

The SBIR Phase II awards for VectraMed's pulmonary hypertension and surgical adhesion technologies facilitated the growth of the company. Dr. Pachence noted that "it was critical that I had that Phase II money, otherwise, I would have never hired employees." The proceeds from these Phase II awards, however, would not have been sufficient by themselves to create and sustain a team of Ph.D. researchers. But the scientific success in Phase I combined with the Phase II award allowed Dr. Pachence to raise some angel financing to complement the SBIR monies. Taken together, VectraMed was allowed to grow its research team from subcontractors to eight full-time scientists.

(4) Simple Mechanism for Early-stage Collaborative Work

In addition to the initial SBIR grants in the areas of pulmonary hypertension and surgical adhesions, VectraMed received another Phase I award in 1999. This grant supported a collaborative effort between VectraMed and a drug delivery firm called MicroDose. MicroDose, a company whose physical location is near VectraMed, has developed a proprietary delivery system using deep lung inhalation. VectraMed needed a delivery system to administer its molecule and the MicroDose technology looked promising. The SBIR award funded a research effort into this possibility. Unfortunately, the combination was not commercially viable and the research was abandoned after Phase I. Nevertheless, the SBIR-funded research facilitated the collaborative effort and helped to resolve the uncertainty about the most effective route of administration for VectraMed's polymer molecule.

ISSUES WITH THE CURRENT SBIR PROGRAM

Dr. Pachence highlighted a few issues of concern about the SBIR Program at NIH. His perspective is comparative since he is able to draw on past experience with the SBIR Program at the Department of Defense and the Advanced Technology Program at the National Institute of Standards and Technology. He made the following four points:

(1) Obtaining the Award Funds is Difficult at the NIH (Including Phase I & Phase II Delays)

Even prior to VectraMed's grants, Dr. Pachence won SBIR awards from the NIH back in the late 1980s. Since those early days, NIH SBIR administration has improved quite a bit. In those days, in order to collect his award funds, he would need to travel to the NIH campus to track down the money. While no longer requiring a dedicated trip to the campus, obtaining funds in the current program is still somewhat difficult because it requires an active pursuit of the funds, typically in the form of a phone call to the Study Section leader. This problem extends to the delays between Phase I and Phase II awards. The dispersal of the award funds was more efficiently done by both DoD and APT, although setting up the payment protocols at DoD often took a long time.

(2) Program Coordination is Poor Within the NIH

In stark contrast to the SBIR at DoD and the ATP at NIST, the SBIR Program at NIH is poorly coordinated. Responsibility for the administration of the SBIR Program within the NIH is shared by different groups. As far as the awardees are concerned, these groups have poor communication and integration. In contrast, ATP at NIST stands out as a highly successful model. There is a "project insider" within the agency that has a sincere interest in the company's research and actively manages the relationship between the agency and the company. Dr. Pachence says the ATP insider had a real scientific and economic interest, set up quarterly meetings, and produced well thought out reports. SBIR at DoD is also better coordinated than NIH. Perhaps the procurement orientation of the agency's awards necessitated a deeper agency interest in the firms.

(3) NIH Study Section Leaders are Too Busy

A critical underlying problem for the administration of the NIH SBIR Program is that study section leaders are too busy. Dr. Pachence noted that these people are "inundated" with programs and responsibilities. As a consequence, a productive relationship cannot be established.

(4) The NIH Review Process Overemphasizes the Scientific Component of the Proposal

The implementation of the SBIR Program at NIH followed the established organization and procedures traditionally used for purely scientific proposals. As a result, the SBIR proposal review panels are populated with a disproportionate number of academic scientists. Perhaps unintentionally, the science orientation of these individuals created a bias against commercial development grants. While a strong scientific knowledge base is important, understanding medical product development is also a critical component. SBIR proposal review panels should be "balanced" to allow projects with a stronger product development component to be reviewed more favorably.

Appendix E

Bibliography

Acs, Z., and D. Audretsch. 1988. "Innovation in Large and Small Firms: An Empirical Analysis." *The American Economic Review.* 78(4):678-690.

Acs, Z., and D. Audretsch. 1990. *Innovation and Small Firms.* Cambridge, MA: MIT Press.

Advanced Technology Program. 2001. *Performance of 50 Completed ATP Projects, Status Report 2.* National Institute of Standards and Technology Special Publication 950-2. Washington, DC: Advanced Technology Program/National Institute of Standards and Technology/U.S. Department of Commerce.

Alic, John A., Lewis Branscomb, Harvey Brooks, Ashton B. Carter, and Gerald L. Epstein. 1992. *Beyond Spinoff: Military and Commercial Technologies in a Changing World.* Boston: Harvard Business School Press.

American Association for the Advancement of Science. "R&D Funding Update on NSF in the FY2007." Available online at: <*http://www.aaas.org/spp/rd/nsf07hf1.pdf*>.

American Heart Association. "Heart Attack and Angina Statistics." Available online at: <*http://www.americanheart.org/presenter.jhtml?identifier=4591*>.

American Psychological Association. 2002. "Criteria for Evaluating Treatment Guidelines." *American Psychologist.* 57(12):1052-1059.

Archibald, R., and D. Finifter. 2000. "Evaluation of the Department of Defense Small Business Innovation Research Program and the Fast Track Initiative: A Balanced Approach." In National Research Council. *The Small Business Innovation Research Program: An Assessment of the Department of Defense Fast Track Initiative.* Charles W. Wessner, ed. Washington, DC: National Academy Press.

Arrow, Kenneth. 1962. "Economic welfare and the allocation of resources for Invention." Pp. 609-625 in *The Rate and Direction of Inventive Activity: Economic and Social Factors.* Princeton, NJ: Princeton University Press.

Arrow, Kenneth. 1973. "The theory of discrimination." Pp. 3-31 in *Discrimination in Labor Market.* Orley Ashenfelter and Albert Rees, eds. Princeton, NJ: Princeton University Press.

Audretsch, David B. 1995. *Innovation and Industry Evolution.* Cambridge, MA: MIT Press.

Audretsch, David B., and Maryann P. Feldman. 1996. "R&D spillovers and the geography of innovation and production." *American Economic Review* 86(3):630-640.

Audretsch, David B., and Paula E. Stephan. 1996. "Company-scientist locational links: The case of biotechnology." *American Economic Review* 86(3):641-642.

Audretsch, D., and R. Thurik. 1999. *Innovation, Industry Evolution, and Employment.* Cambridge, MA: MIT Press.

Audretsch, D., J. Weigand, and C. Weigand. 1999. "Does the Small Business Innovation Research Program Foster Entrepreneurial Behavior." In National Research Council, *The Small Business Innovation Research Program: An Assessment of the Department of Defense Fast Track Initiative.* Washington, DC: National Academy Press.

Bailey, D. 2004. Process Overview Brief for ONR Partnership Conference. Presentation to Advanced Technology Review Board. August 5.

Baker, Alan. No date. "Commercialization Support at NSF." Draft.

Baker, Alan. 2005. "Incentives and Technology Transition: Improving Commercialization of SBIR Technologies in Major Defense Acquisition Programs." SBTC White Paper. Washington, DC. September 21.

Barfield, C., and W. Schambra, eds. 1986. *The Politics of Industrial Policy.* Washington, DC: American Enterprise Institute for Public Policy Research.

Baron, Jonathan. 1998. "DoD SBIR/STTR Program Manager." Comments at the Methodology Workshop on the Assessment of Current SBIR Program Initiatives,Washington, DC, October.

Barry, C. B. 1994. "New directions in research on venture capital finance." *Financial Management* 23 (Autumn):3-15.

Bator, Francis. 1958. "The anatomy of market failure." *Quarterly Journal of Economics* 72: 351-379.

Bingham, R. 1998. *Industrial Policy American Style: From Hamilton to HDTV.* New York: M.E. Sharpe.

Birch, D. 1981. "Who Creates Jobs." *The Public Interest* 65 (Fall):3-14.

Branscomb, L. M. 2000. *Managing Technical Risk: Understanding Private Sector Decision Making on Early Stage Technology Based Projects.* Washington, DC: Department of Commerce/ National Institute of Standards and Technology.

Branscomb, L. M., and P. E. Auerswald. 2001. *Taking Technical Risks: How Innovators, Managers, and Investors Manage Risk in High-Tech Innovations,* Cambridge, MA: MIT Press.

Branscomb, L. M., and P. E. Auerswald. 2002. *Between Invention and Innovation: An Analysis of Funding for Early-Stage Technology Development.* Gaithersburg, MD: National Institute of Standards and Technology.

Branscomb, L. M., and P. E. Auerswald. 2003. "Valleys of Death and Darwinian Seas: Financing the Invention to Innovation Transition in the United States." *The Journal of Technology Transfer* 28(3-4).

Branscomb, L. M., and J. Keller. 1998. *Investing in Innovation: Creating a Research and Innovation Policy.* Cambridge, MA: MIT Press.

Branscomb, L. M., K. P. Morse, and M. J. Roberts. 2000. *Managing Technical Risk: Understanding Private Sector Decision Making on Early Stage Technology-based Projects.* NIST GCR 00-787. Gaithersburg, MD: National Institute of Standards and Technology.

Brav, A., and P. A. Gompers. 1997. "Myth or reality?: Long-run underperformance of initial public offerings; Evidence from venture capital and nonventure capital-backed IPOs." *Journal of Finance* 52,1791-1821.

Brodd, R. J. 2005. *Factors Affecting U.S. Production Decisions: Why Are There No Volume Lithium-Ion Battery Manufacturers in the United States?* ATP Working Paper No. 05-01, June 2005.

Brown, G., and Turner J. 1999. "Reworking the Federal Role in Small Business Research." *Issues in Science and Technology* XV, no. 4 (Summer).

BRTRC. 1997. "Commercialization of DoD Small Business Innovation Research (SBIR): Summary Report." DoD Contract Number DAAL01-94-C-0050 Mod P00010. October 8.

Bush, Vannevar. 1946. *Science—the Endless Frontier.* Republished in 1960 by U.S. National Science Foundation, Washington, DC.

Carden, S. D., and O. Darragh. 2004. "A Halo for Angel Investors." *The McKinsey Quarterly* 1.

Cassell, G. 2004. "Setting Realistic Expectations for Success." In National Research Council. *SBIR: Program Diversity and Assessment Challenges.* Charles W. Wessner, ed. Washington, DC: The National Academies Press.

Caves, Richard E. 1998. "Industrial organization and new findings on the turnover and mobility of firms." *Journal of Economic Literature* 36(4):1947-1982.

Christensen, C. 1997. *The Innovator's Dilemma.* Boston, MA: Harvard Business School Press.

Clinton, William Jefferson. 1994. *Economic Report of the President.* Washington, DC: U.S. Government Printing Office.

Clinton, William Jefferson. 1994. *The State of Small Business.* Washington, DC: U.S. Government Printing Office.

Coburn, C., and D. Bergland. 1995. *Partnerships: A Compendium of State and Federal Cooperative Technology Programs.* Columbus, OH: Battelle.

Cochrane, J. H. 2005. "The Risk and Return of Venture Capital." *Journal of Financial Economics* 75(1):3-52.

Cohen, L. R., and R. G. Noll. 1991. The Technology Pork Barrel. Washington, DC: The Brookings Institution.

Congressional Commission on the Advancement of Women and Minorities in Science, Engineering, and Technology Development. 2000. *Land of Plenty: Diversity as America's Competitive Edge in Science, Engineering and Technology.* Washington, DC: National Science Foundation/U.S. Government Printing Office.

Cooper, R.G. 2001. *Winning at New Products: Accelerating the process from idea to launch.* In Dawnbreaker, Inc. 2005. "The Phase III Challenge: Commercialization Assistance Programs 1990-2005." White paper. July 15.

Council of Economic Advisers. 1995. *Supporting Research and Development to Promote Economic Growth: The Federal Government's Role.* Washington, DC.

Council on Competitiveness. 2005. *Innovate America: Thriving in a World of Challenge and Change.* Washington, DC: Council on Competitiveness.

Cramer, Reid. 2000. "Patterns of Firm Participation in the Small Business Innovation Research Program in Southwestern and Mountain States." In National Research Council. 2000. *The Small Business Innovation Research Program: An Assessment of the Department of Defense Fast Track Initiative.* Charles W. Wessner, ed. Washington, DC: The National Academies Press.

Davidsson, P. 1996. "Methodological Concerns in the Estimation of Job Creation in Different Firm Size Classes." Working Paper. Jönköping International Business School.

Davis, S. J., J. Haltiwanger, and S. Schuh. 1994. "Small Business and Job Creation: Dissecting the Myth and Reassessing the Facts," *Business Economics* 29(3):113-122.

Dawnbreaker, Inc. 2005. "The Phase III Challenge: Commercialization Assistance Programs 1990-2005." White paper. July 15.

Dertouzos. 1989. *Made in America: The MIT Commission on Industrial Productivity.* Cambridge, MA: The MIT Press.

Dess, G. G., and D. W. Beard. 1984. "Dimensions of Organizational Task Environments." *Administrative Science Quarterly* 29:52-73.

Devenow, A., and I. Welch. 1996. "Rational Herding in Financial Economics. *European Economic Review* 40(April):603-615.

DoE Opportunity Forum. 2005. "Partnering and Investment Opportunities for the Future." Tysons Corner, VA. October 24-25.

Eckstein, 1984. *DRI Report on U.S. Manufacturing Industries.* New York: McGraw Hill.

Eisinger, P. K. 1988. *The Rise of the Entrepreneurial State: State and Local Economic Development Policy in the United State.* Madison, WI: University of Wisconsin Press.

European Commission. 2004. "Entrepreneurship—Flash Eurobarometer Survey." Available online at: <*http://europa.eu.int/comm/enterprise/enterprise_policy/survey/eurobarometer83.htm*>.

Feldman, Maryann P. 1994a. "Knowledge complementarity and innovation." *Small Business Economics* 6(5):363-372.

Feldman, Maryann P. 1994b. *The Geography of Knowledge.* Boston: Kluwer Academic.

Feldman, M.P. and M.R. Kelly. 2001a. "Leveraging Research and Development: The Impact of the Advanced Technology Program." In National Research Council, *The Advanced Technology Program: Assessing Outcomes.* Charles W. Wessner, ed. Washington, DC: National Academy Press.

Feldman, M.P., and M.R. Kelley. 2001b. *Winning an Award from the Advanced Technology Program: Pursuing R&D Strategies in the Public Interest and Benefiting from a Halo Effect.* NISTIR 6577. Washington, DC: Advanced Technology Program/National Institute of Standards and Technology/U.S. Department of Commerce.

Fenn, G. W., N. Liang, and S. Prowse. 1995. *The Economics of the Private Equity Market.* Washington, DC: Board of Governors of the Federal Reserve System.

Fisher, C. 2006. Presentation to SBTC SBIR in Rapid Transition Conference. September 27.

Flamm, K. 1988. *Creating the Computer.* Washington, DC: The Brookings Institution.

Flender, J. O., and R. S. Morse. 1975. *The Role of New Technical Enterprise in the U.S. Economy.* Cambridge, MA: MIT Development Foundation.

Freear, J., and W. E. Wetzel Jr. 1990. "Who bankrolls high-tech entrepreneurs?" *Journal of Business Venturing* 5:77-89.

Freeman, C, and L. Soete. 1997. *The Economics of Industrial Innovation.* Cambridge, MA: MIT Press.

Galbraith, J. K. 1957. *The New Industrial State.* Boston: Houghton Mifflin.

Geroski, P. A. 1995. "What do we know about entry?" *International Journal of Industrial Organization* 13(4):421-440.

Geshwiler, J., J. May, and M. Hudson. 2006. "State of Angel Groups." Kansas City, MO: Kauffman Foundation.

Gompers, P. A. 1995. "Optimal investment, monitoring, and the staging of venture capital." *Journal of Finance* 50:1461-1489.

Gompers, P. A., and J. Lerner. 1977. "Risk and Reward in Private Equity Investments: The Challenge of Performance Assessment." *Journal of Private Equity* 1:5-12.

Gompers, P. A., and J. Lerner. 1996. "The use of covenants: An empirical analysis of venture partnership agreements." *Journal of Law and Economics* 39:463-498.

Gompers, P. A., and J. Lerner. 1998. "What drives venture capital fund-raising?" Unpublished working paper. Harvard University.

Gompers, P. A., and J. Lerner. 1998. "Capital formation and investment in venture markets: A report to the NBER and the Advanced Technology Program." Unpublished working paper. Harvard University.

Gompers, P. A., and J. Lerner. 1999. "An analysis of compensation in the U.S. venture capital partnership." *Journal of Financial Economics* 51(1):3-7.

Gompers, P. A., and J. Lerner. 1999. *The Venture Capital Cycle.* Cambridge, MA: MIT Press.

Good, M. L. 1995. Prepared testimony before the Senate Commerce, Science, and Transportation Committee, Subcommittee on Science, Technology, and Space (photocopy, U.S. Department of Commerce).

Goodnight, J. 2003. Presentation at National Research Council Symposium. "The Small Business Innovation Research Program: Identifying Best Practice." Washington, DC. May 28.

Graham, O. L. 1992. Losing Time: The Industrial Policy Debate. Cambridge, MA: Harvard University Press.

Greenwald, B. C., J. E. Stiglitz, and A. Weiss. 1984. "Information imperfections in the capital market and macroeconomic fluctuations." *American Economic Review Papers and Proceedings* 74:194-199.

Griliches, Z. 1990. *The Search for R&D Spillovers.* Cambridge, MA: Harvard University Press.

Groves, R. M., F. J. Fowler, Jr., M. P. Couper, J. M. Lepkowski, E. Singer, and R. Tourangeau. 2004. *Survey Methodology.* Hoboken, NJ: John Wiley & Sons, Inc.

Haltiwanger, J., and C. J. Krizan. 1999. "Small Businesses and Job Creation in the United States: The Role of New and Young Businesses" in *Are Small Firms Important? Their Role and Impact,* Zoltan J. Acs, ed., Dordrecht: Kluwer.

Hall, Bronwyn H. 1992. "Investment and research and development: Does the source of financing matter?" Working Paper No. 92-194, Department of Economics/University of California at Berkeley.

Hall, Bronwyn H. 1993. "Industrial research during the 1980s: Did the rate of return fall?" Brookings Papers: *Microeconomics* 2:289-343.

Hall, B. H. 2002. *The Financing of Research and Development.* NBER Working Paper 8773, Cambridge, MA: National Bureau of Economic Research.

Hamberg, Dan. 1963. "Invention in the industrial research laboratory." *Journal of Political Economy* (April):95-115.

Hao, K. Y., and A. B. Jaffe. 1993. "Effect of liquidity on firms' R&D spending." *Economics of Innovation and New Technology* 2:275-282.

Hebert, Robert F., and Albert N. Link. 1989. "In search of the meaning of entrepreneurship." *Small Business Economics* 1(1):39-49.

Heilman, C. 2005. "Partnering for Vaccines: The NIAID Perspective" in Charles W. Wessner, ed. *Partnering Against Terrorism: Summary of a Workshop.* Washington, DC: The National Academies Press.

Held, B., T. Edison, S. L. Pfleeger, P. Anton, and J. Clancy. 2006. *Evaluation and Recommendations for Improvement of the Department of Defense Small Business Innovation Research (SBIR) Program.* Arlington, VA: RAND National Defense Research Institute.

Himmelberg, C. P., and B. C. Petersen. 1994. "R&D and internal finance: A panel study of small firms in high-tech industries." *Review of Economics and Statistics* 76:38-51.

Horrobin, D. F. 1990. "The philosophical basis of peer review and the suppression of innovation.," *Journal of the American Medical Association* 263:1438-441.

Hubbard, R. G. 1998. "Capital-market imperfections and investment." *Journal of Economic Literature* 36:193-225.

Huntsman, B., and J. P. Hoban Jr. 1980. "Investment in new enterprise: Some empirical observations on risk, return, and market structure." *Financial Management* 9 (Summer): 44-51.

Institute of Medicine. 1998. "The Urgent Need to Improve Health Care Quality." National Roundtable on Health Care Quality. *Journal of the American Medical Association* 280(11):1003, September 16.

Jacobs, T. 2002. "Biotech Follows Dot.com Boom and Bust." *Nature* 20(10):973.

Jaffe, A. B. 1996. "Economic Analysis of Research Spillovers: Implications for the Advanced Technology Program." Washington, DC: Advanced Technology Program/National Institute of Standards and Technology/U.S. Department of Commerce.

Jaffe, A. B. 1998. "Economic Analysis of Research Spillovers: Implications for the Advanced Technology Program." Washington, DC: Advanced Technology Program/National Institute of Standards and Technology/U.S. Department of Commerce.

Jaffe, A. B. 1998. "The importance of 'spillovers' in the policy mission of the Advanced Technology Program." *Journal of Technology Transfer* (Summer).

Jarboe, K. P., and R. D. Atkinson. 1998. "The Case for Technology in the Knowledge Economy; R&D, Economic Growth and the Role of Government." Washington, DC: Progressive Policy Institute. Available online at: *<http://www.ppionline.org/documents/CaseforTech.pdf>*.

Jewkes, J., D. Sawers, and R. Stillerman. 1958. *The Sources of Invention.* New York: St. Martin's Press.

Jefferson, T., Wager, E., and Davidoff, F. 2002. "Measuring the quality of editorial peer review." *Journal of the American Medical Association* 287:2786-2790.

Judd, C. M., and G. H. McClelland. 1989. *Data Analysis: A Model-Comparison Approach.* San Diego, CA: Harcourt Brace Jovanovich.

Kauffman Foundation. About the Foundation. Available online at: <*http://www.kauffman.org/foundation.cfm*>.

Kleinman, D. L. 1995. *Politics on the Endless Frontier: Postwar Research Policy in the United States.* Durham, NC: Duke University Press.

Kortum, Samuel, and Josh Lerner. 1998. "Does Venture Capital Spur Innovation?" NBER Working Paper No. 6846, National Bureau of Economic Research.

Krugman, P. 1990. Rethinking International Trade. Cambridge, MA: MIT Press.

Krugman, P. 1991. *Geography and Trade.* Cambridge, MA: MIT Press.

Langlois, R. N., and P. L. Robertson. 1996. "Stop Crying over Spilt Knowledge: A Critical Look at the Theory of Spillovers and Technical Change." Paper prepared for the MERIT Conference on Innovation, Evolution, and Technology. Maastricht, Netherlands, August 25-27.

Langlois, R. N. 2001. "Knowledge, Consumption, and Endogenous Growth." *Journal of Evolutionary Economics* 11:77-93.

Lebow, I. 1995. *Information Highways and Byways: From the Telegraph to the 21st Century.* New York: Institute of Electrical and Electronic Engineering.

Lerner, J. 1994. "The syndication of venture capital investments." *Financial Management* 23 (Autumn):16-27.

Lerner, J. 1995. "Venture capital and the oversight of private firms." *Journal of Finance* 50: 301-318.

Lerner, J. 1996. "The government as venture capitalist: The long-run effects of the SBIR program." Working Paper No. 5753, National Bureau of Economic Research.

Lerner, J. 1998. "Angel financing and public policy: An overview." *Journal of Banking and Finance* 22(6-8):773-784.

Lerner, J. 1999. "The government as venture capitalist: The long-run effects of the SBIR program." *Journal of Business* 72(3):285-297.

Lerner, J. 1999. "Public venture capital: Rationales and evaluation." In *The SBIR Program: Challenges and Opportunities.* Washington, DC: National Academy Press.

Levy, D. M., and N. Terleckyk. 1983. "Effects of government R&D on private R&D investment and productivity: A macroeconomic analysis." *Bell Journal of Economics* 14:551-561.

Liles, P. 1977. *Sustaining the Venture Capital Firm.* Cambridge, MA: Management Analysis Center.

Link, Albert N., and John Rees. 1990. "Firm size, university based research and the returns to R&D." *Small Business Economics* 2(1):25-32.

Link, Albert N., and John T. Scott. 1998. "Assessing the infrastructural needs of a technology-based service sector: A new approach to technology policy planning." *STI Review* 22:171-207.

Link, Albert N., and John T. Scott. 1998. *Overcoming Market Failure: A Case Study of the ATP Focused Program on Technologies for the Integration of Manufacturing Applications (TIMA).* Draft final report submitted to the Advanced Technology Program. Gaithersburg, MD: National Institute of Technology. October.

Link, Albert N. 1998. "Public/Private Partnerships as a Tool to Support Industrial R&D: Experiences in the United States." Paper prepared for the working group on Innovation and Technology Policy of the OECD Committee for Science and Technology Policy, Paris.

Link, A. N., and J. T. Scott. 1998. *Public Accountability: Evaluating Technology-Based Institutions.* Norwell, MA: Kluwer Academic.

Link, A. N., and J. T. Scott. 2005. *Evaluating Public Research Institutions: The U.S. Advanced Technology Program's Intramural Research Initiative.* London: Routledge.

Loell, Lani. 2006. Presentation to SBTC SBIR in Rapid Transition Conference. September 27.

Longini, P. 2003. "Hot buttons for NSF SBIR Research Funds," Pittsburgh Technology Council, *TechyVent.* November 27.

Malone, T. 1995. *The Microprocessor: A Biography.* Hamburg, Germany: Springer Verlag/Telos.

Mansfield, E., J. Rapoport, A. Romeo, S. Wagner, and G. Beardsley. 1977. "Social and private rates of return from industrial innovations." *Quarterly Journal of Economics* 91:221-240.

Mansfield, E. 1985. "How Fast Does New Industrial Technology Leak Out?" *Journal of Industrial Economics* 34(2).

Mansfield, E. 1996. *Estimating Social and Private Returns from Innovations Based on the Advanced Technology Program: Problems and Opportunities.* Unpublished report.

Martin, Justin. 2002. "David Birch." *Fortune Small Business* (December 1).

McCraw, T. 1986. "Mercantilism and the Market: Antecedents of American Industrial Policy." In C. Barfield and W. Schambra, eds. *The Politics of Industrial Policy.* Washington, DC: American Enterprise Institute for Public Policy Research.

Mervis, Jeffrey D. 1996. "A $1 Billion 'Tax' on R&D Funds." *Science* 272:942-944.

Morgenthaler, D. 2000. "Assessing Technical Risk," in L. M. Branscomb, K. P. Morse, and M. J. Roberts, eds. *Managing Technical Risk: Understanding Private Sector Decision Making on Early Stage Technology-Based Project.* Gaithersburg, MD: National Institute of Standards and Technology.

Moore, D. 2004. "Turning Failure into Success." In National Research Council. *The Small Business Innovation Research Program: Program Diversity and Assessment Challenges.* Charles W. Wessner, ed. Washington, DC: The National Academies Press.

Mowery, D. 1998. "Collaborative R&D: how effective is it?" *Issues in Science and Technology* (Fall):37-44.

Mowery, D., ed. 1999. *U.S. Industry in 2000: Studies in Competitive Performance.* Washington, DC: National Academy Press.

Mowery, D., and N. Rosenberg. 1989. *Technology and the Pursuit of Economic Growth.* New York: Cambridge University Press.

Mowery, D., and N. Rosenberg. 1998. Paths of Innovation: Technological Change in 20th Century America. New York: Cambridge University Press.

Murphy, L. M. and P. L. Edwards. 2003. *Bridging the Valley of Death—Transitioning from Public to Private Sector Financing.* Golden, CO: National Renewable Energy Laboratory. May.

Myers, S., R. L. Stern, and M. L. Rorke. 1983. A Study of the Small Business Innovation Research Program. Lake Forest, IL: Mohawk Research Corporation.

Myers, S. C., and N. Majluf. 1984. "Corporate financing and investment decisions when firms have information that investors do not have." *Journal of Financial Economics* 13:187-221.

National Aeronautics and Space Administration. 2005. "The NASA SBIR and STTR Programs Participation Guide." Available online at: *<http://sbir.gsfc.nasa.gov/SBIR/zips/guide.pdf>*

National Institute of Allergy and Infectious Diseases. 2005. "Advice on NIH SBIR & STTR Grant Applications." Available online at: *<http://www.niaid.nih.gov/ncn/sbir/advice/advice.pdf>*.

National Institutes of Health. 2003. "National Survey to Evaluate the NIH SBIR Program." Available online at: *<http://grants.nih.gov/grants/funding/sbir_report_2003_07.pdf>*.

National Institutes of Health. 2003. Road Map for Medical Research. Available online at: *<http://nihroadmap.nih.gov/>*.

National Institutes of Health. 2005. *Report on the Second of the 2005 Measures Updates: NIH SBIR Performance Outcomes Data System (PODS).*

National Institutes of Health. 2005. "National Academy of Sciences Study of the Small Business Innovation Research (SBIR) Program-NIH SBIR Program Data." Letter to the National Research Council. 14 February.

National Institutes of Health. "Overview of Peer Review Process." Center for Scientific Review. Available online at: *<http://cms.csr.nih.gov/ResourcesforApplicants/PolicyProcedureReview+Guidelines/OverviewofPeerReviewProcess/>*.

National Research Council. 1986. *The Positive Sum Strategy: Harnessing Technology for Economic Growth.* Washington, DC: National Academy Press.

National Research Council. 1987. *Semiconductor Industry and the National Laboratories: Part of a National Strategy.* Washington, DC: National Academy Press.

National Research Council. 1991. *Mathematical Sciences, Technology, and Economic Competitive-
ness.* James G. Glimm, ed. Washington, DC: National Academy Press.
National Research Council. 1992. *The Government Role in Civilian Technology: Building a New
Alliance.* Washington, DC: National Academy Press.
National Research Council. 1995. *Allocating Federal Funds for R&D.* Washington, DC: National
Academy Press.
National Research Council. 1996. *Conflict and Cooperation in National Competition for High-
Technology Industry.* Washington, DC: National Academy Press.
National Research Council. 1997. *Review of the Research Program of the Partnership for a New
Generation of Vehicles: Third Report.* Washington, DC: National Academy Press.
National Research Council. 1999. *The Advanced Technology Program: Challenges and Opportunities.*
Charles W. Wessner, ed. Washington, DC: National Academy Press.
National Research Council. 1999. *Funding a Revolution: Government Support for Computing Re-
search.* Washington, DC: National Academy Press.
National Research Council. 1999. *Industry-Laboratory Partnerships: A Review of the Sandia Science
and Technology Park Initiative.* Charles W. Wessner, ed. Washington, DC: National Academy
Press.
National Research Council. 1999. *New Vistas in Transatlantic Science and Technology Cooperation.*
Charles W. Wessner, ed. Washington, DC: National Academy Press.
National Research Council. 1999. *The Small Business Innovation Research Program: Challenges and
Opportunities.* Charles W. Wessner, ed. Washington, DC: National Academy Press.
National Research Council. 2000. *The Small Business Innovation Research Program: An Asssessment
of the Department of Defense Fast Track Initiative.* Charles W. Wessner, ed. Washington, DC:
National Academy Press.
National Research Council. 2000. *U.S. Industry in 2000: Studies in Competitive Performance.* Wash-
ington, DC: National Academy Press.
National Research Council. 2001. *A Review of the New Initiatives at the NASA Ames Research Center.*
Charles W. Wessner, ed. Washington, DC: National Academy Press.
National Research Council. 2001. *The Advanced Technology Program: Assessing Outcomes.* Charles
W. Wessner, ed. Washington, DC: National Academy Press.
National Research Council. 2001. *Attracting Science and Mathematics Ph.Ds to Secondary School
Education.* Washington, DC: National Academy Press.
National Research Council. 2001. *Building a Workforce for the Information Economy.* Washington,
DC: National Academy Press.
National Research Council. 2001. *Capitalizing on New Needs and New Opportunities: Government-
Industry Partnerships in Biotechnology and Information Technologies.* Charles W. Wessner, ed.
Washington, DC: National Academy Press.
National Research Council. 2001. *A Review of the New Initiatives at the NASA Ames Research Center.*
Charles W. Wessner, ed. Washington, DC: National Academy Press.
National Research Council. 2001. *Trends in Federal Support of Research and Graduate Education.*
Washington, DC: National Academy Press.
National Research Council. 2002. *Government-Industry Partnerships for the Development of New
Technologies: Summary Report.* Charles W. Wessner, ed. Washington, DC: The National
Academies Press.
National Research Council. 2002. *Making the Nation Safer: The Role of Science and Technology in
Countering Terrorism.* Washington, DC: The National Academies Press.
National Research Council. 2002. *Measuring and Sustaining the New Economy.* Dale W. Jorgenson
and Charles W. Wessner, eds. Washington, DC: The National Academies Press.
National Research Council. 2002. *Partnerships for Solid-State Lighting.* Charles W. Wessner, ed.
Washington, DC: The National Academies Press.
National Research Council. 2004. *An Assessment of the Small Business Innovation Research Pro-
gram—Project Methodology.* Washington, DC: The National Academies Press.

National Research Council. 2004. Capitalizing on Science, Technology, and Innovation: An Assessment of the Small Business Innovation Research Program/Program Manager Survey. Completed by Dr. Joseph Hennessey.

National Research Council. 2004. *Productivity and Cyclicality in Semiconductors: Trends, Implications, and Questions.* Dale W. Jorgenson and Charles W. Wessner, eds. Washington, DC: The National Academies Press.

National Research Council. 2004. *The Small Business Innovation Research Program: Program Diversity and Assessment Challenges.* Charles W. Wessner, ed. Washington, DC: The National Academies Press.

National Research Council. 2006. *Beyond Bias and Barriers: Fulfilling the Potential of Women in Academic Science and Engineering.*

National Research Council. 2006. *Deconstructing the Computer.* Dale W. Jorgenson and Charles W. Wessner, eds. Washington, DC: The National Academies Press.

National Research Council. 2006. *Software, Growth, and the Future of the U.S. Economy.* Dale W. Jorgenson and Charles W. Wessner, eds. Washington, DC: The National Academies Press.

National Research Council. 2006. *The Telecommunications Challenge: Changing Technologies and Evolving Policies.* Dale W. Jorgenson and Charles W. Wessner, eds. Washington, DC: The National Academies Press.

National Research Council. 2007. *India's Changing Innovation System: Achievements, Challenges, and Opportunities for Cooperation.* Charles W. Wessner and Sujai J. Shivakumar, eds. Washington, DC: The National Academies Press.

National Research Council. 2007. *Innovation Policies for the 21st Century.* Charles W. Wessner, ed. Washington, DC: The National Academies Press.

National Research Council. 2007. *Enhancing Productivity Growth in the Information Age: Measuring and Sustaining the New Economy.* Dale W. Jorgenson and Charles W. Wessner, eds. Washington, DC: The National Academies Press.

National Research Council. 2007. *SBIR and the Phase III Challenge of Commercialization.* Charles W. Wessner, ed. Washington, DC: The National Academies Press.

National Research Council. 2008. *An Assessment of the SBIR Program.* Charles W. Wessner, ed. Washington, DC: The National Academies Press.

National Science Board. 2005. *Science and Engineering Indicators 2005.* Arlington, VA: National Science Foundation.

National Science Board. 2006. *Science and Engineering Indicators 2006.* Arlington, VA: National Science Foundation.

National Science Foundation. Committee of Visitors Reports and Annual Updates. Available online at: *<http://www.nsf.gov/eng/general/cov/>.*

National Science Foundation. Emerging Technologies. Available online at: *<http://www.nsf.gov/eng/sbir/eo.jsp>.*

National Science Foundation. Guidance for Reviewers. Available online at: *<http://www.eng.nsf.gov/sbir/peer_review.htm>.*

National Science Foundation. National Science Foundation at a Glance. Available online at: *<http://www.nsf.gov/about>.*

National Science Foundation. National Science Foundation Manual 14, *NSF Conflicts of Interest and Standards of Ethical Conduct.* Available online at: *<http://www.eng.nsf.gov/sbir/COI_Form.doc>.*

National Science Foundation. The Phase IIB Option. Available online at: *<http://www.nsf.gov/eng/sbir/phase_IIB.jsp#ELIGIBILITY>.*

National Science Foundation. Proposal and Grant Manual. Available online at: *<http://www.inside.nsf.gov/pubs/2002/pam/pamdec02.6html>.*

National Science Foundation. 2004. *Women, Minorities, and Persons with Disabilities in Science and Engineering.* Table 15. Arlington, VA: National Science Foundation. May.

National Science Foundation. 2005. Synopsis of SBIR/STTR Program. Available online at: <*http:// www.nsf.gov/funding/pgm_summ.jsp?Phase Ims_id=13371&org=DMII*>.

National Science Foundation. 2006. "News items from the past year." Press Release. April 10.

National Science Foundation, Office of Industrial Innovation. 2006. "SBIR/STTR Phase II Grantee Conference, Book of Abstracts." Louisville, Kentucky. May 18-20, 2006.

National Science Foundation, Office of Industrial Innovation. Draft Strategic Plan, June 2, 2005.

National Science Foundation, Office of Legislative and Public Affairs. 2003. SBIR Success Story from News Tip. Web's "Best Meta-Search Engine," March 20.

National Science Foundation, Office of Legislative and Public Affairs. 2004. SBIR Success Story: GPRA Fiscal Year 2004 "Nugget." Retrospective Nugget-AuxiGro Crop Yield Enhancers.

Nelson, R. R. 1982. Government and Technological Progress. New York: Pergamon.

Nelson, R. R. 1986. "Institutions supporting technical advances in industry." American Economic Review, Papers and Proceedings 76(2):188.

Nelson, R. R., ed. 1993. National Innovation System: A Comparative Study. New York: Oxford University Press.

O'Hara, D. 1999. "Program Challenges—Operational Views." In National Research Council, *The Small Business Innovation Research Program: Challenges and Opportunities*. Washington, DC: National Academy Press.

Office of Management and Budget. 1996. "Economic analysis of federal regulations under Executive Order 12866."

Office of Management and Budget. 2004. *"What Constitutes Strong Evidence of Program Effectiveness."* Available online at: <*http://www.whitehouse.gov/omb/part/2004_program_eval.pdf*>

Office of the President. 1990. *U.S. Technology Policy*. Washington, DC: Executive Office of the President.

Organization for Economic Cooperation and Development. 1982. *Innovation in Small and Medium Firms*. Paris: Organization for Economic Cooperation and Development.

Organization for Economic Cooperation and Development. 1995. *Venture Capital in OECD Countries*. Paris: Organization for Economic Cooperation and Development.

Organization for Economic Cooperation and Development. 1997. *Small Business Job Creation and Growth: Facts, Obstacles, and Best Practices*. Paris: Organization for Economic Cooperation and Development.

Organization for Economic Cooperation and Development. 1998. *Technology, Productivity and Job Creation: Toward Best Policy Practice*. Paris: Organization for Economic Cooperation and Development.

Organization for Economic Cooperation and Development. 2006. "Evaluation of SME Policies and Programs: Draft OECD Handbook." *OECD Handbook*. CFE/SME 17. Paris: Organization for Economic Cooperation and Development.

Pacific Northwest National Laboratory. SBIR Alerting Service. Available online at: <*http://www. pnl.gov/edo/sbir*>.

Perko, J. S., and F. Narin. 1997. "The Transfer of Public Science to Patented Technology: A Case Study in Agricultural Science." *Journal of Technology Transfer* 22(3):65-72.

Perret, G. 1989. A Country Made by War: From the Revolution to Vietnam—The Story of America's Rise to Power. New York: Random House.

Porter, M. 1998. "Clusters and Competition: New Agendas for Government and Institutions," in *On Competition*, Boston, MA: Harvard Business School Press.

Powell, J.W. 1999. *Business Planning and Progress of Small Firms Engaged in Technology Development through the Advanced Technology Program*. NISTIR 6375. National Institute of Standards and Technology/U.S. Department of Commerce.

Powell, Walter W., and Peter Brantley. 1992. "Competitive cooperation in biotechnology: Learning through networks?" In N. Nohria and R. G. Eccles, eds. *Networks and Organizations: Structure, Form and Action*. Boston: Harvard Business School Press. Pp. 366-394.

Price Waterhouse. 1985. Survey of small high-tech businesses shows Federal SBIR awards spurring job growth, commercial sales. Washington, DC: Small Business High Technology Institute.

Puget Sound Business Journal. 2000. "Phillips to Acquire Optiva Corp." August 22. Available online at: <*http://www.bizjournals.com/seattle/stories/2000/08/21/daily9.html*>

Roberts, Edward B. 1968. "Entrepreneurship and technology." *Research Management* (July): 249-266.

Roberts, E. 1991. *Entrepreneurs in High Technology: Lessons from MIT and Beyond.* Oxford, U.K.: Oxford University Press.

Romer, P. 1990. "Endogenous technological change." *Journal of Political Economy* 98:71-102.

Rosa, P. and A. Dawson. 2006. "Gender and the commercialization of university science: academic founders of spinout companies." *Entrepreneurship & Regional Development.* 18(4):341-366.

Rosenbloom, R. and W. Spencer. 1996. *Engines of Innovation: U.S. Industrial Research at the End of an Era.* Boston: Harvard Business School Press.

Rubenstein, A. H. 1958. *Problems Financing New Research-Based Enterprises in New England.* Boston, MA: Federal Reserve Bank.

Ruegg, Rosalie. 2001. "Taking a Step Back: An Early Results Overview of Fifty ATP Awards." In National Research Council. *The Advanced Technology Program: Assessing Outcomes.* Charles W. Wessner, ed. Washington, DC: National Academy Press.

Ruegg, Rosalie, and Irwin Feller. 2003. *A Toolkit for Evaluating Public R&D Investment Models, Methods, and Findings from ATP's First Decade.* NIST GCR 03-857.

Ruegg, Rosalie, and Patrick Thomas. 2007. *Linkages from DoE's Vehicle Technologies R&D in Advanced Energy Storage to Hybrid Electric Vehicles, Plug-in Hybrid Electric Vehicles, and Electric Vehicles.* U.S. Department of Energy/Office of Energy Efficiency and Renewable Energy.

Sahlman, W. A. 1990. "The structure and governance of venture capital organizations." *Journal of Financial Economics* 27:473-521.

Salomo, S., R. Leifer, and H. G. Gemünden, eds. 2007. "Research on Corporate Radical Innovation Systems-A Dynamic Capabilities Perspective." *Journal of Engineering and Technology Management,* 24(1-2):1-166.

Saxenian, Annalee. 1994. *Regional Advantage: Culture and Competition in Silicon Valley and Route 128.* Cambridge, MA: Harvard University Press.

SBIR World. SBIR World: A World of Opportunities. Available online at: <*http://www.sbirworld.com*>.

Scarpa, Toni. 2006. "Research Funding: Peer Review at NIH." *Science.* 311(5757):41. January 6.

Scherer, F. M. 1970. *Industrial Market Structure and Economic Performance.* New York: Rand McNally College Publishing.

Schumpeter, J. 1950. *Capitalism, Socialism, and Democracy.* New York: Harper and Row.

Scotchmer, S. 2004. *Innovation and Incentives.* Cambridge MA: The MIT Press.

Scott, John T. 1998. "Financing and Leveraging PublicPrivate Partnerships: The Hurdle-lowering Auction." *STI Review* 23:67-84.

Siegel, D., D. Waldman, and A. Link. 2004. "Toward a Model of the Effective Transfer of Scientific Knowledge from Academicians to Practitioners: Qualitative Evidence from the Commercialization of University Technologies." *Journal of Engineering and Technology Management* 21(1-2).

Society for Prevention Research. 2004. *Standards of Evidence: Criteria for Efficacy, Effectiveness and Dissemination.* Avilable online at: <*http://www.preventionresearch.org/softext.php*>.

Sohl, Jeffrey. 1999. *Venture Capital* 1(2).

Sohl, J., J. Freear, and W.E. Wetzel Jr. 2002. "Angles on Angels: Financing Technology-Based Ventures—A Historical Perspective." *Venture Capital: An International Journal of Entrepreneurial Finance* 4 (4).

Solow, R. S. 1957. "Technical Change and the Aggregate Production Function." *Review of Economics and Statistics* 39:312-320.

Stiglitz, J. E., and A. Weiss. 1981. "Credit Rationing in Markets with Incomplete Information." *American Economic Review* 71:393-409.

Stokes, D.E. 1997. *Pasteur's Quadrant: Basic Science and Technological Innovation.* Washington, DC: Brookings Institution Press.

Stowsky, J. 1996. "Politics and Policy: The Technology Reinvestment Program and the Dilemmas of Dual Use." Mimeo. University of California.

Tassey, Gregory. 1997. *The Economics of R&D Policy.* Westport, CT: Quorum Books.

Tibbetts, R. 1997. "The Role of Small Firms in Developing and Commercializing New Scientific Instrumentation: Lessons from the U.S. Small Business Innovation Research Program," in J. Irvine, B. Martin, D. Griffiths, and R. Gathier, eds. *Equipping Science for the 21st Century.* Cheltenham UK: Edward Elgar Press.

Tirman, John. 1984. *The Militarization of High Technology.* Cambridge, MA: Ballinger.

Turner, J., and G. Brown. 1999. "The Federal Role in Small Business Research." *Issues in Science and Technology.*

Tyson, Laura, Tea Petrin, and Halsey Rogers. 1994. "Promoting entrepreneurship in Eastern Europe." *Small Business Economics* 6:165-184.

University of New Hampshire Center for Venture Research. 2007. *The Angel Market in 2006.* Available online at: <*http://wsbe2.unh.edu/files/Full%20Year%202006%20Analysis%20Report%20-%20March%202007.pdf*>

U.S. Census Bureau. 2007. "Table 130: Consumer Price Indexes of Medical Care Prices: 1980 to 2005." In *Statistical Abstract of the United States.* Washington, DC: U.S. Census Bureau.

U.S. Congress. House Committee on Science, Space, and Technology. 1992. *SBIR and Commercialization: Hearing Before the Subcommittee on Technology and Competitiveness of the House Committee on Science, Space, and Technology, on the Small Business Innovation Research [SBIR] Program.* Testimony of James A. Block, President of Creare Inc. Pp. 356-361.

U.S. Congress. House Committee on Science, Space, and Technology. 1998. *Unlocking Our Future: Toward a New National Science Policy: A Report to Congress by the House Committee on Science, Space, and Technology.* Washington, DC: Government Printing Office. Available online at: <*http://www.access.gpo.gov/congress/house/science/cp105-b/science105b.pdf*>.

U.S. Congress. House Committee on Science, Space, and Technology. Subcommittee on Environment, Technology, and Standards. 2005. Hearing on "Small Business Innovation Research: What is the Optimal Role of Venture Capital." Hearing Charter. June 28.

U.S. Congress. House Committee on Science, Space, and Technology. Subcommittee on Environment, Technology, and Standards. 2005. Hearing on "Small Business Innovation Research: What is the Optimal Role of Venture Capital." Testimony by Frederic Abramson. June 28.

U.S. Congress. House Committee on Science, Space, and Technology. Subcommittee on Environment, Technology, and Standards. 2005. Hearing on "Small Business Innovation Research: What is the Optimal Role of Venture Capital." Testimony by Jonathan Cohen. June 28.

U.S. Congress. House Committee on Science, Space, and Technology. Subcommittee on Environment, Technology, and Standards. 2005. Hearing on "Small Business Innovation Research: What is the Optimal Role of Venture Capital." Testimony by Ron Cohen. June 28.

U.S. Congress. House Committee on Science, Space, and Technology. Subcommittee on Environment, Technology, and Standards. 2005. Hearing on "Small Business Innovation Research: What is the Optimal Role of Venture Capital." Testimony by Carol Nacy. June 28.

U.S. Congress. House Committee on Science, Space, and Technology. Subcommittee on Technology and Innovation. 2007. Hearing on "Small Business Innovation Research Authorization on the 25th Program Anniversary." Testimony by Robert Schmidt. April 26.

U.S. Congress. House Committee on Small Business. Subcommittee on Workforce, Empowerment, and Government Programs. 2005. *The Small Business Innovation Research Program: Opening Doors to New Technology.* Testimony by Joseph Hennessey. 109th Cong., 1st sess., November 8.

U.S. Congress. Senate Committee on Small Business. 1981. Senate Report 97-194. *Small Business Research Act of 1981.* September 25. Washington, DC: U.S. Government Printing Office.

U.S. Congress. Senate Committee on Small Business. 1999. Senate Report 106-330. *Small Business Innovation Research (SBIR) Program.* August 4. Washington, DC: U.S. Government Printing Office.

U.S. Congress. Senate Committee on Small Business. 2006. *Strengthening the Participation of Small Businesses in Federal Contracting and Innovation Research Programs.* Testimony by Thomas Bigger. 109th Cong., 2nd sess., July 12.

U.S. Congress. Senate Committee on Small Business. 2006. *Strengthening the Participation of Small Businesses in Federal Contracting and Innovation Research Programs.* Testimony by Michael Squillante. 109th Cong., 2nd sess., July 12.

U.S. Congressional Budget Office. 1985. Federal financial support for high-technology industries. Washington, DC: U.S. Congressional Budget Office.

U.S. Court of Appeals for the Federal Circuit. 2006. *Night Vision Corp. v. The United States of America.* No. 06-5048. November 22.

U.S. Court of Federal Claims. 2005. *Night Vision Corp. v. The United States of America.* No. 03-1214C. 25 May.

U.S. Department of Defense, Under Secretary for Acquisition, Technology, and Logistics. 2004. *Interim Defense Acquisition Guidebook.* Available online at: <*https://akss.dau.mil/dag/DoD5000. asp?view=document*>.

U.S. Department of Education. 2005. "Scientifically-Based Evaluation Methods: Notice of Final Priority." *Federal Register.* 70(15):3586-3589. January 25.

U.S. Food and Drug Administration. 1981. Protecting Human Subjects: Untrue Statements in Application. 21 C.F.R. §314.12

U.S. Food and Drug Administration. *Critical Path Initiative.* Available online at: <*http://www.fda. gov/oc/initiatives/criticalpath/*>

U.S. General Accounting Office. 1987. Federal research: Small Business Innovation Research participants give program high marks. Washington, DC: U.S. General Accounting Office.

U.S. General Accounting Office. 1989. *Federal Research: Assessment of Small Business Innovation Research Program.* Washington, DC: U.S. General Accounting Office.

U.S. General Accounting Office. 1992. *Small Business Innovation Research Program Shows Success but Can Be Strengthened.* RCED-92-32. Washington, DC: U.S. General Accounting Office.

U.S. General Accounting Office. 1997. *Federal Research: DoD's Small Business Innovation Research Program.* RCED-97-122, Washington, DC: U.S. General Accounting Office.

U. S. General Accounting Office. 1998. *Federal Research: Observations on the Small Business Innovation Research Program.* RCED-98-132. Washington, DC: U.S. General Accounting Office.

U.S. General Accounting Office. 1999. *Federal Research: Evaluations of Small Business Innovation Research Can Be Strengthened.* RCED-99-198, Washington, DC: U.S. General Accounting Office.

U.S. Government Accountability Office. 2005. *Defense Technology Development: Management Process Can Be Strengthened for New Technology Transition Programs.* GAO-05-480. Washington, DC: U.S. Government Accountability Office. June.

U.S. Government Accountability Office. 2006. *Small Business Innovation Research: Agencies Need to Strengthen Efforts to Improve the Completeness, Consistency, and Accuracy of Awards Data,* GAO-07-38, Washington, DC: U.S. Government Accountability Office.

U.S. Government Accountability Office. 2006. *Small Business Innovation Research: Information on Awards made by NIH and DoD in Fiscal years 2001-2004.* GAO-06-565. Washington, DC: U.S. Government Accountability Office.

U.S. Small Business Administration. 1992. *Results of Three-Year Commercialization Study of the SBIR Program.* Washington, DC: U.S. Government Printing Office.

U.S. Small Business Administration. 1994. *Small Business Innovation Development Act: Tenth-Year Results.* Washington, DC: U.S. Government Printing Office (and earlier years).